112 Topics in Current Chemistry

Fortschritte der Chemischen Forschung

Managing Editor: F. L. Boschke

Preparative Organic Chemistry

With Contributions by
J. Káš, E. M. Kosower, P. Rauch,
A. Suzuki, I. Szele, H. Zollinger

With 41 Figures and 21 Tables

Springer-Verlag Berlin Heidelberg GmbH
1983

This series presents critical reviews of the present position and future trends in modern chemical research. It is addressed to all research and industrial chemists who wish to keep abreast of advances in their subject.

As a rule, contributions are specially commissioned. The editors and publishers will, however, always be pleased to receive suggestions and supplementary information. Papers are accepted for "Topics in Current Chemistry" in English.

ISBN 978-3-662-15302-4 ISBN 978-3-540-40976-2 (eBook)
DOI 10.1007/978-3-540-40976-2

Library of Congress Cataloging in Publication Data. Main entry under title: Preparative organic chemistry.

(Topics in current chemistry = Fortschritte der chemischen Forschung; 112)
Includes bibliographies and index.
1. Chemistry, Organic—Addresses, essays, lectures.
I. Káš, J. II. Series: Topics in current chemistry; 112.
QD1.F58 vol. 112 [QD255] 540s [547] 83-660

Softcover reprint of the hardcover 1st edition 1983
Originally published by Springer-Verlag Berlin Heidelberg New York in 1983.

2152/3020-543210

Table of Contents

Azo Coupling Reactions
Structures and Mechanisms

Ivanka Szele and Heinrich Zollinger

Technisch-Chemisches Laboratorium, Eidgenössische Technische Hochschule (ETH), CH-8092 Zürich, Switzerland

Table of Contents

1 Introduction

Diazonium ions are Lewis acids in which the β-nitrogen atom is the centre of electrophilic character. The addition of nucleophiles at the β-nitrogen is called an azo (or diazo) coupling reaction and depending on the atom which provides the lone pair of electrons, C-, N-, O-, P- or S-coupling can occur.

In this paper reactions of aromatic, heteroaromatic and related diazonium ions with nucleophiles are discussed. In such reactions substitution by the diazonium ion of an electrofugic atom or group bonded to carbon takes place. Occasionally reference is made to N- and P-coupling. In Section 4 the respective substitution at nitrogen (formation of diazoamino compounds) is included for comparative purposes.

The products of these substitution reactions at carbon are azo compounds; azo coupling is the reaction by which about 50% of all industrial dyes manufactured in the last 100 years have been produced.

The industrial aspects of this reaction will, however, not be discussed in this paper. We will concentrate on structural and mechanistic problems which, to our knowledge, have not been summarized in detail in the last two decades. In the 1950's work on azo coupling reactions with aromatic substrates, e.g. wih phenols, naphthols, naphthylamines, as well as with activated methylene compounds such as enols, demonstrated that azo coupling reactions are, in many respects, ideal systems for mechanistic investigations of electrophilic aromatic substitutions: Azo coupling reactions can be run in dilute aqueous buffer solutions, and therefore acid-base phenomena can be studied better than with other electrophilic aromatic substitutions, which often take place in less familiar systems such as concentrated sulfuric acid for nitrations, or apolar solvents for Friedel-Crafts reactions. A clear differentiation between specific and general base catalysis in electrophilic aromatic substitutions was made possible for the first time using an azo coupling reaction. The same is true for a quantitative evaluation of kinetic hydrogen isotope effects.

These results were previously discussed in a monograph published in 1961 [1], but were later dealt with only briefly by various authors. In the last twenty years some effects, found many decades ago in coupling and other substitution reactions, could be elucidated with respect to their structural and mechanistic basis, e.g. product ratios due to mixing effects — a phenomenon which was not understandable at all only a few years ago.

This review summarizes such investigations. We hope that it will give suggestions for further work on the understanding and the expansion of the scope of the azo coupling reaction as well as catalyze the transfer of methods used successfully in investigations of the azo coupling to other reactions in organic chemistry.

In order to keep this review to a manageable size we do not discuss in detail related reactions, e.g. additions of diazonium ions to simple anions like OH^-, CN^-, and N_3^-, or the so-called oxidative azo coupling reaction, discovered by Hünig in which electrophilic reagents comparable but not identical with diazonium ions are obtained by oxidation of heteroaromatic hydrazones.

2 Structure of Reagents in Azo Coupling Reactions

2.1 Diazo Components

By far the most important azo coupling reaction, i.e. the reaction of arenediazonium ions with aromatic coupling components, has been known for many decades and has been extensively reviewed [1]. Therefore, in this section we shall concentrate on reactions of aliphatic diazonium ions, diazoalkanes and diazocarbonyl compounds which have not been known previously as well as on coupling reactions which have gained in importance more recently, specifically those of heterocyclic diazonium ions.

2.1.1 Aliphatic Diazonium Ions

Aliphatic diazonium ions are formed as intermediates in deamination reactions of aliphatic amines. Due to the extremely good leaving group ability of nitrogen, however, they usually cannot be trapped by an azo coupling reaction, but rather undergo rapid dediazoniation to give carbocation intermediates [2]. Recently, however, it has been found that in some alkanediazonium ions azo coupling successfully competes with dediazoniation. All the examples known up to now concern cyclopropanediazonium ions [3-5] and bridgehead diazonium ions [6-9] in which the loss of nitrogen would lead to a very unstable carbocation.

Kirmse and coworkers have studied the reaction of alkanediazonium ions with amines [3,4] and with lithium azide [3,5]. Cyclopropanediazonium ions [2] give azo coupling products *6* and *7* with dimethylamine and with ethylamine, respectively (1) [3,4]. However, no azo coupling of *1* with phenols was observed. In the reaction

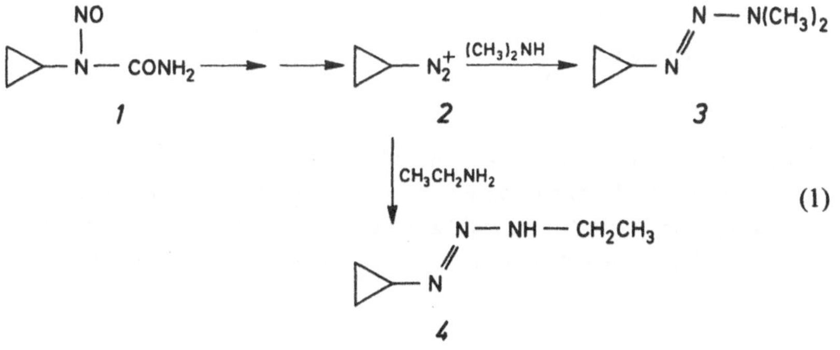

NO
N — $CONH_2$ → → N_2^+ $\xrightarrow{(CH_3)_2NH}$ N=N — $N(CH_3)_2$

1 *2* *3*

↓ $CH_3CH_2NH_2$ (1)

N=N — NH — CH_2CH_3

4

with lithium azide Kirmse and coworkers [3,5] determined by ^{15}N-labelling of the α-nitrogen atom in *1* that only 3% of the product (cyclopropyl azide) was formed by direct substitution of the diazonio group by azide ions, while the major pathway (97%) was an addition of N_3^- to the β-nitrogen of the intermediate diazonium ion *2* and dediazoniation of the pentazene formed.

Curtin and coworkers have observed the azo coupling of the bridgehead diazonium ion *6* with 2-naphthol at low temperature (2) in 50% yield.

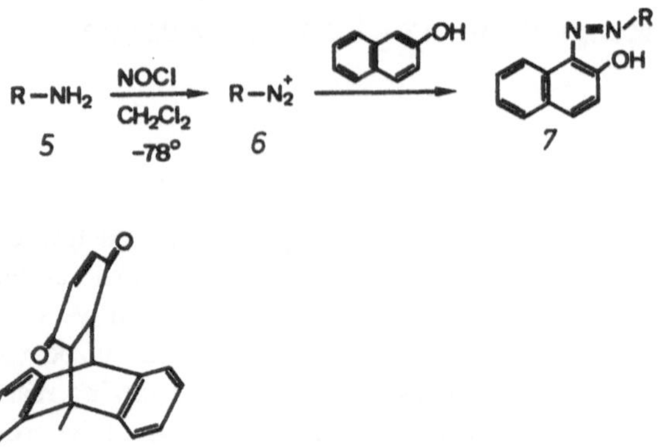

$$R{-}NH_2 \xrightarrow[CH_2Cl_2,\ -78°]{NOCl} R{-}N_2^+ \longrightarrow 7 \qquad (2)$$

5 6 7

Also Scherer and Lunt [7–9] could demonstrate the azo coupling of the extremely electrophilic bridgehead polychlorinated homocubanediazonium ion *8* with several aromatic coupling components (3). The dediazoniation of *8* surprisingly yields radical intermediates [9], thus reflecting the high energy content of the bridgehead perchlorinated homocubyl cation. The hydrogenated analogue of *8*, on the other hand, yields 'normal' ionic intermediates under the same reaction conditions [10].

$$R{-}NH_2 \xrightarrow{HNO_2} R{-}N_2^+ \rightarrow R\cdot$$

8

$$\downarrow ArH$$

$$Ar{-}N{=}N{-}R \quad \sim 90\% \qquad (3)$$

R–NH₂ = [structure], ArH = [structures]

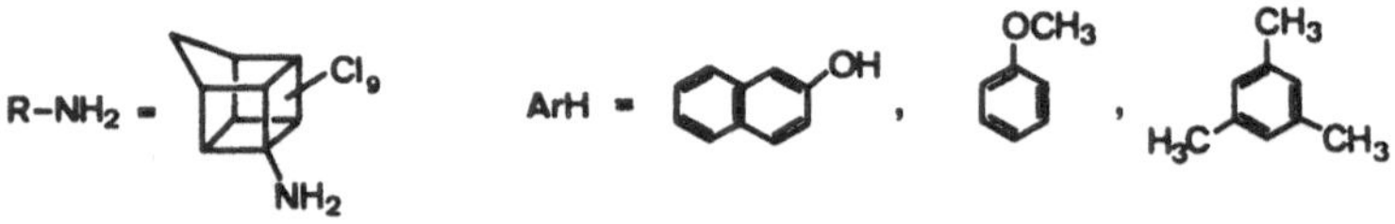

Other alkanediazonium ions, including the bridgehead bicyclo[2.2.1]heptane-1-diazonium ion *9*, do not undergo azo coupling reactions but give only nucleophilic substitution products [4,5]. In addition to the examples given above, some alkene-diazonium ions generated from nitrosooxazolidones can also add azide ions [5,11].

9

It is interesting to compare the behavior of short-lived diazonium ion intermediates with that of the relatively stable and isolable alkenediazonium ions first prepared by Reimlinger [12] and by Bott [13,14]. Until now there has been no conclusive evidence that the latter compounds undergo an azo coupling reaction. The reaction of alkene-diazonium ions *10* with primary amines, which gives triazoles *11* in reasonable yield [15,16], can be rationalized in terms of an azo coupling reaction (Path A), as well as of a nucleophilic attack on the β-carbon atom of the C—C double bond (Path B) (4).

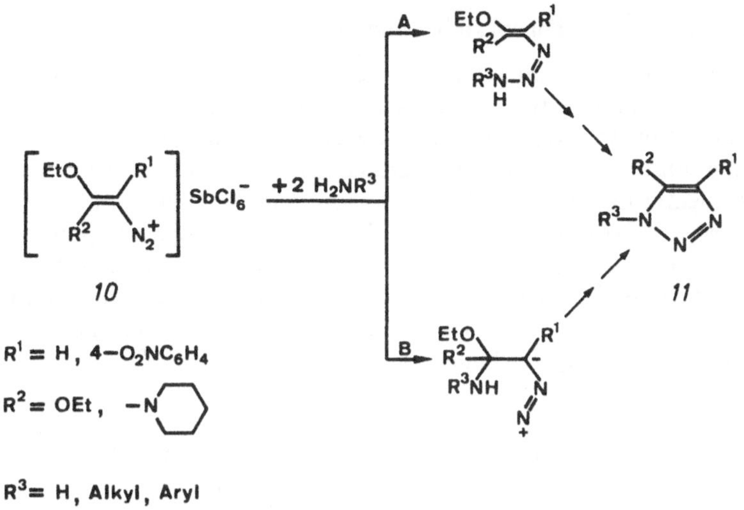
A
B
SbCl$_6^-$
+2 H$_2$NR^3
10
11
R^1 = H, 4—O$_2$NC$_6$H$_4$
R^2 = OEt, —N
R^3= H, Alkyl, Aryl

(4)

The extremely electrophilic alkenediazonium salt *12* was shown to react with nucleophiles, e.g. with anisole, at the β-carbon atom (5) [14]. Compound *12* and other

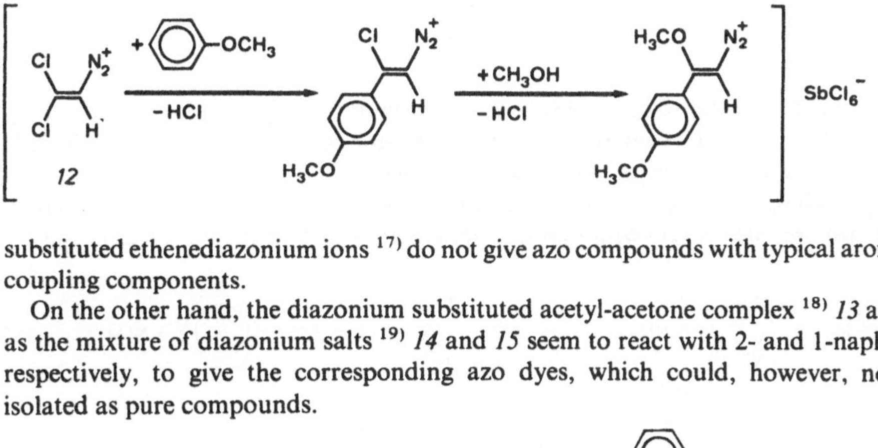
+CH$_3$OH
-HCl
-HCl
SbCl$_6^-$
12

(5)

substituted ethenediazonium ions [17] do not give azo compounds with typical aromatic coupling components.

On the other hand, the diazonium substituted acetyl-acetone complex [18] *13* as well as the mixture of diazonium salts [19] *14* and *15* seem to react with 2- and 1-naphthol, respectively, to give the corresponding azo dyes, which could, however, not be isolated as pure compounds.

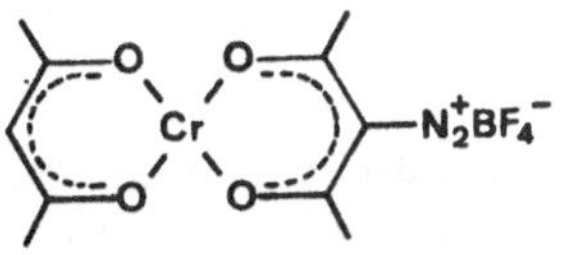

13

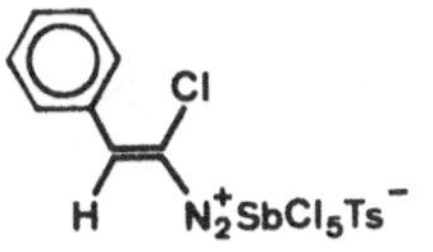

14

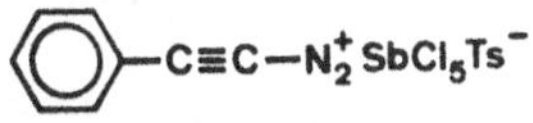

15

In conclusion it can be said that alkane- and alkenediazonium ions react with nucleophiles by a variety of pathways, one of them, in certain cases, being the azo coupling reaction. Small changes in the substrate structure, as well as in the reaction conditions, can drastically change the reaction pathway, indicating that the energy requirements for the competing reactions are rather similar.

2.1.2 Diazoalkanes, Diazoketones and Diazoesters

Compared to diazonium salts, diazo compounds are generally much less reactive towards nucleophiles than towards electrophiles. As a result of this azo coupling reactions of diazo compounds are the exception rather than the rule. Electron withdrawing substituents on the diazo carbon increase the reactivity towards nucleophiles. Consequently the ability to undergo azo coupling reactions increases from diazomethane to diazocarbonyl- and 2-diazo-1,3-dicarbonyl compounds. Among the earliest reactions known were those with cyanide and sulfite ions [1,20].

Tertiary phosphines, as opposed to amines, can form stable addition complexes with diazoalkanes [21,22], probably due to the ability of phosphorus to stabilize the betaine with its empty d orbitals (6).

$$R_2CN_2 + Ph_3P \rightarrow \left[R_2\bar{C}-N{=}N-\overset{+}{P}Ph_3 \leftrightarrow R_2C=N-N{=}PPh_3\right] \quad (6)$$

Carbanions derived from organometallic reagents react with aryldiazoalkanes [23 to 25], diazoketones [26] and diazoesters [27] to yield hydrazones after hydrolysis (7).

$$Ph_2C=N_2 + PhMgBr \rightarrow Ph_2\bar{C}-N{=}N-Ph \rightarrow Ph_2C=N-NHPh \quad (7)$$

2-Diazo-1,3-dicarbonyl compounds are electrophilic enough to give azo coupling products with reactive aromatic azo coupling components [28,29], as well as with CH acidic compounds [30] such as β-diketones and β-keto-esters.

Dicyanodiazomethane [31] even gives an azo-coupling product with dimethylaniline (8) and also reacts with diaryldiazomethanes to give the mixed azines (9).

$$(NC)_2C=N_2 + C_6H_5-NMe_2 \longrightarrow (NC)_2C=N-NH-C_6H_4-NMe_2 \quad (8)$$

$$(NC)_2C=N_2 + Ph_2CN_2 \rightarrow Ph_2C\begin{cases}\overset{+}{N}\equiv N\\ N=N-\underline{C}(CN)_2\end{cases} \rightarrow Ph_2C=N-N=C(CN)_2 \quad (9)$$

Bisarylsulfonyldiazomethanes can also react with diphenyldiazomethane and the azine is formed without passing through a carbene intermediate (10) [32].

$$Ph_2CN_2 + N_2C(SO_2Ph)_2 \rightarrow Ph_2C=N-N=C(SO_2Ph)_2 \quad (10)$$

In 1969 the first example was reported of a coupling reaction of diazomethane with

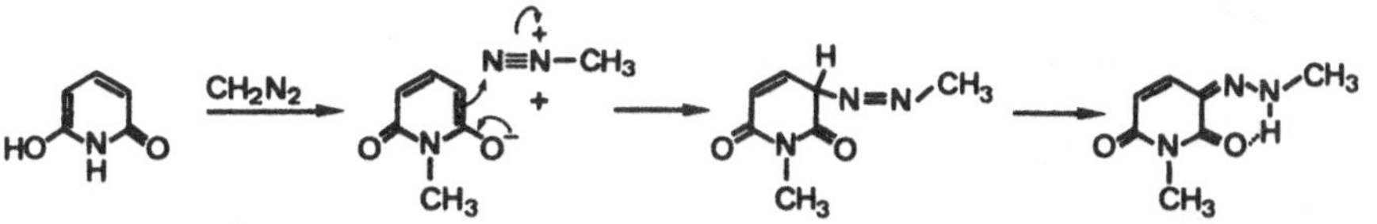

(11)

a heterocyclic coupling component [33] (11). Compounds *16* to *18* can be synthesized in a similar manner [34–36].

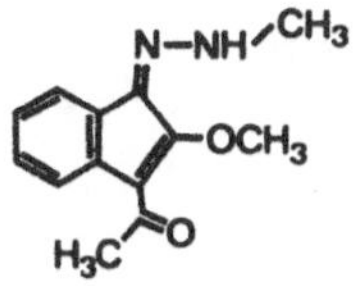

16 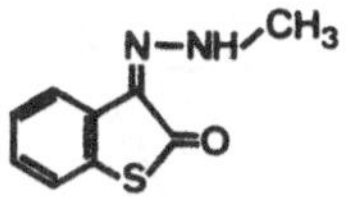 *17* 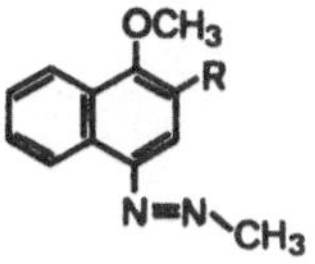 *18*

An interesting dichotomy of reaction paths was recently observed by Huisgen and coworkers [37] in reactions of diazocarbonyl compounds with enamines: While diazomonocarbonyl compounds react with 2,5-dimethyl-1-pyrrolidinocyclopentene *19* in a cycloaddition reaction to give 2-pyrazolines *20*, dimethyl diazomalonate undergoes an azo coupling reaction and the hydrazone *21* is formed (12). This nicely

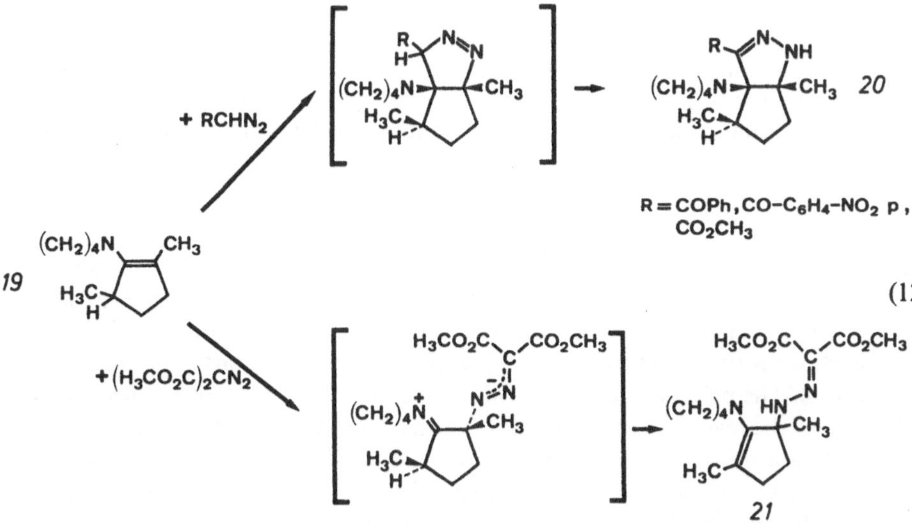

$R = COPh, CO{-}C_6H_4{-}NO_2\ p, CO_2CH_3$

(12)

substantiates the statement that the reactivity of substituted diazo compounds in 1,3-dipolar additions decreases with increasing delocalization of the negative charge on the carbon atom into the substituents. The opposite trend is observed for the reactivity in azo coupling reactions.

Another interesting coupling reaction with enamines was recently observed with bis(methylsulfonyl)diazomethane *22* [40].

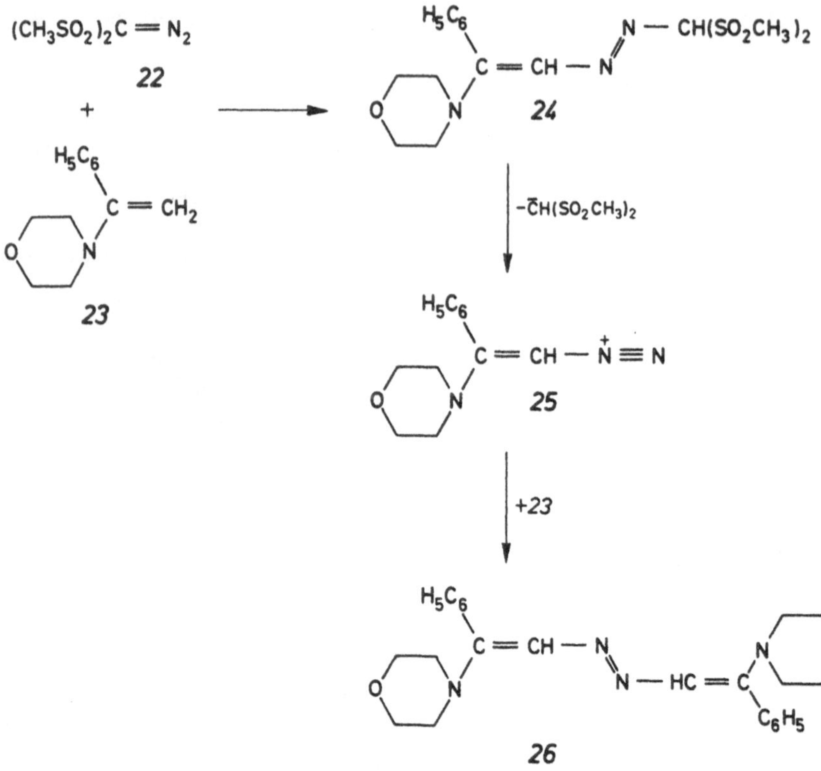

$(CH_3SO_2)_2C{=}N_2$ **22** + **23** → **24** $\xrightarrow{-\bar{C}H(SO_2CH_3)_2}$ **25** $\xrightarrow{+23}$ **26**

The azo compound *24* is probably formed initially. It decomposes to give the vinyldiazonium ion *25* that in its turn couples with another molecule of *23* to give the final product *26* (13) [40]. Actually the diazodiester *22* acts as a diazo-transfer agent.

2.1.3 Diazooxides[1]

p-Diazooxides *27* and their o-isomers simultaneously display the properties of both aliphatic and aromatic diazo components. They can be considered as analogues of conjugated diazoketones. On the other hand, a specific feature of many of their reactions is their conversion to hydroxyarenediazonium ions *28* in the presence of acids (14). The pK_a value of the p-hydroxybenzenediazonium ion *28*, for example, is 3.19 [42], so the reactivity of compounds of this type will depend a lot on the acidity of the reaction medium. *28* is much more electrophilic than *27*, and the measured rate therefore depends on the position of equilibrium (14). Recently a comprehensive review on diazooxides has appeared [43], also including azo coupling reactions, and therefore only a few selected reactions will be mentioned here.

1 In this review the name diazooxides will be used for compounds of the type *27*, since it seems to be in best agreement with the IUPAC [41] Rules (C10.1 and C84.2). Other names also encountered in the literature are: diazoquinones, quinonediazides and diazophenols.

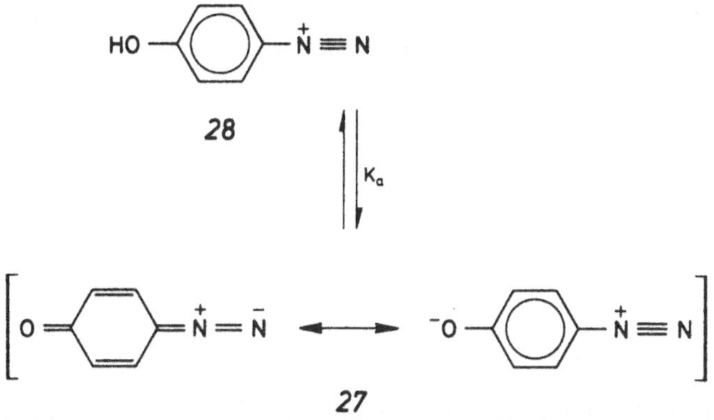

(14)

Diazooxides react with hydroxyaromatic coupling components in the same manner as diazonium salts, giving dihydroxysubstituted azo compounds. An interesting feature of this reaction is the fact that the reaction rate increases with medium acidity, reaches a maximum (between 50–80% H_2SO_4 in reactions with resorcinol, depending on the diazooxide under study) and then decreases [44,45].

Basically the reactivity in azo coupling reactions of diazonium salts is larger than that of diazooxides [46] (1,4-diazooxides again are more reactive than the 1,2-species), and therefore the concentration of the more reactive form is increasing with the medium acidity. Up to a certain point this compensates for the parallel decrease in the concentration of the more reactive coupling component, the phenolate ion. In principle, however, there are no qualitative differences in the behavior of diazooxides and arenediazonium salts in azo coupling reactions with hydroxyaromatic compounds. This is also supported by a thorough study of the coupling with 2-naphthol [46]: in a Hammett plot of log k vs. σ^+, data for diazonium salts and diazooxides fall on one straight line ($\varrho = 2.55$) [42], indicating a uniform mechanism of the azo coupling process.

Substituted 1,4-diazooxides react with secondary aliphatic amines to give the corresponding triazenes (diazoamino compounds) in high yield, while the products with primary amines are surprisingly unstable and decompose by a radical mechanism [47]. Reactions with diazoalkanes yield asymmetrical azines *29* (15) in the case of 1,4-diazooxides, and cyclic benzoxadiazines *30* with 1,2-diazooxides (16) [48].

(15)

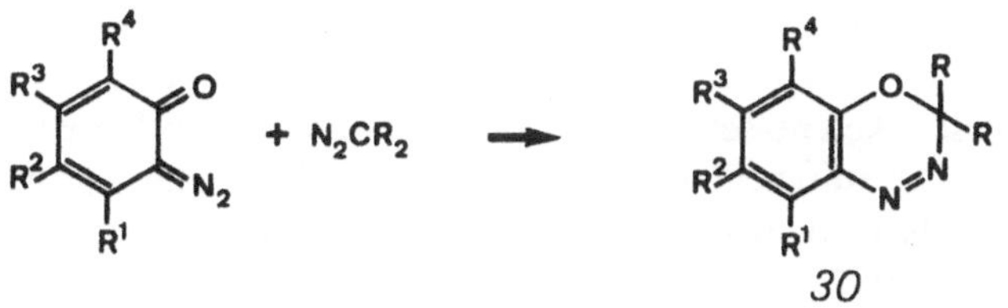

(16)

An interesting reaction of the 1,2-diazooxide *31* was recently discovered in the gas phase[49] (17), i.e. the intramolecular interaction of the nucleophilic carbonyl oxygen atom with the electrophilic diazo group and formation of the benz-1,2,3-oxadiazole *32*. At 40 °C the equilibrium mixture consists of 10–20% *31* and 80–90% *32*. *32* is energetically more favorable by only 4 kJ/mole in the gas phase, while in solution and in the solid state the equilibrium is shifted completely to the diazooxide form *31*[49]. The respective thioanalogue, however, forms the heterocycle benzothiodiazole in solution, also[50].

31 *32*

1,2-Diazooxides react easily with ketenes at low temperature[51–53]. Depending on the reaction conditions, both monoadducts *33* and/or di-adducts *34* can be obtained (18). When the reaction is run in the presence of alcohols, the product *36* observed is that of azo coupling of the 1,2-diazooxide with esters of substituted acetic acids (19, 20)[54]. Since the latter do not react directly with 1,2-diazooxides, the authors concluded that first a ketene semiacetal *35* is formed in which the electron density of the double bond is sufficient to result in azo coupling. The alternative process proceeding via an intermediate cyclic adduct of type *37* and subsequent ring opening to *36* could be excluded[54].

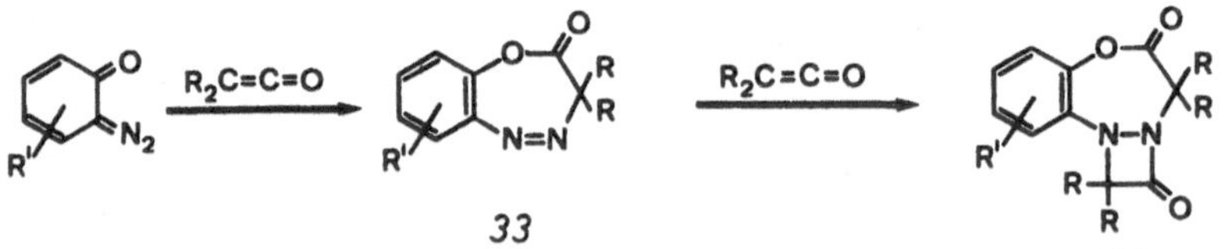

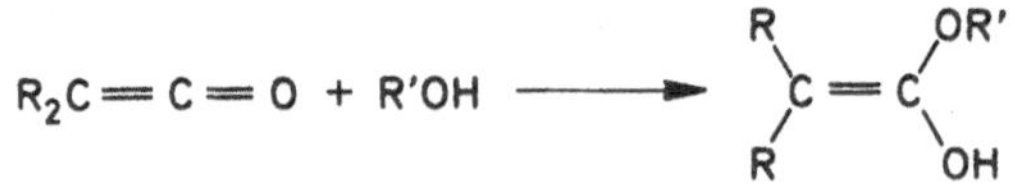

$R_2C{=}C{=}O$ *33* $R_2C{=}C{=}O$ *34*

(18)

$$R_2C{=}C{=}O + R'OH \longrightarrow R_2C{=}C(OR')(OH) \quad (19)$$

35

(20)

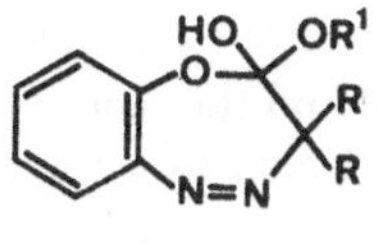

HO OR[1] / O / R / R / N=N

37

2.1.4 Heterocyclic Diazo and Diazonium Compounds

Even though azo coupling reactions of arenediazonium ions are among the oldest and most widely studied reactions in organic chemistry, those of the heterocyclic diazonium compounds have only become known in more detail fairly recently. In the comprehensive review by Zollinger [55] about 20 years ago they are only briefly mentioned, and it is pointed out that, due to the poor stability of heterocyclic diazonium salts, the desired azo dyes are often prepared by oxidative coupling of the corresponding hydrazones. The situation has changed drastically since then, and nowadays dyes with heterocyclic diazo components are prepared on a large scale and are very important because of their excellent brightness and high tinctorial power. The amount of patent literature on that subject bears witness to it [56]. Heterocyclic diazonium salts are also interesting as synthetic intermediates for the preparation of compounds with potential pharmaceutical value [57].

$$\mathrm{HArNH_2} \xrightarrow{\mathrm{HNO_2}} \mathrm{HAr\overset{+}{N}H_2NO} \underset{}{\overset{-\mathrm{H}^+}{\rightleftharpoons}} \mathrm{HArNHNO} \rightleftharpoons \mathrm{HArN{=}NOH}$$

$$\overset{+\mathrm{H}^+}{\rightleftharpoons} \mathrm{HArN_2^+} \overset{-\mathrm{H}^+}{\rightleftharpoons} \mathrm{\bar{A}rN_2^+} \leftrightarrow \mathrm{Ar{=}N_2} \qquad (21)$$

Ar: heteroaromatic ring minus two hydrogen atoms

A specific feature of heterocyclic diazonium compounds is illustrated by the diazotization of the corresponding primary amines (21) [58]. If the ring contains an acidic hydrogen atom, the main reaction products are the diazonium ion and the corresponding diazo compound. On the other hand, if no acidic hydrogen atom in the ring is present, the product of nitrosation of the amine is usually a mixture of the nitrosamine, the diazohydroxide and the diazonium cation, the concentration of the diazonium species increasing with the acitity of the medium [58]. The pK_a values of a series of azoldiazonium salts in aqueous solutions at 0 °C indicate that the acidity of the diazonium ions increases with the number of nitrogen atoms in the ring [58]. In addition, the structure and the yield of the isolated diazotization product also depend on the solubility of the products, the rates of the substitution reactions

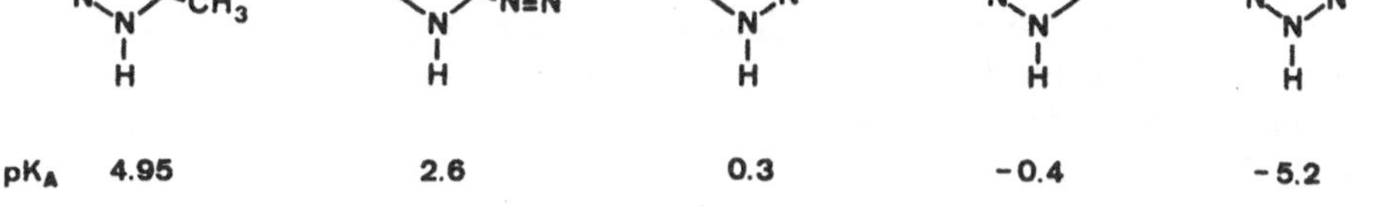

	H_3C, $\overset{+}{N}{\equiv}N$, CH_3, N, N–H (pyrazole)	N, $\overset{+}{N}{\equiv}N$, N–H (imidazole)	N, N, $\overset{+}{N}{\equiv}N$, N–H (triazole)	N, N, $\overset{+}{N}{\equiv}N$, N–H (triazole)	N, N, N, $\overset{+}{N}{\equiv}N$, N–H (tetrazole)
pK_A	4.95	2.6	0.3	−0.4	−5.2

with the nucleophiles present and of decomposition reactions. In view of this it is not surprising that, depending on the heterocyclic amine and the reaction conditions, different products are obtained upon diazotization, and also that the azo coupling reaction competes more or less successfully with the other reactions of the diazonium ions produced [57].

Few quantitative studies of the azo coupling reactions of heterocyclic diazo compounds are known. The most conclusive one is by Sawaguchi, Hashida and Matsui [59], who measured the coupling rates of seventeen heterocyclic diazonium ions with R-acid (2-naphthol-3,6-disulfonic acid), and compared them with those of benzenediazonium ion with the same coupling component. All the heterocyclic diazonium ions investigated react faster with R-acid than does benzenediazonium salt. This is in accordance with the higher electronegativity of the ring nitrogen atoms relative to carbon. A detailed discussion of the individual effects is beyond the scope of this review and is given in the original paper [59].

The high reactivity of heterocyclic diazonium ions in azo coupling reactions is the reason why in some cases the primary diazotization products cannot be isolated. 2-Methyl-5-aminotetrazole *38*, for example, yields directly the diazoamino compound *39* on diazotization, the intermediate diazonium ion being reactive enough to give the N-coupling product with the parent amine, even under strongly acidic conditions (22) [60,61].

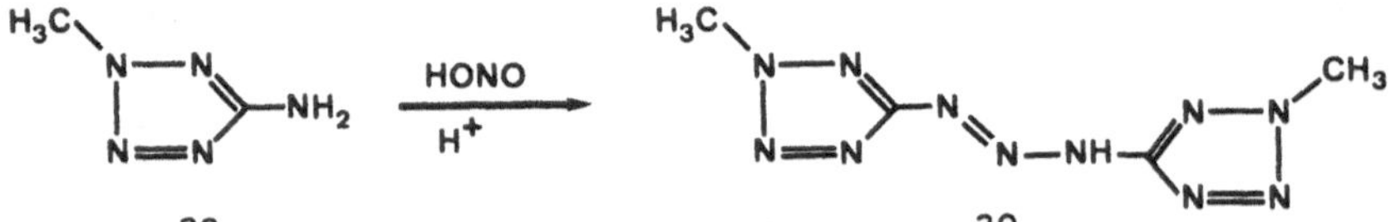

(22)

The unsubstituted diazotetrazol *40* even couples with hydrazine in a manner not observed with benezenediazonium ions to form the hexazadiene *41* (23) [62].

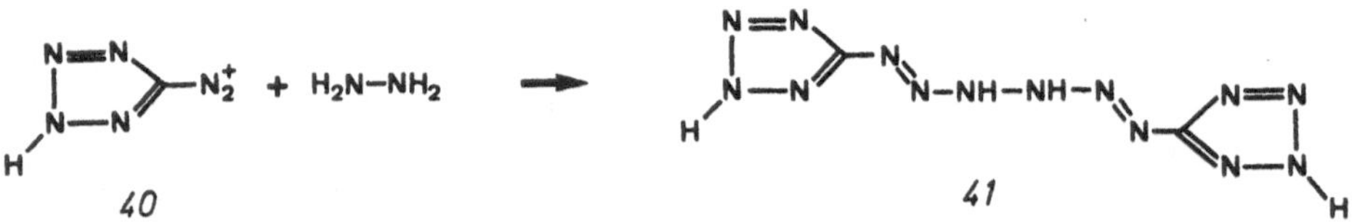

(23)

An interesting example of how the reaction conditions can influence the structure of the product is given in (24). Depending on the acidity of the reaction medium and on the reaction time, the diazotization of aminotriazoles *42* yields the nitrosamines *43*, the chloro compounds *44*, or the azo coupling product, i.e. the diazoamino compounds *45* [63].

42 → 43: $NaNO_2/HCl$ (18%), 30 min

42 → 44: $NaNO_2/HCl$ (20%), 12 hr

42 → 45: $NaNO_2/HCl$ (5–10%)

(24)

$R = n\text{-alkyl}$; $Ar = p\text{-}X\text{-}C_6H_4$

Most diazotized heterocyclic amines undergo normal coupling reactions with common aromatic coupling components, as well as with some C—H acidic compounds [57]. This pertains for five-membered ring systems such as pyrrols *46* [64], pyrazoles *47* [65–67], imidazoles *48* [68], 1,2,4-(*49*) [69,70] and 1,2,3-triazoles *50* [71], tetrazoles *51* [59], thiazoles *52* [72], thiadiazoles *53* [72], furans *54* [73] and thiophanes *55* [74], as well as for six-membered ring systems such as pyridines *56* [75,76] and pyrimidines *57* [76].

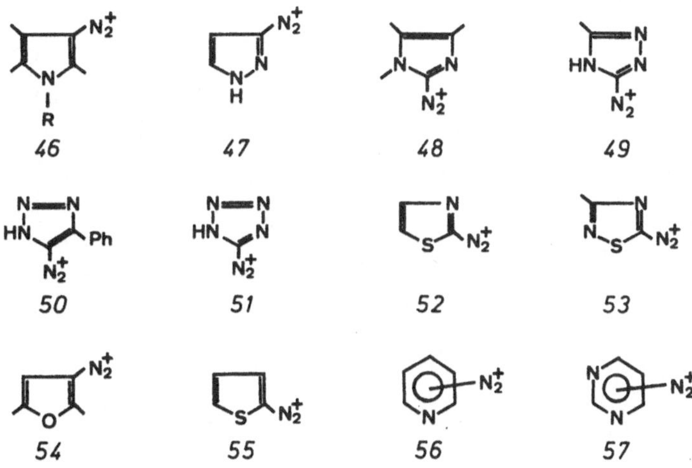

Reactions of potentially high synthetic utility are intramolecular azo coupling reactions of heterocyclic diazonium ions to give new fused-ring heterocycles. Some examples are given in (25–27) [77–79]. The use of heterocyclic diazo compounds in organic synthesis has recently been reviewed by Tišler and Stanovink [80].

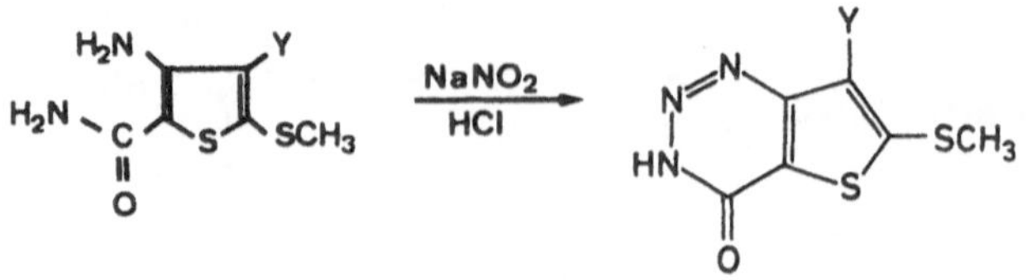

(25)

Y = RCO, CN

(26)

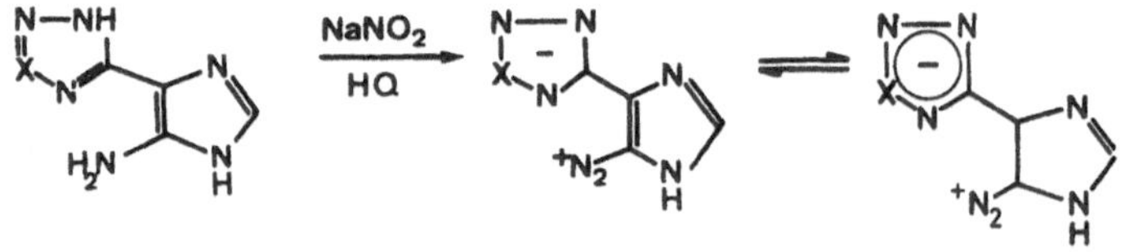

(27)

An interesting cyclization reaction following azo coupling of the diazoimidazole *58* with reactive methylene substrates to finally yield the 1,2,4-triazine *59* has recently been reported [81] (28).

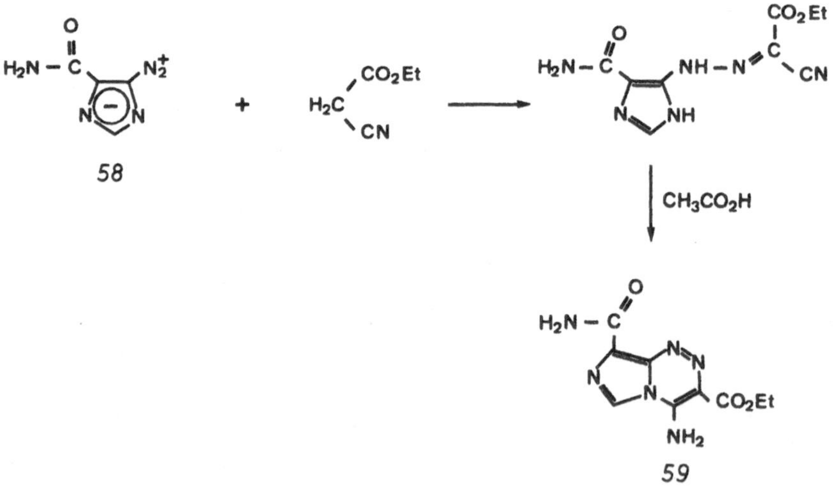

(28)

Diazodicyanoimidazole *60* undergoes a variety of reactions, two of which proceed without loss of nitrogen [81], and which seem to be an azo coupling reaction and a dipolar cycloaddition (29).

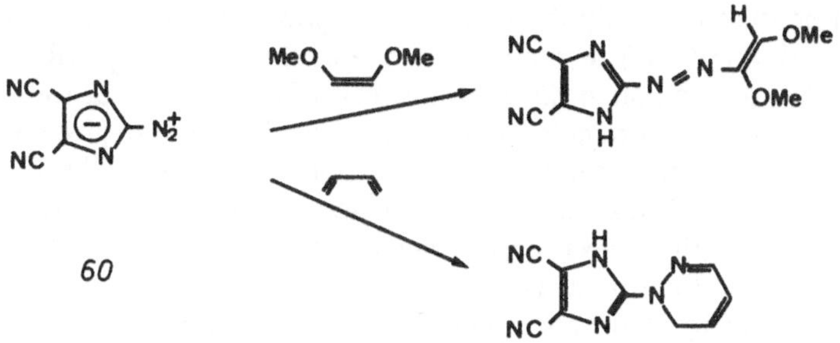

(29)

60

2.2 Coupling Components

2.2.1 Compounds with Activated Methyl, Methylene and Methine Groups

In azo coupling reactions diazonium and diazo compounds act as electrophilic reagents and will therefore attack the atom of the coupling component which exhibits the greatest nucleophilicity. Since diazo and diazonium species are relatively weak electrophiles, in order to be suitable as coupling components, compounds must have a structure where quite a high electron density can be built up on the reacting carbon atom. Therefore diazonium ions only react with aliphatic carbon atoms activated by neighboring electron withdrawing groups (usually acyl or nitro). This is not surprising since it is known that the conjugate base of the coupling component is the actual reactive species [82-87] (see Sect. 4.1).

Azo coupling reactions with activated C—H groups have been reviewed previously [88-90]. Generally the coupling of activated aliphatic carbon atoms with benzenediazonium ions is known as the Japp-Klingemann synthesis [91-93] (30).

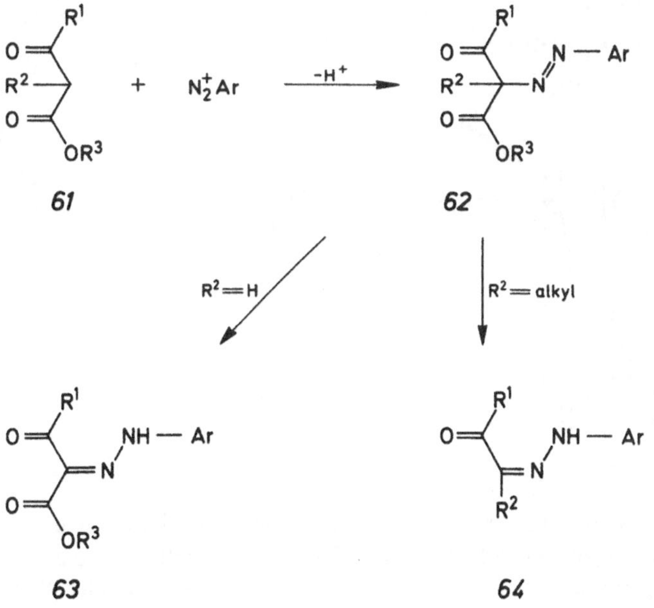

61 62

(30)

63 64

The final product is usually the more stable hydrazone *63* or *64*, formed either on tautomerization ($R^2 = H$) or removal of one of the acyl groups (R^2 = alkyl). The intermediate azo compound *62*, however, can often be isolated, too.

Acetone reacts with two moles of diazonium ion at pH 6–9 to give the formazan *66* (the hydrazone *65* is the final product under acidic conditions) [85] (31). The reaction with the first mole of diazonium ion is rate determining, and the observed coupling rate with the ω-methylglyoxal arylhydrazone *65* is, due to its higher acidity, faster than that with acetone by about seven orders of magnitude [87]. A similar observation was made by Wahl and Lebris [94–96] for the azo coupling with 2-methylbenzthiazole.

$$CH_3{-}\overset{O}{\overset{\|}{C}}{-}CH_3 \xrightleftharpoons{-H^+} CH_3{-}\overset{O^-}{\overset{|}{C}}{=}CH_2 \xrightarrow{ArN_2^+} CH_3{-}\overset{O}{\overset{\|}{C}}{-}CH_2{-}N{=}N{-}Ar$$

$$\underset{66}{CH_3{-}\overset{O}{\overset{\|}{C}}{-}\underset{N=N-Ar}{\underset{|}{C}}{=}N{-}NH{-}Ar} \xleftarrow[\text{fast}]{ArN_2^+} \underset{65}{CH_3{-}\overset{O}{\overset{\|}{C}}{-}CH{=}N{-}NH{-}Ar} \qquad (31)$$

An interesting case of intramolecular azo coupling to an activated methyl group is the well-known synthesis of indazoles *72* starting from o-toluenediazonium salts *69* [97] or from N-nitroso-N-acetyl-o-toluidine *67* [97,98]. Recently the reaction mechanism (32) was elucidated [99]. When the reaction is run in the presence of D_2O or

(32)

CH$_3$COOD, no deuterium is incorporated in the diazonium salt *69*. On the other hand, the synthesis of indazol *74* starting from the optically active diazonium salt *73* results in complete racemization (33). Therefore it can be concluded that the reaction proceeds via intermediate *70* by an intramolecular azo coupling, which is faster than reprotonation to give *69*. The direct route from *68* to *71* can be excluded [99].

$$\text{73} \longrightarrow \text{74 (racemic)} \qquad (33)$$

Some of the examples where a diazo or a diazonium compound reacts with activated double bonds, specifically with enamines [37,40] and with 1,2-dimethoxyethene [81] have been mentioned in sections 2.1.2 and 2.1.4, respectively (12, 29). A similar reaction is the azo coupling of arenediazonium salts with tropolones (34) [100]. The diazonio group reacts at the C-5 atom, so the reaction is used as a diagnostic tool for the presence of substituents at C-5, as well as for the synthesis of 5-amino tropolones by reduction of the azo compound.

(34)

In view of the many coupling reactions of conjugate bases of activated C—H compounds with diazonium and diazo species, it is not surprising that carbanions from other sources can act as coupling components, too, e.g. in (35–37) [101–103].

$$ArN_2^+ + R_2Zn \xrightarrow{DMF} Ar{-}N{=}N{-}R \qquad (35)$$

$$R{-}C{\equiv}C^- + ArN_2^+ \rightarrow R{-}C{\equiv}C{-}N{=}N{-}Ar \xrightarrow{H_2O} R{-}\overset{O}{\overset{\|}{C}}{-}CH{=}N{-}NH{-}Ar \qquad (36)$$

$$(CF_3)_2C = CF_2 \overset{F^-}{\rightleftarrows} (CF_3)_3C^- \xrightarrow{ArN_2^+} (CF_3)_3C{-}N{=}N{-}Ar \qquad (37)$$

2.2.2 Heterocyclic Compounds

Heterocyclic coupling components have, in a similar manner as heterocyclic diazo components (sect. 2.1.4), recently become very important and are used in large scale industrial production of dyestuffs and chemicals. A recent review [104] deals with the industrial use of pyrazolones *75*, iminopyrazoles *76*, pyridones *77*, aminopyrimidines *78*, aminopyridines *79*, hydrochinolines *80* and aminothiazoles *81* as coupling components.

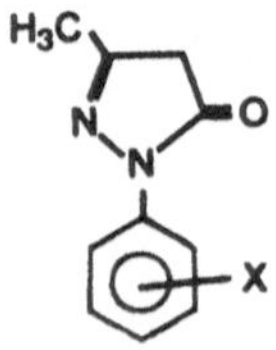

75 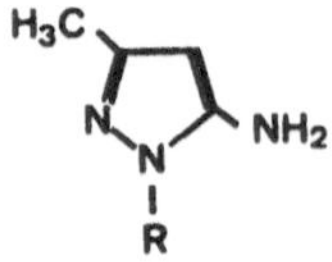 76 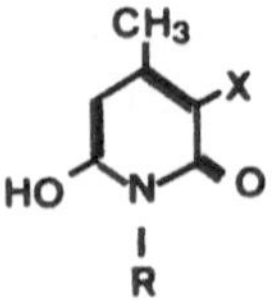 77 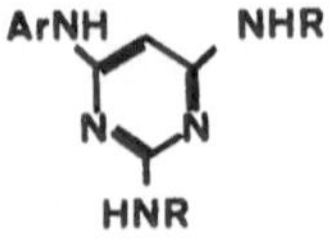 78

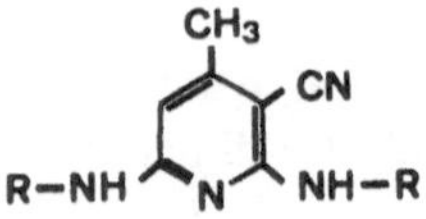

79 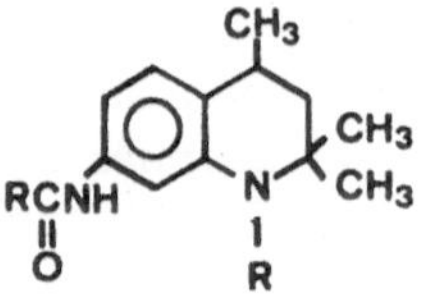 80 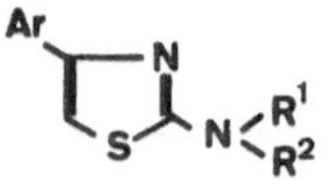 81

Recently the interaction of arenediazonium ions with some purine bases, nucleosides and nucleotides was investigated [105,106]. Adenine, adenosine and 5′-adenylic acid react with arenediazonium ions in neutral and basic solution to yield N-6 triazenes *82*, which readily decompose in basic solution to give 8-aryladenines *83* via intermolecular free-radical substitution reactions in which the ribose moiety is removed from the purine (38) [106]. Under the same conditions guanine (*84*, R=H) reacts with diazonium ions to yield the 8-azo coupling products *85* in 60–85% yield,

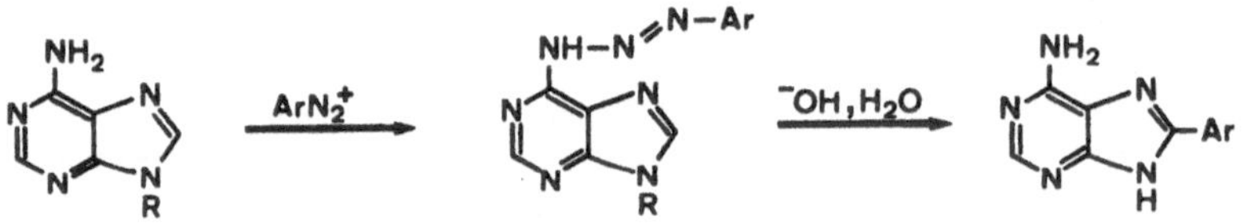

ArN_2^+ → *82* —$^-OH, H_2O$→ *83* (38)

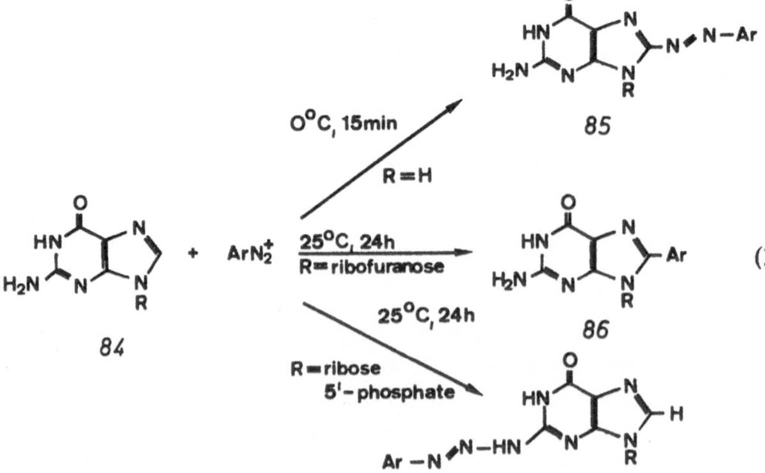

84 + ArN_2^+ —0°C, 15min, R=H→ *85*; —25°C, 24h, R=ribofuranose→ *86*; —25°C, 24h, R=ribose 5′-phosphate→ *87* (39)

guanosine (*84*, R = ribofuranose) reacts much more slowly to give 8-arylguanosines *86* in 20–30% yield, while 5'-guanylic acid (*84*, R = ribose-5'-phosphate) reacts even more slowly to yield about 10% of the triazenes *87* with very electronegative diazonium ions (39) [105]. In all cases the reactive heterocyclic species is the conjugate anion. The reasons for the different reactivity of the various coupling components are discussed [105]. Reactions (38) and (39) are of biological interest in view of the known carcinogenic activity of nitrosamines and alkylnitrosoureas, which can yield unstable diazonium ions that react with DNA and RNA [107].

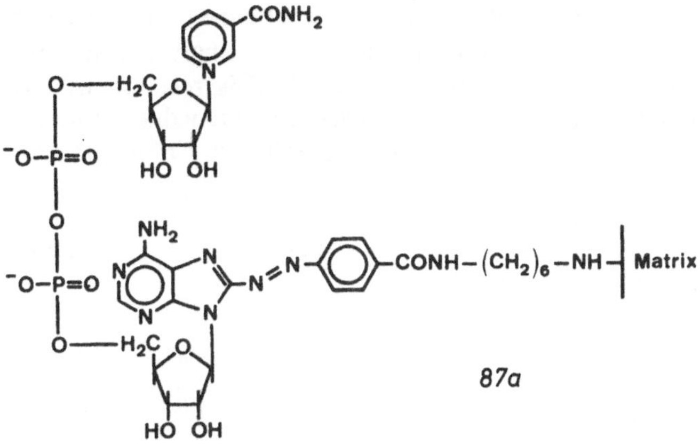

87a

The fact that adenosine and its derivatives are azo coupling components is used for immobilizing nicotinamide-adenine nucleotide (NAD^+) for affinity chromatography purposes. In *87a* NAD^+ is bonded to a matrix through an azo bond. *87a* is used for the purification of dehydrogenase enzymes [107a].

Butler and coworkers have studied the reactions of arenediazonium ions with pyrroles [108–110]. Pyrroles are known to react readily with diazonium ions under acid conditions to give the monoazo compound [111]. 1-Methyl- and 2,5-dimethylpyrrole react with diazotized sulfanilic acid to give *88* and *89*, respectively [108], in

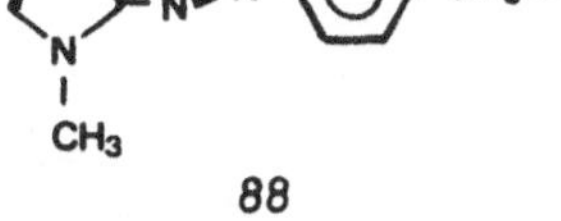

88

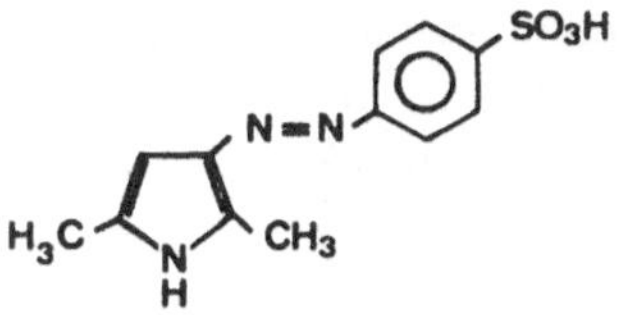

89

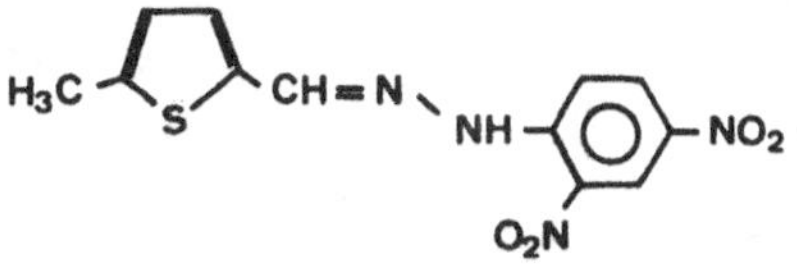

90

accordance with the α-positions in pyrrole being more susceptible to electrophilic attack than the β-positions [112]. When the α-positions in pyrrole are blocked by methyl groups, the coupling occurs at the β-position to give *89*. For 2,5-dimethylthiophene, on the other hand, the reaction with 2,4-dinitrobenzenediazonium ion results in partial coupling at the methyl group to yield the hydrazone *90* [113]. A detailed kinetic investigation [108] of the azo coupling of substituted benzenediazonium ions with eleven different pyrroles is consistent with the S_E2 mechanism of electrophilic aromatic substitution (see Sect. 4.3). The reactive species are the unprotonated pyrroles. Linear Hammett plots were obtained and the ϱ values for the different pyrrole systems varied in the range 1.2–1.6 [108], belonging to the lowest values observed for azo coupling reactions (see Sect. 4.6). Analytically pure samples of the azo compounds of type *88* and *89* could not be obtained from reactions in aqueous solution, but were easily prepared in chloroform where the diazonium salts were solubilized with dicyclohexano-18-crown-6 [109]. This reaction will be discussed in Section 2.3. Dipyrrylmethanes *91* also react with arenediazonium ions in aqueous acid to give azopyrroles *92* and formaldehyde (40) [110]. At low pH a mechanism with

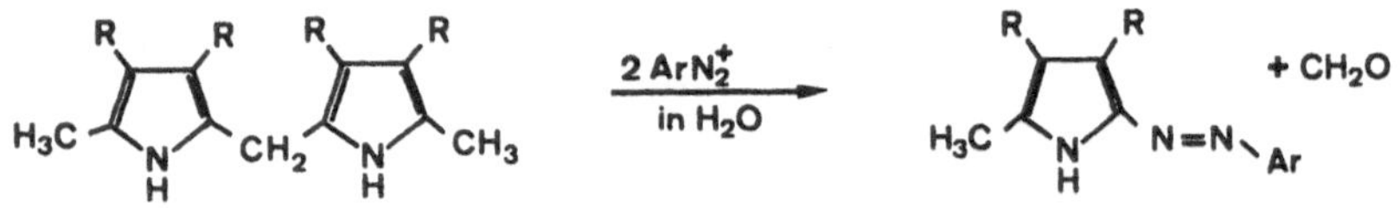

$$91 \xrightarrow[\text{in } H_2O]{2\,ArN_2^+} 92 + CH_2O \qquad (40)$$

R = H or CH_3

fast cleavage of the methylene bridge brought about by attack of H^+ (to give CH_2O and two pyrrole molecules), followed by slow coupling with pyrrole monomer is proposed. At higher pH the methylene bridge is cleaved by diazonium ions and there is a change in mechanism [110].

Several papers have appeared recently dealing with azo coupling reactions with various indole systems as coupling components [114–118].

The rates of azo coupling at the 3-position for a series of 2-, 2,4- and 2,4,6-substituted indoles *93* (41) with p-substituted arenediazonium ions in mixed aqueous and some aprotic solvents are reported [114]. The pH-dependence of these azo coupling reactions will be discussed in Section 4.1 and their kinetic hydrogen isotope effects in Section 4.3.

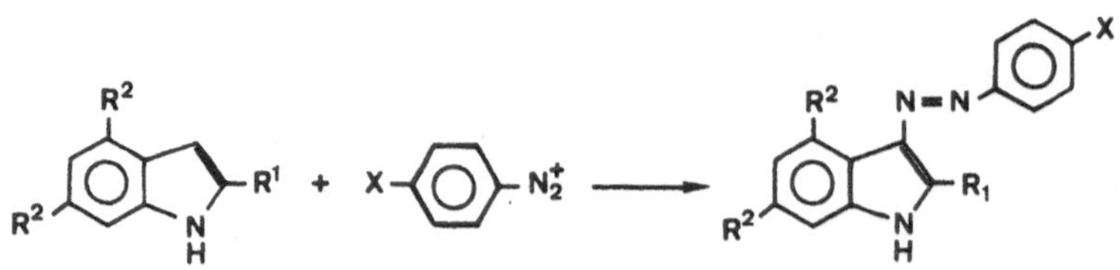

$$93 + X\text{-}C_6H_4\text{-}N_2^+ \longrightarrow \text{3-(4-X-phenylazo)indole} \qquad (41)$$

R^1 = Me, ^{t}Bu

R^2 = H, Me, ^{t}Bu

Genkina and coworkers [115,116] investigated the comparative reactivity of isomeric benzindoles and of 4,5,6,7-dibenzindole *94* in azo coupling reactions. They found that the dibenzindole *94* was more reactive toward azo coupling at C-3 than indole, 4,5-benzindole or 6,7-benzindole, contrary to the predictions based on electron density at C-3. This is another example that it may be dangerous making conclusions about the relative reactivities of compounds solely on the basis of the respective electron densities at the reactive center.

Recently the formation of an unusual triazene derivative *99* was reported in the coupling reaction of skatole *95* (R=CH_3) (42) [118]. Based on the change of the UV spectrum of the reaction mixture with time, a mechanism involving primary formation of the azo compound *97* followed by the hydrolytic cleavage of the indolenine system and reclosing of the ring to finally give *99* was proposed (42) [117].

94 95 96 97 (42) 99 98

2.2.3 Isocyclic Aromatic Substrates

The term 'coupling component' (as well as 'diazo component') originated in the azo dye technology. It refers to the nucleophilic reagent in an azo coupling reaction. Originally it referred to substituted aromatic hydrocarbons, particularly benzene and naphtalene derivatives, substituted by a hydroxy or an amino group. This reaction has been studied in great detail (see Chapter 4). The aromatic coupling component must be activated (X=NR_2 or OR). If X=OH, the reactive form of the substrate is usually the conjugate base (see Sect. 4.1). Therefore the control of pH in azo coupling reactions is important, even more so since the concentration of the active diazonium form decreases rapidly in basic solution. For coupling with phenols the maximum rate is observed at pH $\sim$ 10. In more acidic medium the measured rate constant is inversely proportional to $[H^+]$ because of the depletion of the reactive phenolate ion, while in base it is inversely proportional to $[HO^-]^2$ due to the conversion of the diazonium ion to the unreactive diazotate [119].

Normally, the preferred site of coupling is para to the activating group. This will be discussed in more detail in Sections 4.3 and 4.5. The reactivity order for benzene or naphthalene coupling components is $O^- > NR_2 > NHR > OR \sim OH \gg Me$.

Thus the phenolates and enolates are 7 to 10 orders of magnitude more reactive than their conjugate acids (enols, phenols)[120], as discussed in more detail in Section 4.1. In addition to H^+, other groups such as SO_3, Cl, Br and I can also be displaced in azo coupling reactions with aromatic substrates[121,122].

In coupling reactions with aromatic amines as coupling components, two reaction pathways are possible: N-coupling to give diazoamino compounds *100*, and C-coupling with the formation of aminoazo compounds *101* (43). Usually primary

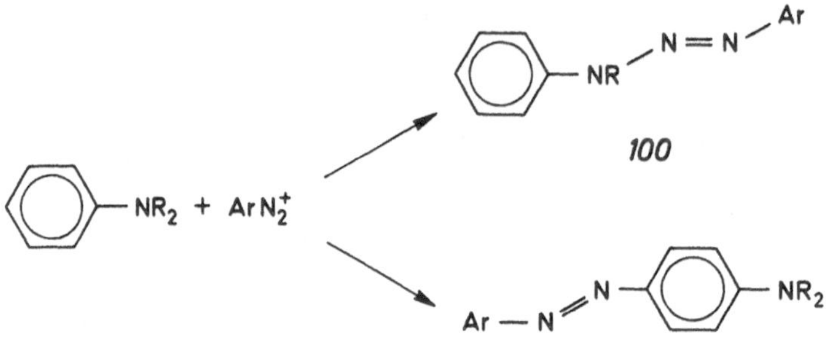

(43)

aromatic amines form diazoamino compounds *100*, except when the nucleophilic reactivity of the aryl residue is raised by substituents or fused rings, when aminoazo formation takes place[123]. In aminoazo formation, however, the initial attack of the diazonium ion may still be at the amino nitrogen atom, but the decomposition of the σ_N-complex might be too rapid to allow its identification. This will be discussed in more detail in Section 4.4.

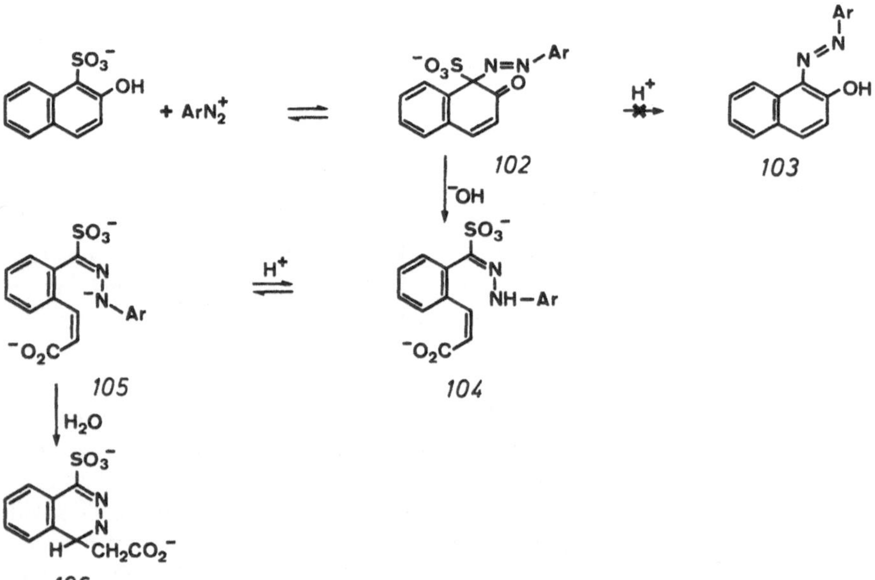

An interesting reaction of diazonium salts with 2-naphthol-1-sulfonic acid was observed[124] in the pH range 10–15. Instead of the usual diazodesulfonation reaction to give the azo dyes *103*, addition of a hydroxide ion to the σ-complex *102* takes place and the cinnamic acids *104* are formed, which finally undergo an intramolecular cyclization to give the heterocycles *106*. A simplified mechanism is shown in (44) [124]. In this reaction the fragmentation of the the σ-complex *102* competes successfully with the release of the electrofugic leaving group SO_3. When the leaving group is a proton (2-naphthol as coupling component), no fragmentation is observed.

3 Influence of the Reaction Medium on Azo Coupling Reactions

3.1 Azo Coupling Reactions in Organic Solvents

Compared with the abundance of data dealing with azo coupling reactions in aqueous solutions, relatively few investigations deal with reactions in organic solvents. The first reports were simply stating that specific diazonium salts, which are soluble in benzene [125] and in dipolar aprotic solvents [126], can react with coupling components in these media.

The first investigation on the role of solvent in azo coupling reactions was performed by Penton and Zollinger [127], who measured the rates for the azo coupling of 4-toluenediazonium salts with N,N-dimethylaniline in water, sulfolane, acetonitrile and nitromethane. They found that the rates only varied within a factor of 5 and did not correlate with the standard solvent parameters. A change of the diazonium salt counterion (HSO_4^- or BF_4^-) did not influence the rate. After the Gutmann donor numbers for solvents [128] became known, Bagal and coworkers [129] performed a more detailed investigation of various p- and m-substituted benzenediazonium salts with N,N-dimethylaniline using a larger variety of solvents. They showed that a linear correlation existed between the logarithms of the rate constants and the Gutmann solvent donor numbers. The sensitivity of the coupling rates to a change of solvent increased with the reactivity of the diazonium ion, ranging from a factor of 30 for p-toluenediazonium ion to as high as 4700 for p-nitrobenzenediazonium ion. Hammett's reaction constants ϱ, as can be expected, also varied with the solvent (from ca. 2.5 in dimethylacetamide to ca. 4.3 in nitromethane), as well as with the reaction temperature.

The solvent effects on azo coupling reactions with different coupling components were studied by Hashida and coworkers [130–133]. No large solvent effect was found in reactions of m-nitrobenzenediazonium ion with 1,3,5-trimethoxybenzene [130]. On the other hand, the rate for the coupling of benzenediazonium tetrafluoroborate with 2-naphthol increased by a factor of ca. 10^8 when the solvent was changed from water to acetonitrile! The reactivity sequence in the solvents under study followed the increase of the pK_a values of 2-naphthol in these solvents. The observed results were interpreted in terms of the difference in solvation of the naphtholate ion [130]. Similar solvent effects to those observed for coupling reactions with 2-naphthol were also found in reactions of p-methoxybenzenediazonium tetrafluoroborate with acetylacetone and imidazole and were rationalized accordingly, i.e. by different solvation of the carbanionic coupling component [132].

The same conclusion was made in a kinetic study of solvent effects in reactions of benzenediazonium tetrafluoroborate with substituted phenols [131]. As expected due to the difference in solvation, the effect of the p-substituents is smaller in protic than in dipolar aprotic solvents. Ortho-alkyl substitution of phenol was found to increase the coupling rate, again as would be expected for electron releasing substituents. However, this rate acceleration was larger in protic than in dipolar aprotic solvents, since in the former case the anion solvation is much stronger to begin with, and therefore steric hindrance to solvation will have a larger effect [131].

Fischer and Zollinger [122] have studied the reaction of p-chlorobenzenediazonium chloride with 2-naphthol-1-sulfonate. A very interesting dependence of the reaction products and reaction rate on the solvent was observed: The initially formed complex *102* decomposes under ring fragmentation in aqueous base [124] (see (44) in Sect. 2.2.3); it does not react in water, ethanol or dimethylsulfoxide, reacts very slowly in acetone or glacial acetic acid to give a small amount of azo compound *103*, while in dichloromethane, chloroform, benzene or pyridine quantitative, though slow, formation of the azo compound is observed. Based on kinetic measurements an S_N1-type mechanism was postulated for the formation of the azo dye *103* from the σ-complex *102* [122].

In several investigations of the azo coupling of arenediazonium ions with aromatic amines in organic solvents general base catalysis and consequently a primary deuterium isotope effect were observed [133–135].

Messmer and Szimán [135] measured the coupling rates of p-methoxybenzenediazonium tetrafluoroborate with N,N-dimethylaniline, m-toluidine and 1- and 2-naphthylamine in nitrobenzene as solvent. Except for reactions with m-toluidine, the reaction rates were reduced significantly if a deuterium ion had to be displaced instead of a proton. The kinetic isotope effect varied from $k_H/k_D = 1.5$ for 2,4,6-d_3-

$$Ar{-}N_2^+ + C_6H_5{-}NR_2 \underset{k_{-1}}{\overset{k_1}{\rightleftharpoons}} \underset{107}{[\text{Ar}{-}N{=}N(H)C_6H_5^{+}{-}NR_2]} \xrightarrow{k_2\,[\text{Base}]} Ar{-}N{=}N{-}C_6H_4{-}NR_2 + \text{BaseH}^+ \qquad (45)$$

N,N-dimethylaniline to $k_H/k_D = 4.4$ for 1-d-2-naphthylamine. The azo coupling of p-tert-butylbenzenediazonium ion with 4-d-N,N-dimethylaniline in 1,2-dichloroethane showed an isotope effect $k_H/k_D = 1.96$ [134]. The results were consistent with the S_E2 mechanism if the rate limiting step is proton transfer from the σ-complex

107 to a second molecule of amine acting as a base (45)[2] [134,135]. The similarity of the deuterium isotope effect values indicates transition states with approximately the same structure in the two aprotic solvents of differing polarity.

Hashida and coworkers could nicely establish the importance of different factors which can influence the mechanism shown in (45). They investigated the base catalysis (by pyridine) in azo coupling reactions of benzenediazonium ion with N,N-dialkylanilines in six organic solvents [133], and could demonstrate the role of solvent as a proton acceptor from the σ-complex intermediate *107*. In solvents of relatively high basicity such as methanol and dimethylsulfoxide, neither significant base catalysis nor an isotope effect was observed. On the other hand, in less basic solvents such as nitromethane, the reaction was strongly accelerated by base and showed a considerable deuterium isotope effect (up to $k_H/k_D = 2.8$). Thus a change in solvent can induce a change in the rate limiting step of the reaction. Reducing the basicity of the reaction medium makes the second step in (45) slower relative to the first step. A reduction of the concentration of the amine coupling component has, not surprisingly, the same effect. This is manifested by the observation that the kinetic deuterium isotope effect for the azo coupling of benzenediazonium ion to N.N-bis-(2-hydroxyethyl)aniline in nitromethane changes from $k_H/k_D = 0.99$ to $k_H/k_D = 2.8$ if the concentration of the aniline is reduced from 10^{-1} to 10^{-3} molar.

In conclusion it can be said that a change of the solvent can influence different azo coupling reactions in very different ways, depending on which step of the S_E2 mechanism is rate limiting, whether the coupling component is an anion or a neutral species etc.

3.2 Reactions Catalyzed by Micelles

Micellar catalysis of azo coupling reactions is possible; it has been studied recently by Poindexter and McKay [136] and in more detail by Rufer [137]. The first mentioned authors studied the reaction of 4-nitrobenzenediazonium salt with 2-naphthol-6-sulfonic and 2-naphthol-3,6-disulfonic acids in the presence of sodium dodecylsulfate and hexadecyltrimethylammonium bromide. With the anionic as well as with the cationic additive an inhibition (up to 15-fold) was observed. This result was to be expected based on the principles of micellar catalysis, since the charges of the two reacting species are opposite to one another.

Therefore Rufer studied an azo coupling reaction with uncharged coupling components, namely that of p-methoxybenzenediazonium ions with 1-naphthylamine and 1-amino-2-methylnaphthalene, buffered to pH 7 at 25 °C. Both reactions are strongly catalyzed by sodium dodecylsulfate (up to 1100 times with 1-amino-2-methylnaphthalene) in cases where the surfactant concentration was higher than the critical micelle concentration (CMC). For the coupling reaction with 1-naphthylamine, both the o/p-ratio and the amount of the 2,4-bisphenylazo product are dependent on the sodium dodecylsulfate concentration. The formation of the 2,4-bisphenylazo com-

2 For a detailed discussion of general base catalysis and kinetic hydrogen isotope effects in azo coupling reactions, see Section 4.3. The problem of competitive N-coupling of diazonium ions with aromatic amines and its implications on the mechanism shown in (45) is treated in Section 4.4.

pound indicates that mixing effects are present (see Sect. 4.2). Reactions with 4-sulfobenzenediazonium zwitterions are only slightly influenced by added surfactant.

The catalytic effect can be explained by the increased concentrations of both reagents in the neighborhood of the micelles: The diazonium ion is attracted by the opposite charge of the micelles and the naphthylamine by solubilization in the micelles. Rufer succeeded in treating these effects not only qualitatively, but also on a quantitative basis.

Hashida and coworkers [138] have studied the effects of cationic, anionic and nonionic surfactants on azo coupling reactions. By using diazonium ions (*108–110*) as well as coupling components (*111*, *112*) bearing a different overall charge, they could confirm the statement that micellar effects are governed by the formal charge of both reactants.

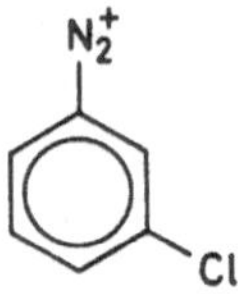

108  *109* 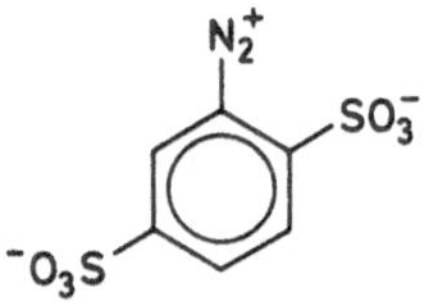 *110* 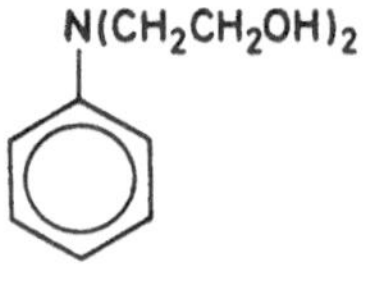 *111* 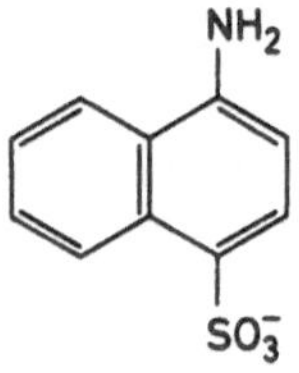 *112*

If the two reacting species have opposite formal charge, the reaction rate is usually depressed by cationic as well as by anionic micellar reagents, as was shown previously [136]. This is due to the fact that either of the reagents will, for electrostatic reasons, be excluded from the micelle. This electrostatic impasse was successfully circumvented by functionalizing a cationic surfactant with an arenediazonium ion moiety [139]. The azo coupling with the micellar diazonium ion surfactant *113* was faster than the azo coupling with the model diazonium ion *114* by factors of 122 and 244, in reactions with 2-naphthol and with 2-naphthol-6-sulfonate, respectively.

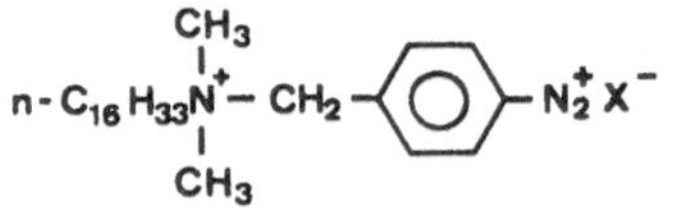

113 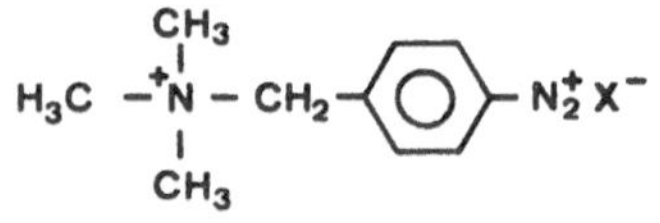 *114*

In the context of Hashida's work with diazonium ions *108–110*, i.e. compounds with net charge $Z_D = +1, 0$ and -1 respectively, which react with coupling components *111* and *112* of net charge $Z_C = 0$ and -1 in the presence of surfactants, it should be mentioned that the question of the contribution of Coulombic attraction ($Z_D = +1$ reacts with $Z_C = -1$) versus Coulombic repulsion ($Z_D = -1$ reacts with $Z_C = -1$) to the measured rate in water without additives has not been answered yet. The problem is, however, quite complicated, since a change in charge also implies a change of substituents. The contribution of the Coulombic interaction to the total effect of substituents (mesomeric, inductive, steric, hyperconjugation etc.) has, as far as we know, not been investigated in detail. However, the importance of Z_D and Z_C for Brönsted salt effects on azo coupling reactions has been well known for many years [140].

3.3 Phase Transfer Catalysis

Phase transfer catalysis is nowadays a well-established technique in preparative chemistry. However, it has been confined almost exclusively to nucleophilic substitution reactions. Nevertheless some examples of phase transfer catalyzed azo coupling reactions are known.

Cram, Gokel and coworkers [141–144] have shown that diazonium tetrafluoroborates can be solubilized in non-polar solvents with crown ethers and in some cases also by simple anion exchange using a equimolar amount of tetramethylammonium chloride. In this way they could obtain azo dyes [141–143] and arenediazocyanides [144] in good yield in various non-polar solvents. In a similar way analytically pure azopyrroles, which could not be prepared by other methods, were made in chloroform using dicyclohexano-18-crown-6 as solid-liquid phase transfer catalyst [109]. The first products formed were the tetrafluoroborate salts *115* and *116* of azopyrroles, which were easily deprotonated with aqueous ammonia.

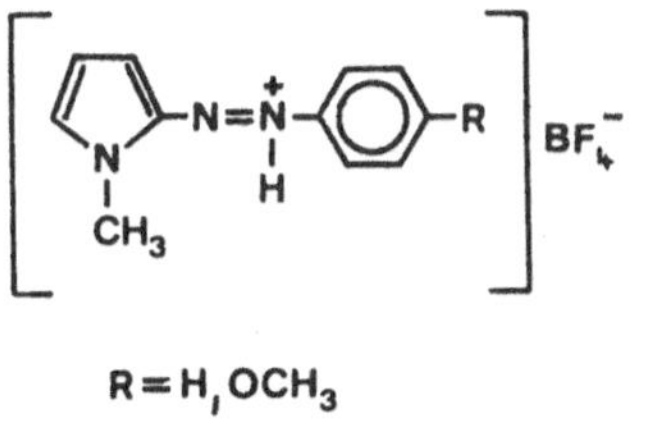

R = H, OCH_3

115

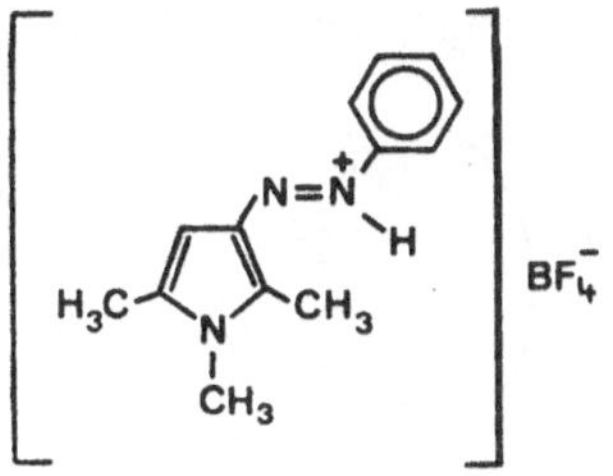

116

Juri and Bartsch [134] found that azo coupling reactions in the presence of crown ethers are slower than the reaction without this additive. This result indicates that the reactivity of crown-complexed diazonium ions in azo coupling reactions is much lower than that of the free diazonium ion. It may even be zero since Cram et al. [141] found that no rotaxane-type azo product, i.e. azo compound with a 'collar-like' crown ether ring around the azo bridge, was formed. The observation of better yields of azo products, or the formation of azopyrroles (above) in the presence of crown ethers, seems to be a true phase-transfer effect of the two reaction components into the organic phase. In water the azo coupling is probably less favourable in these cases, either because of low solubility of one of the reagents, or because of dominant competitive decomposition of the diazonium ion (homolytic and heterolytic dediazonation [145]).

Recently Griffiths and coworkers reported the first example of a liquid-liquid phase transfer catalysis in an electrophilic substitution reaction [146,147]. They showed that the coupling reactions between 4-nitrobenzenediazonium chloride and N-ethylcarbazole *117* (46) or N,N-diphenylamine in aqueous media are accelerated by using a two-phase water-dichloromethane system containing sodium 4-dodecylbenzene-sulfonate as a transfer catalyst for the diazonium ion. The greatest rate acceleration

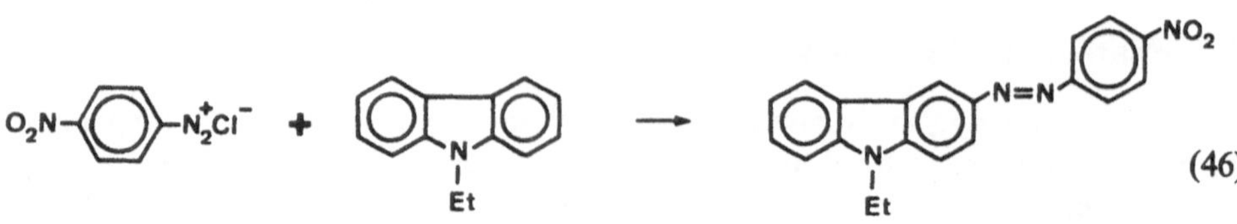

(46)

117

by a factor of ca. 50 was observed for the reaction shown in (46). Several amine coupling components were tried out, but phase transfer catalysis was not a general phenomenon and was only observed with amines that have a low rate of N-coupling, as well as a low rate of C-coupling. Different water-organic solvent systems were used, and it was shown that the polarity of the organic phase is important, dichloromethane or nitrobenzene being much better solvents than toluene or ethyl acetate [146)].

While the phase transfer catalyst used by Griffith's group is a well known surfactant, which may complicate the reactions by micelle formation, Kobayashi, Sonoda and Iwamoto [148)] used sodium tetrakis[3,5-di(F-methyl)phenyl]borate *118*, which, due to its spherical shape, does not form micelles.

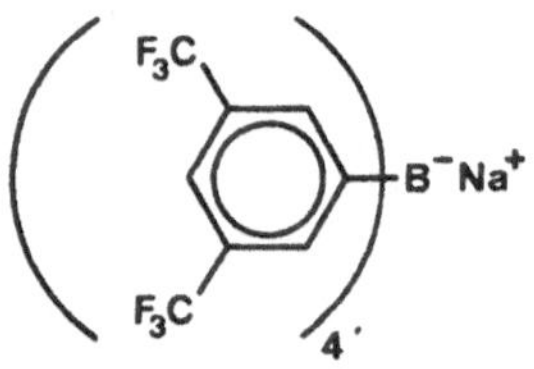

118

Again a remarkable rate acceleration was observed by use of the boron compound *118* as a phase transfer catalyst for arenediazonium tetrafluoroborates (Ar=Ph, p-$NO_2C_6H_4$) with a range of coupling components (aromatic amines and phenol) in liquid-solid (benzene-ArN_2^+) and liquid-liquid (H_2O—CH_2Cl_2) two-phase systems. As opposed to the results of Griffiths and coworkers [146)], the reactions in the presence of borate *118* were accelerated irrespective of the reactivity of the coupling component. Amine coupling components gave N-coupling products only in the absence of appropriate deprotonating agents, while in most cases C-coupling products were isolated in good yields (75÷85%) [148)].

4 Mechanisms

4.1 Acid-base Pre-equilibria

It is well known that the rates of all azo coupling reactions in aqueous or partly aqueous solutions are highly dependent on acidity. Conant and Peterson [149)] made the first quantitative investigation of this problem. They demonstrated that the rate of coupling of a series of naphtholsulfonic acids is proportional to the ^{-}OH concentration in the range pH 4.50 to 9.15. They concluded that the substitution proper is preceded by an acid-base equilibrium in *one* of the two reactants which was assumed to be the diazohydroxide-diazonium ion equilibrium, or in other words, that the reacting equilibrium forms are the undissociated naphthol and the diazohydroxide.

We mention Conant and Peterson's classical work more than 50 years later for two reasons. First, because of their correct conclusion with regard to the involvement of *one* acid-base equilibium; second, because of their incorrect interpretation. As discussed later by Wistar and Bartlett [150], the same pH-dependence is obtained assuming the naphtholate-naphthol equilibrium. Pütter [151] as well as Zollinger and Wittwer [119, 152] extended the pH-range of rate measurements and Zollinger [140, 153] investigated the influence of ionic strength (salt effects) on the rates of azo coupling reactions. All their results showed clearly that the diazonium ion reacts with naphtholate ions (phenolate, enolate) and with (unprotonated) aromatic amines.

Nowadays it is of course common knowledge that in all electrophilic substitutions, *it is* the most acidic equilibrium form of the electrophilic reagent and the most basic form of the nucleophilic substrate which give rise to the highest rate of the substitution proper. The first positive evidence for this rule came from the investigations of azo coupling mentioned above. Azo coupling reactions were particularly suitable for the development of this rule because these reactions could be studied in aqueous solutions within conventional pH ranges. Respective investigations on other electrophilic aromatic substitutions, e.g. nitrations, sulfonations or Friedel-Crafts-reactions can only be made in much more complex systems.

We call [152] the measured, second-order rate constant k_s. The subscript 's' refers to the fact that this constant was calculated on the basis of the analytical ('stoichiometric') concentrations of reagents, the diazo and the coupling components, not taking into account the pre-equilibria. If, however, we refer to the intrinsic rate constant of the substitution proper, we call that constant k. For azo coupling of a diazo component with 1-naphthol, for example, k refers to the fractions of the two reagents which are present as diazonium ion and naphtholate ion, respectively. These are calculated with the help of the equilibrium constants $(pK_1 + pK_2)/2$ and pK_a for the pH value of the solution under investigation. Intrinsic rates for various steps are designated with respective subscripts, e.g. k_1, k_{-1}, k_2 etc.

The investigations summarized earlier do not eliminate the possibility that *other* equilibrium forms than the diazonium ion as well as the conjugate bases of phenols, naphthols and enols also react. Until now no evidence is available for the reactivity of the *cis*- and *trans*-diazohydroxides and the respective diazotates.

For azo coupling reactions of naphthols, however, semiquantitative work made in the fifties [154, 155] suggested that undissociated naphthol does react with diazonium ions, but at rates which are several orders of magnitude lower than those of naphtholates.

Quantitative evaluations of kinetic measurements using strongly electrophilic diazonium ions and, as coupling components, 1-naphthol, 2-naphthol-6-sulfonic acid and resorcinol in aqueous acid were made in the seventies by Štěrba and coworkers [120, 156–158]. In a typical case (2,6-dichloro-4-nitrobenzenediazonium ion and 1-naphthol) the dependence of the logarithm of the measured rate constant (k_s) on pH was linear with a slope of 1. At pH < 1, however, a practically constant value for k_s was obtained. The measured rate constants correspond therefore to equation 47 in which the first term relates to the reaction of the naphtholate ion and the second to that of the undissociated naphthol. K_a is the acidity constant of 1-naphthol.

$$k_s = k^{O^-} \frac{K_a}{K_a + [H^+]} + k^{OH} \tag{47}$$

Table 1 gives some of Štěrba's results for 1-naphthol, resorcinol, 1-methoxynaphthalene, 3-methoxyphenol and 1,3-dimethoxybenzene.

Table 1. Rate constants ($l \cdot mol^{-1} \cdot s^{-1}$) for azo coupling reactions of l-naphthol and resorcinol, the corresponding anions and some methoxy derivatives in water, 20 °C. Data from Štěrba et al. [120,156–158]

	3-Nitrobenzene-diazonium ion	2-Chloro-4-nitrobenzene-diazonium ion
Resorcinol dianion	2.83×10^9	—
Resorcinol monoanion	3.20×10^6	—
3-Methoxyphenolate anion	3.47×10^6	—
Resorcinol	5.78×10^{-2}	1.75
3-Methoxyphenol	2.63×10^{-2}	5.78×10^{-1}
1,3-Dimethoxybenzene	—	9.58×10^{-2}
1-Naphtholate anion	3.73×10^8	2.47×10^9
1-Naphthol	4.70×10^{-1}	1.92×10^1
1-Methoxynaphthalene	—	2.30×10^{-1} [a]

a in 32% (vol.) acetic acid

The data in Table 1 show that 1-naphtholate anion is 10^8 times more reactive than the undissociated naphthol, which is 10^2 times more reactive than 1-methoxynaphthalene. The rate ratios for the monoanion of resorcinol relative to resorcinol, 3-methoxyphenol and 1,3-dimethoxybenzene are of similar magnitude. The dissociation of both OH groups of resorcinol gives rise to a rate constant ($2.83 \times 10^9\ l\ mol^{-1}\ s^{-1}$) which, in our opinion, is probably mixing or diffusion controlled (see Sect. 4.2).

In a recent paper Cox et al. [159] found, as expected, that the triple anion of 2,3-dihydroxynaphthalene-6-sulfonic acid has a higher reactivity than the double anion. They observed a specific catalytic effect of borate buffers, but unfortunately they did not investigate it further.

There is some evidence that also with acetylacetone as a coupling component, the neutral compound may be reactive in addition to the conjugate anion. The latter has, however, a reactivity which is many powers of ten higher. The evidence for the reactivity of the neutral acetylacetone comes from rate measurements with 2,6-dichloro-4-nitrobenzenediazonium ion at pH less than zero. Štěrba et al. [86] plotted the logarithms of the measured rate constants (k_s) against pH and, below pH 0, against H_0. The plot is linear at pH $\geq$ 0, as expected for the anion acting as substrate. The curve levels off below pH 0. For an anion as nucleophile, however, H_- should be used rather than H_0. Štěrba's data are therefore only qualitatively correct. Unfortunately the incorrect plot does not allow calculation of the rate constant of the neutral molecule.

Štěrba et al. [160] were successful, however, in solving that problem for another azo coupling component, namely 1-phenyl-3-methyl-5-pyrazolone (*119*), which in solution exists as a mixture of tautomeric forms A–C. The conjugate base *120* reacts 10^9 times faster than the neutral compound.

A B C

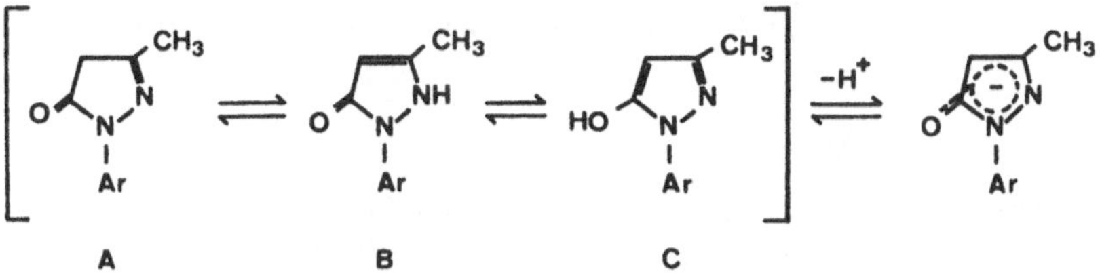

119 *120*

Analogous investigations on the reactivities of the acid and base forms were made for 2-methylindole (*121*), its 4,6-di-t-butyl derivative [114], and for pyrrole (*122*) [161]. In the cases of the indoles, the anion is ca. 2×10^8 times more reactive than the neutral compound. In the case of pyrrole the undissocciated compound was the dominant nucleophile in the pH range from 4.7 to 8.2, whereas in the pH region >10.0 the reactive species was the conjugate base.

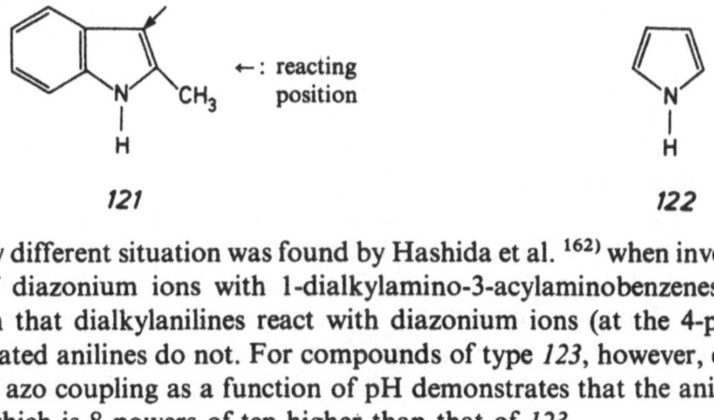

←: reacting position

121 *122*

A slightly different situation was found by Hashida et al. [162] when investigating the reaction of diazonium ions with 1-dialkylamino-3-acylaminobenzenes (*123*). It is well known that dialkylanilines react with diazonium ions (at the 4-position), but that N-acylated anilines do not. For compounds of type *123*, however, evaluation of the rates of azo coupling as a function of pH demonstrates that the anion *124* has a reactivity which is 8 powers of ten higher than that of *123*.

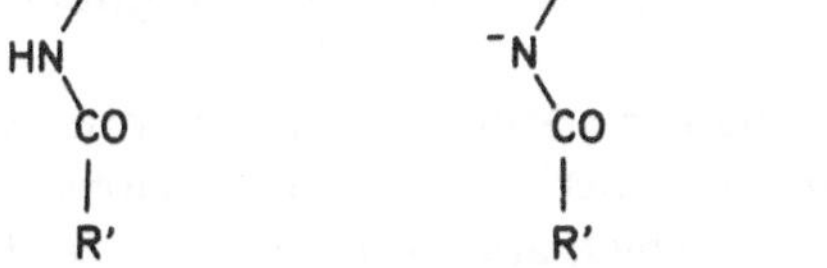

123 *124*

We think that an additional conclusion follows from the work of Hashida et al. It is known that derivatives of aniline (and naphthylamines) of the type *125* react with diazonium ions [163], which is surprising since the groups X are electron-withdrawing and therefore decrease the nucleophilicity of the aromatic ring. On the other hand, however, they increase the acidity of the —NH-group and thus the concentration of the respective anion, which is probably the reactive species. It would be easy to test this hypothesis by investigating the coupling rate of compounds of type *125* as a function of the acidity of the medium.

C_6H_5–NH—X X = $-SO_2C_6H_4CH_3$, -COR, -COAr, -CN, $-NO_2$, $-SO_3^-$

125

The investigations of acid-base pre-equilibria of active methylene compounds (C-acids) as coupling components began in 1968 [83], i.e. about two to three decades later than those on phenols (naphthols) and aromatic amines. The most extensive and comprehensive paper on pre-equilibria in azo coupling of active methylene compounds was published by Hashida's group [164] in 1971. They investigated no fewer than 28 coupling components with respect to the rate dependence on acidity! This series of compounds includes acetoacetanilides substituted in the benzene ring, malonic acid ethyl ester and malonitrile, acetyl- and benzoylacetone, alicyclic β-dicarbonyl compounds (e.g. dimedone), bis-(alkylsulfonyl)methanes (e.g. CH_3SO_2-$CH_2SO_2CH_3$), nitroacetic acid ethyl ester, 1-phenyl-3-methyl-5-pyrazolone and other 5-pyrazolones.

Based on the pH-dependence of the coupling rates (in most cases measured over about one pH unit in the pH range 2.0 to 7.2) it becomes clear that in all these compounds it is the conjugate base which enters the substitution proper. This result actually removes the problem, in the case of mono- and particularly of β-di-carbonyl compounds, of whether it is the keto or the enol form which enters into an electrophilic substitution by diazonium ions, by halogenating agents and many other reagents. The keto *and* the enol form are distinct species, but they have *one* (common) conjugate base! This was made clear quite early by chemists working with quantitative methods in this field, e.g. by Schwarzenbach [165] in 1944, but even today there are many chemists who seem not to be aware of it. Štěrba's work [86] on the reactivity of acetylacetone in its neutral form (besides the highly reactive anion) indicates that it is likely that the reacting neutral species is the enol, not the ketone.

In addition, Hashida et al. [164] found a linear correlation between log k and the pK_a of 15 β-dicarbonyl compounds. This indicates that in these cases the nucleophilicity and the basicity of the anions are closely related. The same result was obtained by Hashida et al. [166] for the azo coupling reactivity of p-substituted phenolate ions.

However, it is well known that this correlation does not always exist. Indeed, Hashida et al. [164] found that two trifluoro derivatives of acylacetones and *all* other active methylene compounds, i.e. those which are not β-dicarbonyls, show no such correlation between reactivity and basicity.

The rates of azo coupling reactions with β-dicarbonyl compounds as a function of pH were also studied by Štěrba's group [84,86,160,167]. Their results are consistent with those of Hashida but cover a smaller range of compounds.

From the dependence of the measured coupling rate (k_s) of acetylactone on the solvent acidity (H_2O, MeOH, Me_2SO, DMF and H_3CCN) it was also found that in all these solvents it is the carbanion which is the most reactive species [132].

The only really different case is the azo coupling reaction of nitroethane investigated by Štěrba et al. [83,168]. With p-nitrobenzenediazonium ion the reaction is zero order with respect to diazonium ion and first order in both nitroethane and base. Obviously the rate-limiting step is the dissociation of nitroethane; the rate of formation of the anion is slower than its subsequent rate of reaction with this diazonium ion. For reactions with diazonium ions of lower reactivity [168] the reaction system (48, 49) with the nitroethane anion as steady state intermediate had to be applied.

$$CH_3CH_2NO_2 + B \underset{k_{-1}}{\overset{k_1}{\rightleftharpoons}} CH_3CHNO_2^- + BH^+ \tag{48}$$

$$CH_3CHNO_2^- + ArN_2^+ \xrightarrow{k_2} \text{Azo product} \tag{49}$$

4.2 Mixing and Diffusion Effects

As early as 1891, Noelting and Grandmougin [169,170] reported that in the azo coupling reaction of equimolar amounts of 1-naphthol and 4-chlorobenzenediazonium ion in water, an appreciable amount of 2,4-bisazo compound is obtained besides the expected 2- and 4-monoarylazo-1-naphthol. The percentage of 2,4-bisazo compound is particularly high under alkaline conditions.

The formation of this product of a consecutive disubstitution is apparently not easy to understand. It is known that an arylazo residue *de*creases the reactivity of a phenol or a naphthol for substitution by a second diazonium ion by 3 to 5 orders of ten. In phenols and naphthol the favorable effect of promoting the dissociation of a proton from the substrate does not work; it was found only for C-acids (see discussion in the preceding Sect. 4.1).

An explanation for this strange effect could be given only 86 years later! It is based on the concept that intrinsic rates of consecutive chemical reactions can be disguised by microdiffusion effects if molecular diffusion is slower than the intrinsic rates of chemical reaction.

This concept was first tested with another electrophilic substitution, which also yields unexpectedly large amounts of the disubstitution product, namely the nitration of durene (1,2,4,5-tetramethylbenzene). For this reaction it has been known since 1870 that an equimolar mixture of a nitrating reagent (e.g. HNO_3 in H_2SO_4) and durene yield 35–40% 3,6-dinitrodurene, 2–5% 3-mononitrodurene, 35–40% unreacted durene and some byproducts. Many chemists investigated this surprising reaction over almost a century. As recently as 1969–71 Zollinger et al. [171,172] found that mixing influenced the ratio of mono- to dinitrodurene. A qualitative explanation was possible with a so-called droplet model [173]:

In mixing two solutions (in the same solvent or in two completely miscible solvents) one has to differentiate between macro- and micromixing. Macromixing refers to the convection by stirring. Mixing with the help of stirring does, however, not lead directly to complete mixing on a molecular level. It is known from hydrodynamics that in macromixing liquid elements, so-called eddies, are formed, whose characteristic radius R lies in the range 10^{-3} to 10^{-2} cm. Their degradation to a micro-homogeneous solution (micromixing) is determined by the rate of *molecular* diffusion. If the chemical reaction proper is faster than this process, diffusion will disguise the rate of the chemical reaction.

These processes are shown qualitatively for the nitration of durene in the droplet model of Figure 1. At time $t = t_0$ it is assumed that the eddy of the solution which contains the nitrating reagent (the reactive equilibrium form: $NO_2^+ = +$ in Figure 1) is present in the durene solution (durene = ○). After the time intervall Δt (less than 10^{-4} s) all nitronium ions at the surface of this eddy have reacted with neighbouring durene molecules, and since the rate of mononitration is much faster than microdiffusion, ideally we expect the eddy to be surrounded by a monolayer of mononitrodurene (M). If now the rate of *di*nitration is also faster than the random diffusion of molecules in the micromixing process, this layer of mononitrodurene will react with the adjacent layer of nitronium ions before a significant number of mononitrodurene molecules diffuses into the durene solution. After the second time interval a layer of dinitrodurene molecules (D) with only few mononitrodurene molecules on its surface formed. Thus a much higher yield of dinitrodurene is found.

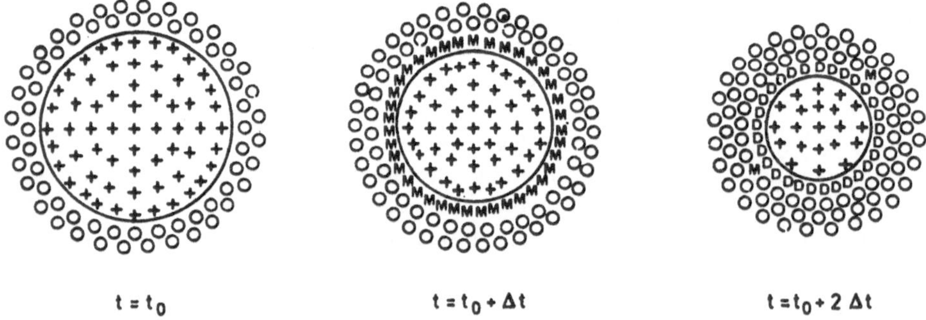

Fig. 1. Droplet model for nitration of durene [173]: + = nitronium ion; ○ = durene; M = mononitrodurene; D = dinitrodurene

This qualitative explanation could be verified by Rys et al. [174, 175] for such competititive, consecutive second-order reactions with the aid of a computer model.

This model predicts that in these reactions the relative yields of the secondary products are dependent on four parameters, namely the stoichiometric ratio of the reagents A and B at time zero (A is the reagent which can be substituted twice), the volume ratio of the solutions of A and B, and the two moduli $\Phi_{B,1}$ and $\Phi_{B,2}$. These moduli are characterized by the mean radius $\bar{R}$ of the eddies, the initial concentration

$[B]_0$ of the reagent B, the intrinsic rate constants k_1 and k_2 and the diffusion coefficient of molecules in the solvent used:

$$\Phi^2_{B,1} = \frac{\bar{R}^2 k_1 [B]_0}{D} \qquad \Phi^2_{B,2} = \frac{\bar{R}^2 k_2 [B]_0}{D}$$

The computation results say that if $k_1 \geqq 10^3 k_2$ and the term $\Phi^2_{B,2}$ has a numerical value >1, mixing effects are probable. Thus with an initial volume ratio of 1, initial concentrations $[A]_0 = [B]_0 = 1M$, $D = 10^{-5}\ cm^2 \cdot s^{-1}$, disguised kinetics are likely to occur if k_2 has values of $1\ l\ mol^{-1}\ s^{-1}$ or more.

This model was verified by Rys et al. [176] first for nitrations of six different di-, tri- and tetramethylbenzenes and subsequently for the azo coupling reaction investigated by Noelting and Grandmougin [169,170] (see above). The theoretical results fitted the analytical results of coupling experiments of 4-chlorobenzenediazonium ion with 1-naphthol very well. The intrinsic rate constants show clearly that mixing effects are to be expected ($k_{2,p} = 51$ to $58\ l\ mol^{-1}\ s^{-1}$ for the reaction of 2-arylazo-1-naphthol at the 4-position; $k_{2,o} = 180$ to $200\ l\ mol^{-1}\ s^{-1}$ for the reaction of 4-arylazo-1-naphthol at the 2-position).

More recently Rys et al. [178,179] started investigations on the influence of pH and buffer concentration on the mixing model. Since in azo coupling reactions a proton is liberated in the substitution proper, the local pH at the edges of eddies will disguise the product ratio further. In addition, chemical engineering aspects, namely mixing-disguised azo coupling reactions in a continuous stirred tank reactor, were investigated recently [180].

Does this model give us a practical solution for the synthesis of monosubstitution products in high yields? The model taught us that reactions are not disguised by micromixing if the intrinsic rate constant k_2 is significantly lower than $1\ l\ mol^{-1}\ s^{-1}$. As discussed in Section 4.1, the intrinsic rate constant refers to unit concentrations of the acid-base equilibrium species, which enter the substitution proper, but not to analytical concentrations. If therefore the azo coupling reaction mentioned above is not run in the range of maximal measured rates (i.e. equilibria on the side of 1-naphtholate ion and diazonium ion) but at lower or higher pH, less bisazo product is obtained. At lower pH the equilibrium concentration of 1-naphtholate ion decreases, at higher pH that of the diazonium ion[3] is reduced.

In several papers from Štěrba's group azo coupling reactions are described for which intrinsic rate constants close to diffusion control ($>10^7\ l\ mol^{-1}\ s^{-1}$) were measured. The coupling components used can, however, not form bisazo compounds. Therefore no mixing effects of the type described in this section were detectable.

4.3 The Substitution Proper of C-Coupling Reactions

For a long time azo coupling was considered to belong to the group of electrophilic aromatic substitutions.

3 Higher pH (pH > 9) is, however, not to be recommended because of the competing homolytic dediazoniation (i.e. decomposition) of diazonium ions [145,181–183].

The discussions in the 50's on the mechanism of these substitutions concentrated mainly on the problem as to whether they were one-step reactions in which the release of hydrogen is associated with the rate-determining part of the reaction (the so-called S_E3 mechanism), or if they were two-step reactions (S_E2) in which the initial attack of the electrophile on the aromatic substrate is followed by the release of the hydrogen ion.

Unequivocal evidence for the S_E2 mechanism came in 1955 in three papers by Zollinger[185–187], in which general base catalysis and primary kinetic hydrogen isotope effects of azo coupling reactions of various types were treated. In particular, such reactions with no isotope effects ($k_H/k_D \cong 1.0$) and no general base catalysis, others with large isotope effects ($k_H/k_D = 6.5$) and (practically) linear base catalysis, and intermediate cases with isotope effects around 3.0 and 'less-than-linear' base catalysis were found.

All these reactions fit the steady-state equation (52) for the S_E2 mechanism (50, 51).

$$Ar'N_2^+ + Ar{-}H \underset{k_{-1}}{\overset{k_1}{\rightleftharpoons}} \overset{+}{Ar}\!\begin{smallmatrix}\diagup H \\ \diagdown N{=}N\diagdown Ar'\end{smallmatrix} \tag{50}$$

σ-complex

$$\overset{+}{Ar}\!\begin{smallmatrix}\diagup H \\ \diagdown N{=}N\diagdown Ar'\end{smallmatrix} + B \xrightarrow{k_2} Ar\diagdown N{=}N\diagdown Ar' + HB^+ \tag{51}$$

$$\text{rate} = k[Ar'{-}N_2^+]\,[Ar{-}H] = \frac{k_1 k_2[B]}{k_{-1} + k_2[B]}[Ar'{-}N_2^+]\,[Ar{-}H] \tag{52}$$

The first case mentioned corresponds in (52) to $k_2[B] \gg k_{-1}$, the second to $k_2[B] \ll k_{-1}$ and the third to $k_2[B] \cong k_{-1}$.

Earlier work by Melander and coworkers[188,189] on nitrations, brominations and sulfonations did not exclude the S_E3 mechanism unambiguously, as was discussed by Hammond[190].

It is interesting to note today that a large number of textbooks of organic chemistry and practically all books on physical organic chemistry mention those investigations made in 1955. Yet most of them say that the *occurrence* of an isotope effect is evidence of the S_E2 mechanism. This statement is wrong, since one expects an isotope effect for an S_E3 reaction, too.

Only few authors discuss the really essential part of that work, namely:

a) the dependence of the measured 'overall' values of k_H/k_D on the structure of the σ-complex (steric crowding by neighbouring groups, C—N bond and C—H bond strength changes influenced by the electrophilicity of the benzenediazonium ion) and on the base B (concentration and structure of B). The intrinsic isotope effect, i.e. k_{2H}/k_{2D} (isotope effect of the second step, i.e. equation 51) is, however, almost independent of these parameters.

b) the most direct evidence for the general occurrence of the S_E2 mechanism is the non-linear dependence of the rate constant k on [B]. It seems likely, however, that a paper describing only that dependence of the rate on [B], but not the isotope effects, would not have received the same worldwide interest as the results including the isotope effects.

For the reactions studied in the original papers [185-187], the position of sulfonic groups in positions peri-, ortho- and meta- to the reacting carbon was decisive for the magnitude of steric crowding in the σ-complex. Later, Snyckers and Zollinger [191] studied the kinetics of the reaction of the 4-sulfobenzenediazonium zwitterion with 9 derivatives of 2-naphthol, each containing a substituent of varying size in the 8-position. The rate ratios k_2/k_{-1} (52) depend on a distance R^f (*126*). This factor corresponds to the distance between the center A of the hindering group (or atom) in the 8-position and a point N to which the azo nitrogen attached to the sp^3-carbon in the σ-complex will 'swing' during conversion to products in the second step of the substitution, *minus* the van der Waals radius R^w of the peri group.

126

The free radius R^f varied from 0.23 to 9.65 Å and k_2/k_{-1} varied from 1.07 to 3.93×10^{-2} $l\,mol^{-1}$. There is a fairly good linear relationship between $\log R^f$ and $\log(k_2/k_{-1})$ for all 2-naphthol derivatives with substituents in the 8-position, but not for the 7,8-benzo derivative (3-phenanthrol).

On the other hand, the rate constant k_1 does not depend on the changing steric influence of substituents in the 8-position, but correlates surprisingly well with the Hammett-Brown constant σ_m^+. This result indicates that the formation of a sp^3-hybridized carbon atom (at the 1-position of the σ-complex) leads to a compound without significant steric interaction of the electrophile with substituents in the 8-position. The σ-complex cannot be planar and is asymmetric. The preferred conformation of a σ-complex of this type is illustrated in Figure 2: The pseudo-axial position of the electrophile E^+ decreases the steric interaction between this group and the peri substituent R.

Fig. 2. Diagrammatic representation of the preferred conformation of a σ-complex in substitutions of 8-substituted 2-naphtholate anions by the electrophile E^+

Challis et al. [114] observed kinetic deuterium isotope effects in the azo coupling of 2-methyl-4,6-di-tert.-butylindole (*127*) and its anion. The origin of this effect must also be attributed to steric hindrance of the proton transfer step in the substitution proper, since 2-deuterated methylindole and unsubstituted indole [192] do not give isotope effects.

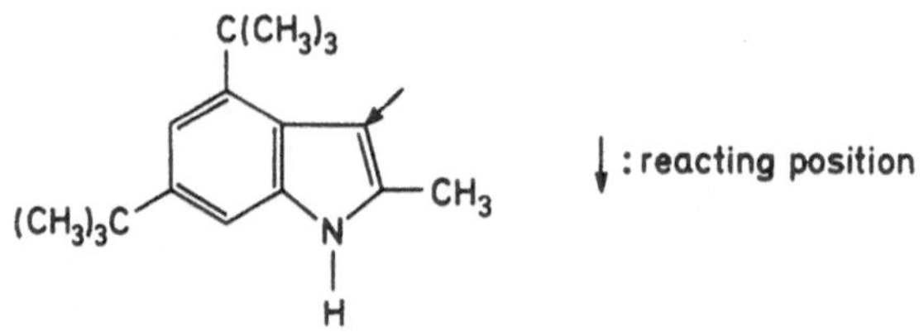

$C(CH_3)_3$ $(CH_3)_3C$ N H CH_3

↓ : reacting position

127

The subject of the stereochemistry of intermediates in electrophilic aromatic substitutions, including azo coupling reactions, was reviewed some years ago [193].

Verification of structures like that in Figure 2 by X-ray crystallography is unfortunately not possible since such σ-complexes are steady-state intermediates. To the best of our knowledge, only one group of compounds is known, where σ-complexes containing diazo groups could be observed. Ershov et al. [194] added an equimolar amount of p-benzoquinone diazide to four different 2,6-dialkylphenols in acetonitrile/pyridine and kept the mixture in the dark at 0–3 °C for 2 to 3 days. After evaporation and acidification they obtained 3′,5′-dialkyl-4,4′-dihydroxyazobenzenes. Kinetic data indicate that intermediates of type (*128*) were formed rapidly and then rearranged slowly into the tautomeric dialkyl-dihydroxyazobenzenes. Unfortunately, these authors neither isolated the intermediates, nor characterized them by instrumental analysis.

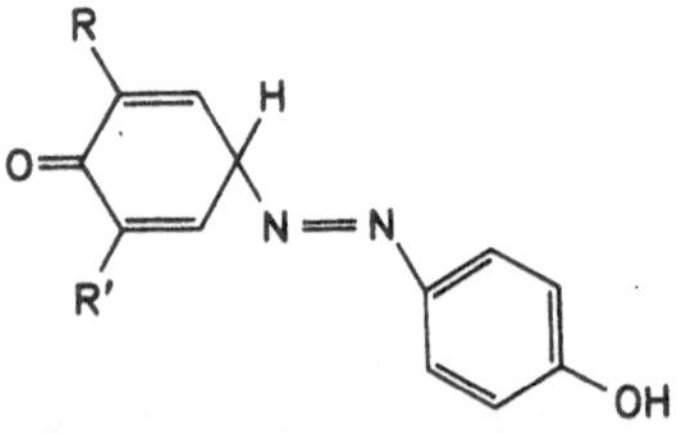

	R	R′
128a	Me	Me
128b	iso-Pr	iso-Pr
128c	Me	tert-Bu
128d	tert-Bu	tert-Bu

It is likely that in azo coupling reactions the diazonium ion is added to the nucleophilic center of the substrate in such a way that the azo group in the σ-complex has a trans-(E)-configuration. Although there is no strict evidence against primary formation of a compound with cis-(Z)-configuration, Zollinger [195] showed that all additions of 'hard' nucleophiles to the β-nitrogen of a diazonium ion (e.g. ^{-}OH) have an early transition state and the cis-configuration is the primary product. With 'soft' nucleophiles (C- and N-coupling) a late transition state is feasible; trans-isomers are formed and cis-isomers could either not be obtained at all until today (diazoamino compounds, i.e. N-coupling) or are available only by photochemical isomerization of the trans-compounds. For the hard/soft-acid/base concept in such additions to diazonium ions see also Section 4.6.

An additional argument in favour of the primary formation of trans-isomers in azo coupling reactions was brought forward recently by Schwarz and Zollinger [182]. As shown in (53), proton release from the trans-isomer of the σ-complex is facilitated either by hydrogen bonding with the lone electron pair of the N_α-atom of the diazo group, or by the easier approach of an external proton acceptor to the proton (or by both these effects). In the first effect a water molecule may be involved in a transition state similar to *131* with proton transfer in a six-membered ring system instead of the four-membered ring system of (53) (see later in this Section).

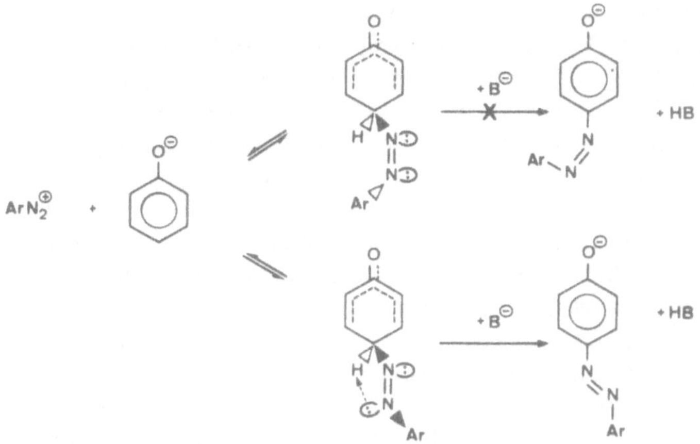

(53)

As mentioned above, the original investigations on the general base catalysis of azo coupling reactions [185-187] were made with nucleophilic substrates in which the reacting carbon was sterically hindered to different degrees by sulfonic groups in various positions. This resulted in relatively large values of k_{-1} (large relative to $k_2[B]$) in (52). Later work with Jermini and Koller [196] demonstrated that k_{-1} can be large even without significant steric hindrance if diazonium ions of very low electrophilicity are used. This is the case for benzenediazonium ions with hydroxy groups in the o- or p-position[4]. Since the diazonio group is a very strongly electron withdrawing group (it is actually by far the strongest such group known), the pK_a-value of the hydroxy groups is very low, for example for 2-diazophenol-4-sulfonic acid $pK_a = -0.04 \pm 0.10$ [196]. In the pH range usual for coupling reactions with such diazo components (pH 9–12), Jermini et al. [196] showed that it is the anion *130* and not the zwitterion *129* which enters the substitution proper — in spite of the argument brought forward in Section 4.1 for the higher reactivity of the zwitterion *129*.

4 Benzenediazonium ions with hydroxy groups in the o-position are industrially important diazo components for the production of metal complex dyes.

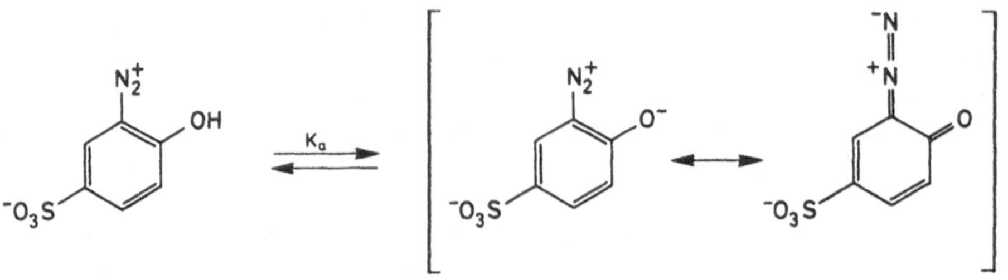

129 $\underset{}{\overset{K_a}{\rightleftharpoons}}$ *130* (54)

In a preliminary communication Szimán and Messmer [135] reported kinetic deuterium isotope effects in C-coupling reactions of 4-methoxybenzenediazonium tetrafluoroborate with N,N-dimethylaniline, m-toluidine, 1- and 2-naphthylamine in nitrobenzene solution (k_H/k_D = 1.5, 1.0, 3.3 and 4.4, respectively). As it has been mentioned that all such reactions are base catalyzed, it is difficult to understand why the reaction with m-toluidine does not display an isotope effect. As shown in Section 4.4, azo coupling reactions with aromatic amine may be rather complex with respect to mechanism. A critical discussion of Szimán and Messmer's paper is therefore impossible. Although announced in their communication, a full paper has to the best of our knowledge, never been published.

Recent ^{13}C- and ^{15}N-NMR studies of J. D. Roberts et al. [197] show that for 4-hydroxybenzenediazonium ion the compound with an ionized OH group (corresponding to *130* above) has a C—N bond with significantly more double-bond character than the compound with an undissociated OH group; the latter, however, has more C—N double bond character than the unsubstituted benzenediazonium ion.

As azo coupling reactions with rate-limiting proton transfers are catalyzed by general bases, they follow Brönsted's catalysis law. This was previously demonstrated in 1955 [187] and studied in more detail by Zollinger et al. in 1969–1974 [198, 199]. Four coupling reactions were investigated with respect to catalysis by seven pyridine derivatives, 1,4-diazabicyclo[2.2.2]octane, N,N′-dimethylpiperazine, N-methylmorpholine, water and hydroxide ion. Brönsted's β-values are different for heteroaromatic bases (β = 0.072 to 0.30) and alicyclic tertiary amines (β = 0.079 to 0.60). Plotting the logarithms of the intrinsic kinetic isotope effects k_{2H}/k_{2D} for the reactions with all N- and O-bases against the pK_a of the conjugate acids of the catalysts gives astonishingly smooth curves, with maximum isotope effects (k_{2H}/k_{2D} up to 8.4) at pK_a 1 to 2. Such maxima are expected [200] for that base whose conjugate acid has the same acidity as the substrate.

In the case of reaction (51) it is the σ-complex which is undergoing deprotonation. The pK_a values of these ketone-like σ-complexes can be derived from the constant K_T for the tautomeric keto-enol equilibrium of the phenol and the acid dissociation constant K_a for the phenol. One gets $pK_{a(keto)} \approx 1$.

2-Methyl- and 2,6-dimethylpyridine as catalysts yield, with sterically hindered σ-complexes, higher isotope effects (k_{2H}/k_{2D} up to 10.8). Such values are qualitatively understandable, since the basic center of these pyridine derivatives cannot approach the C—H group easily. The possibility of tunneling can be excluded for these reactions, as the ratio of the frequency factors A_H/A_D and the difference in activation energies $E_D - S_H$ (Arrhenius equation) do not have abnormal values.

The large number of values of k_{2H}/k_{2D} and of β for these azo coupling reactions allowed a critical comparative evaluation of these indices with respect to the problem of the 'symmetry' of transition states[5] in proton transfer reactions [199]. It is concluded that, while kinetic isotope effects are much more sensitive than Brönsted exponents to variations in pK_a, the use of either quantity as an index of transition state symmetry may be doubtful.

The secondary α-deuterium isotope effects of azo coupling reactions are very close to unity. For the reaction of 4-nitrobenzenediazonium ion with the triple anion of 1-d-2-naphthol-6,8-disulfonic acid catalyzed by pyridine $k_{1H}/k_{1D} = 1.057 \pm 0.042$ [199]. The t-test showed that at the 98% confidence level $k_{1H}/k_{1D} > 1$. This is probably due to hyperconjugation in the σ-complex which is not fully compensated for by the sp^2-sp^3 change in hybridization at the reaction center.

An interesting problem related to the substitution proper and base catalysis is the o/p-ratio of azo coupling reactions with 1-naphthol, 1-naphthylamine and their derivatives with sulfonic groups in the 3- or 5-position. Stamm and Zollinger [201] demonstrated in 1957 that with 1-naphthol-3-sulfonic acid, reaction at the 2-position is less sensitive to base catalysis by acetate ions than that at the 4-position. The σ-complexes for both positions are sterically hindered by the neighboring sulfonic group to the same extent. The complex leading to the substitution at the 4-position is, however, in addition sterically influenced by the C—H group in the peri- (i.e. 5-) position.

On the other hand, the catalytic effect of water as a base is stronger at the 2-position. This result can be explained by a proton transfer by a water molecule which solvates the O^--group in the reagent 3-sulfo-1-naphtholate ion. As shown by *131*, the base is already in the optimum position when the stage of the σ-complex is reached. This explanation is supported by a comparison of the entropies of activation for reaction at the 2- and the 4-positions, respectively.

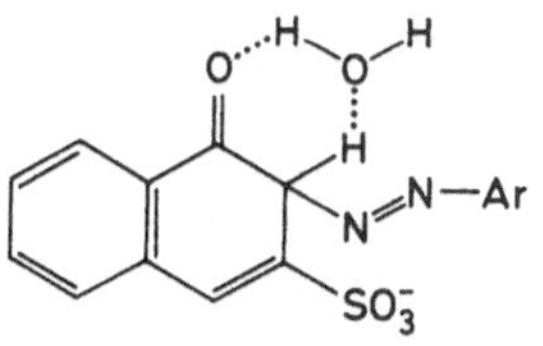

131

This explanation was given in 1957. It was very unconventional at that time for various reasons.

a) A direct intramolecular proton transfer in the σ-complex *132* or a mechanism in which the diazonium ion first attacks the naphtholate oxygen and then moves to the ortho position (*133*, *134*) was considered to be unlikely. Today we know that such

5 We write symmetry in quotation marks because it is not strictly a symmetry, as the proton donor and proton acceptor particles are not identical.

processes would involve symmetry-forbidden four-center transition states in contrast to the more favorable six-membered transition states in *131*. This is remarkable because in 1957 the symmetry rules of Woodward and Hoffmann were not yet known.

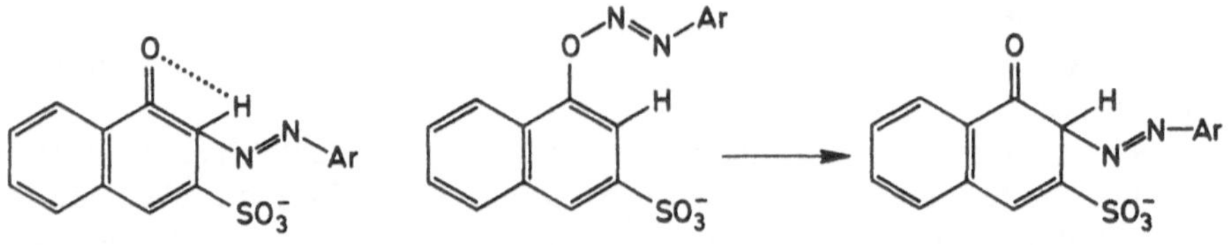

132 *133* *134*

b) In *131* the water molecule has bifunctional character; it is a proton donor to the carbonyl oxygen and proton acceptor for the C—H group. The general importance of the bifunctionality of water was recognized clearly by Eigen [202], but only in 1965, i.e. eight years after this example. At about the same time (1966), however, Bell [203, 204] proposed that the methanediol dehydration to formaldehyde and water proceeds by means of a concerted mechanism involving one or more water molecules in a cyclic hydrogen-bonded structure *135*, *136* (55). These structures are clearly analogous to *131*. Several further examples for such 'pseudointermolecular' proton transfers have been found since that time [205, 206]. Very recently ab initio MO calculations at the STO-3G level made by Maggiora et al. [207, 208] confirm that the activation barrier for the hydration of formaldehyde is dramatically lowered if an additional water molecule is involved in the reaction.

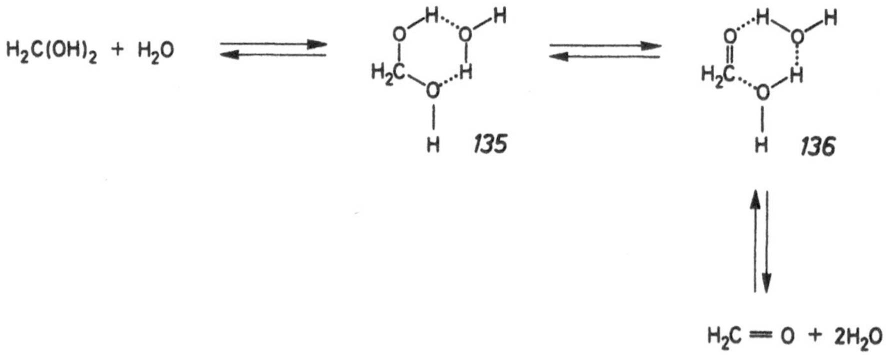

$$H_2C(OH)_2 + H_2O \rightleftharpoons 135 \rightleftharpoons 136 \rightleftharpoons H_2C{=}O + 2H_2O \qquad (55)$$

Although the mechanisms with six-membered cyclic structures *131* and *135*, *136* have striking similarities, it should be emphasized that *135* → *136* involve only proton transfers between oxygen atoms (but also the formation/dissociation of a carbon-oxygen bond), whereas in *131* a proton transfer from carbon to oxygen occurs.

In a broader context, but based on his own investigations of general acid- and based catalyzed additions to carbonyl groups, Jencks [211a] classifies a reaction like the azo coupling of 1-naphthol-3-sulfonic acid dianion catalyzed by water molecule(s) hydrogen-bonded at the naphtholate oxygen as a pre-association mechanism.

In the 70's the postulate of structures of type *131* was supported by activation entropy studies of Demian [209,210] and investigations on the o/p-ratio of coupling of 1-naphthylamine made by Hashida et al. [211].

Kishimoto and coworkers[212] showed in the 70's that for 1-naphtholate ion the reaction at the 4-position is catalyzed by a variety of bases ($H_2PO_4^-$, HPO_4^{2-}, AcO^-, pyridine and water), whereas the rate of reaction at the 2-position is almost independent of the concentration of these bases, except water. Their results therefore coincide well with previous work.

In addition, these authors [213,214] found a general acid catalysis by protonated pyridines in coupling reactions of 1-naphtholate ion if weakly electrophilic diazonium ions were used. In this case it is likely that the general acid protonates the carbonyl oxygen of the σ-complex, with a concerted or stepwise deprotonation at the 4-position (transition state *137*).

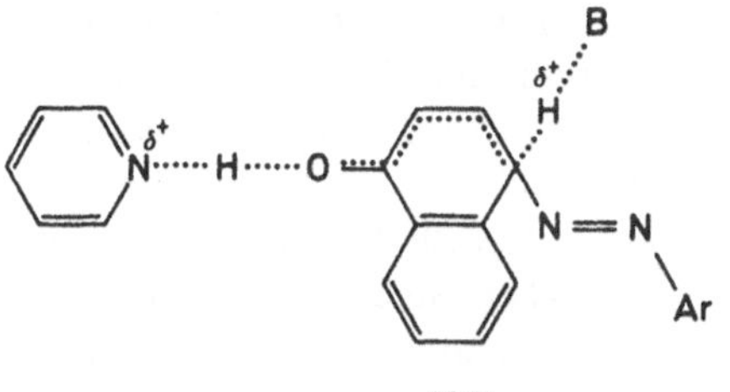

137

Proper intramolecular proton transfer in the second step of an azo coupling reaction was found by Snyckers and Zollinger [191,215] in the reaction of 8-(2′-pyridyl)-2-naphtholate ion (with transition state *138*). This compound shows neither a kinetic deuterium isotope effect nor general base catalysis, in contrast to the sterically similar 8-phenyl-2-naphtholate ion. Obviously the heterocyclic nitrogen is the proton acceptor.

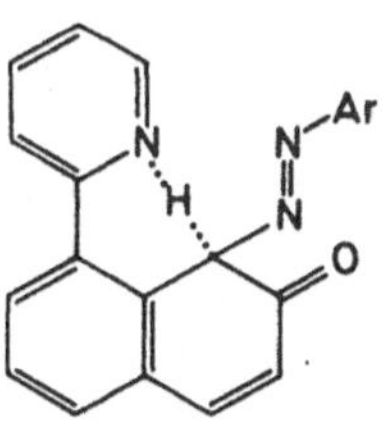

138

Besides pyridine derivatives and bases used as buffers (mentioned above), some other compounds catalyze azo coupling reactions. The mechanism of catalysis by urea was investigated by Gloor and Zollinger [216,217] for the second substitution of a bis-diazonium ion, i.e. for the formation of the bisazo compound from biphenyl-4,4′-bis-diazonium ion (bisdiazotized benzidine) with 2-naphthol-3,6-disulfonic acid. The catalytic activity of urea is due to general base catalysis (rather surprinsingly as urea is a very weak base!), but in addition to two other effects, namely:

a) A dielectric medium effect, as urea increases the dielectric constant of the aqueous solution. This effect facilitates the approach of the diazo component, i.e. the monoazo compound formed from biphenyl-4,4′-bis-diazonium ion and the triple anion of 2-naphthol-3,6-disulfonic acid (net charge: —1) to the triple anion of 2-naphthol-3,6-disulfonic acid. The Coulombic repulsion of the two reacting anionic species is reduced by the increased dielectric constant of the solvent.

b) Since the diazo component for the second substitution is a large, planar and dipolar compound (one positive charge at one end of the compound, two negative charges at the other), it easily forms aggregates (dimers and oligomers). These aggregates have a lower electrophilic reactivity, as shown by the observation that the apparent kinetic order n with respect to the diazo component is not 1.0 but only 0.65. Urea is known to have a disaggregating effect on azo compounds [218, 219].

Due to this aggregation effect, the measured rate constants of the second coupling reaction are not truly first order with respect to the diazo component. Gloor and Zollinger [216] calculated the actual rate constant referring to the monomeric diazo component from the (overall) measured rates and the dimerisation equilibrium constant. The rate of reaction of the monomeric diazo component is significantly larger than the overall rate. The ratio of rates of the first to the second azo coupling reaction of biphenyl-4,4′-bis-diazonium ion with the triple anion of 2-naphthol-3,6-disulfonic acid at 15 °C, k_1/k_2, is 80. This value is significantly lower than that communicated for the same sequence of reactions by Hashida et al. [220] two years before ($k_1/k_2 = 647$ at 20 °C). The value of Hashida et al. for k_2 is not correct as those authors did not recognize that the second coupling reaction is subject to the aggregation effect mentioned.

The mechanistic background of other compounds which catalyze azo coupling reactions [221] has not been investigated yet.

There are two cases where the general base catalysis observed for an azo coupling reaction is not due to a rate-limiting proton transfer from the σ-complex (51) but to deprotonations of the coupling component when the species which enters the substitution is formed. These reactions follow the general scheme (56–58).

$$\mathrm{H{-}R{-}H} + \mathrm{B} \underset{k_{-1}}{\overset{k_1}{\rightleftharpoons}} \mathrm{H{-}R^-} + \mathrm{HB^+} \tag{56}$$

$$\mathrm{Ar{-}N_2^+} + \mathrm{H{-}R^-} \underset{k_{-2}}{\overset{k_2}{\rightleftharpoons}} \mathrm{Ar{-}N{=}N{-}R{-}H} \tag{57}$$

$$\mathrm{Ar{-}N{=}N{-}R{-}H} + \mathrm{B} \underset{k_{-3}}{\overset{k_3}{\rightleftharpoons}} \mathrm{Ar{-}N{=}N{-}R^-} + \mathrm{HB^+} \tag{58}$$

$$\mathrm{Ar{-}N{=}N{-}R^-} \overset{\text{fast}}{\rightleftharpoons} \mathrm{Ar{-}N{=}N{-}R{-}H}$$

In this scheme $\overset{\overset{\large H}{|}}{R}$—H refers to the coupling component. The hydrogen above the residue R is that atom which is substituted in the azo coupling proper.

In the azo coupling of nitroethane [83, 168] (already discussed in Section 4.1) the forward step of (56) refers to the formation of the nitroethane anion. It is a rate-limiting deprotonation and therefore general base-catalyzed.

In the azo coupling reaction of acetoacetanilide [160] reaction steps (56) and (57) are a steady-state system, i.e. $k_1[B] \ll k_{-1}[HB^+] \sim k_2[Ar{-}N_2^+]$; $k_{-2} \cong 0$ with a fast consecutive deprotonation (58). As with nitroethane, this reaction is general base-catalyzed because of the rate-determining formation of the anion of acetanilide (56). In contrast to the coupling of nitroethane, however, the addition of the diazonium ion (57) is rate-limiting. The overall kinetics are therefore between zero and first order with respect to diazonium ion and not strictly independent of $[ArN_2^+]$ as in the nitroethane coupling reaction.

As seen from (58), the primary product of arylazo substitution is the anion $Ar{-}N_2{-}R^-$. It is protonated in a subsequent step either at a basic center of the residue R (preferentially at an oxygen atom, for example, in coupling with phenol), or at the azo nitrogen next to the group Ar forming a hydrazone. The problem as to whether the final product is a hydroxyazo- or a quinonehydrazone-type compound is therefore *not* a question of the substitution proper but of the consecutive tautomeric equilibrium.

The overall azo coupling at carbon does in practice not run in the reverse direction, although reversibility has been found in rare cases under rigorous conditions [222].

For many years the question has been discussed as to whether other intermediates are involved in electrophilic aromatic substitutions in addition to σ-complexes. Most claims that π-complexes or radical pairs are intermediates are ambiguous. It is not possible to differentiate between an intermediate on the direct way from reagents to products and an adduct of the reagents in a side equilibrium, if in the formation or dissociation of such a compound no additional particle is added or transferred to another particle. Only in such a case can the steady-state equations be tested by checking the dependence of the overall rate constant on the concentration of such particles.

To be more specific, the two mechanisms (59) and (60) cannot be differentiated, since in steps 1–2–3 of (59) and 1′–2′–3′ of (60) no additional particles are added or transferred. In steps 4 and 4′, however, the non-linear rate-dependence on [B] provides unambiguous evidence for the involvement of a σ-complex (see beginning of this Section and equation 52).

$$Ar{-}N_2^+ + Ar'{-}H \overset{①}{\rightleftarrows} \left|\begin{matrix} Ar{-}N_2^+ \\ \uparrow \\ Ar'{-}H \end{matrix}\right| \overset{②}{\rightleftarrows} \left| Ar{-}\dot{N}_2 \quad \overset{+}{\dot{A}}r'{-}H \right| \overset{③}{\rightleftarrows} Ar{-}N_2{-}\overset{+}{A}r'{-}H$$

π-(or charge transfer) complex — Radical pair — σ-complex

$$\text{σ-complex} \xrightarrow[+B]{④} Ar{-}N_2{-}Ar' + HB^+$$

(59)

$$\begin{array}{l} Ar{-}N_2^+ + Ar'{-}H \overset{③}{\rightleftarrows} Ar{-}N_2{-}\overset{+,}{Ar'}{-}H \overset{④}{\longrightarrow} Ar{-}N_2{-}Ar' + HB^+ \\ \quad \downarrow\uparrow ① \\ \left| \begin{array}{c} Ar{-}N_2^+ \\ \uparrow \\ Ar'{-}H \end{array} \right| \\ \quad \downarrow\uparrow ② \\ |Ar{-}\dot{N}_2 + \dot{A}r'\overset{+}{-}H| \end{array} \tag{60}$$

In the reaction of the strongly electrophilic 4-nitrobenzenediazonium ion with 2-naphthol-6,8-disulfonic acid, which yields a sterically hindered σ-complex, Koller and Zollinger [223] actually observed the rapid formation of a π-complex spectrophotometrically at low pH-values (i.e. $k_{-1} > k_2[B]$ in (52)). Its concentration decreases slowly and at the same rate as that of the formation of the azo product. ^{1}H-NMR indicates that the π-complex is not localized. All π-electrons of the benzene and the naphthalene system are involved in the complex formation to a similar degree, in contrast to the respective complex of the same naphthol derivative with the iodonium ion, where a localized complex is formed [224]. Bagal et al. [225] investigated the role of donor-acceptor complexes in the azo coupling reaction of 4-nitrobenzenediazonium ion with 2-naphthylamine-3,6-disulfonic acid and of the 4-chlorobenzenediazonium ion with 2-naphthol-6-sulfonic acid in more detail. Their kinetic results are, as would be expected, compatible with mechanisms (59) *or* (60).

Evidence for the formation of radicals in azo coupling systems was given by Bubnov et al. [226] by ^{15}N-CIDNP signals when mixing an acetone solution of $^{15}N_\alpha,^{15}N_\beta$-benzenediazonium tetrafluoroborate with methanolic sodium phenolate. It may be, in the opinion of the present authors, that these signals — probably corresponding to a considerably high concentration of radicals — are either specific for a water-free system or may be the result of the competitive homolytic protodediazoniation, a reaction which is known to occur in ethanol solutions of diazonium ions.

Whereas the work of Koller and Zollinger [223], Bagal et al. [225] and that of Bubnov et al. [226] discussed above does not allow us to differentiate with certainty whether π-complexes or radical pairs, respectively, are intermediates (59) or side-equilibrium products (60), a completely different method of synthesis of azo compounds from diazonium salts involving radical intermediates was found recently by Citterio and coworkers [227,228]. It is a new general synthesis of arylazoalkanes on the basis of equations (61, 62), i.e. addition of alkyl radicals to arenediazonium ions (61) and reduction of the intermediate azo radical cation adducts by metal salts (62).

$$R^{\cdot} + Ar{-}N_2^+ \rightarrow R{-}N{=}\dot{N}^{+}{-}Ar \tag{61}$$

$$R{-}N{=}\dot{N}^{+}{-}Ar + M^{n+} \rightarrow R{-}N{=}N{-}Ar + M^{(n+1)+} \tag{62}$$

The preferred source for alkyl radicals $R^{\cdot}$ in this reaction are alkyl iodides, which give rise to alkyl radicals cleanly in the presence of arenediazonium salts, and

Ti^{3+} or Fe^{2+} salts following equations (63, 64). The overall stoichiometric equation is therefore (65). The yields vary between 36 and 79%.

$$Ar{-}N_2^+ + Ti^{3+} \rightarrow Ar^{\cdot} + N_2 + Ti^{4+} \tag{63}$$

$$Ar^{\cdot} + R{-}I \rightarrow Ar{-}I + R^{\cdot} \tag{64}$$

$$2Ar{-}N_2^+ + R{-}I + 2M^{n+} \rightarrow Ar{-}N{=}N{-}R + Ar{-}I + 2M^{(2n+1)^+} + N_2 \tag{65}$$

The new reaction appears to be a simple one-step procedure particularly for tertiary alkyl-aryldiazenes for which alternative synthetic routes are less convenient. Aryl radicals or alkyl radicals, in which the carbon-centered radical is bonded to an electron-withdrawing group (i.e. COOR, COR, $CONR_2$, CN, SO_2R, etc.), do not, however, add to diazonium salts or give only poor results [229]. This result indicates that the radical must have a relatively high nucleophilicity in order to be able to react with a diazonium ion.

Colonna et al. [230] gave experimental evidence that radical pairs, probably as intermediates (59) and not as side-equilibrium products (60), can be detected in coupling reactions in which the electrofugic leaving-group is not a proton, i.e. in the so-called ipso-substitutions. Colonna and coworkers found that derivatives of 1-methyl-2-phenylindole with 11 different substituents in the 3-position (*139*) react with 4-nitrobenzenediazonium ions to form 3-(4'-nitrophenylazo)-1-methyl-2-phenylindole. In one case ($X{=}SCH_3$) an ESR spectrum was observed which corresponds to the radical cation *140*. The authors interpreted this result in terms of electron transfer process (59) preceding the formation of the σ-complex. This seems feasible since *139* ($X{=}SCH_3$) has the lowest oxidation potential in the indole series studied, and 4-nitrobenzenediazonium ion has the highest electron affinity of the mono-substituted arenediazonium ions.

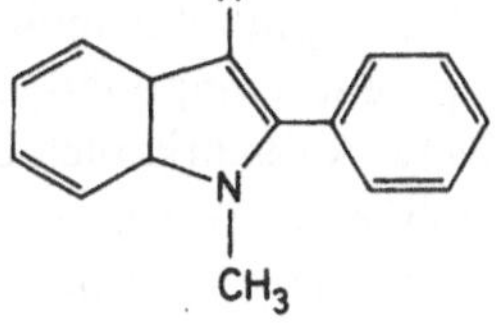

139 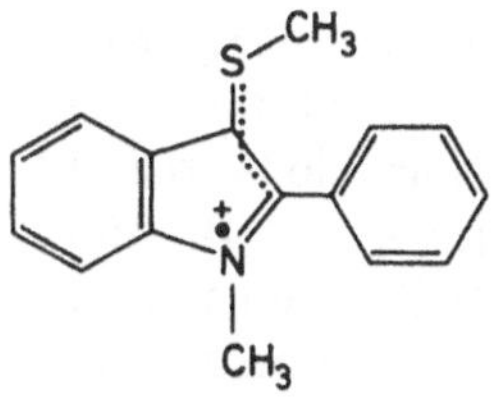 *140*

Azo coupling reactions with similar compounds giving rise to ipso-substitutions have been described before, namely with dipyrrylmethanes [110] and with di-indolylmethanes [231]. They were interpreted by assuming the formation of a σ-complex only. On the basis of Colonna's work with *139* ($X{=}SCH_3$) it is obvious that radical pair formation is observable only if the nucleophilic substrate has an extremely low oxidation potential.

A similar case seems to be that of the reactions of arenediazonium ions with phenothiazine described by Bisson, Hanson and Slocum [232]. Colonna et al. also reported an azo coupling reaction with nine 4-substituted N,N-dimethylanilines (e.g. *141*). With $X = COCH_3$ the substrate was monodemethylated and N-coupling took place. The authors interpreted the result in terms of scheme (66).

$$N(CH_3)_2C_6H_4COCH_3 + Ar{-}N_2^+ \longrightarrow {\cdot}\overset{+}{N}(CH_3)_2C_6H_4COCH_3 + Ar{-}N_2^{\bullet}$$

$$\xrightarrow{H_2O} NHCH_3C_6H_4COCH_3 \xrightarrow{ArN_2^+} H_3C{-}N(N{=}N{-}Ar)C_6H_4COCH_3 \qquad (66)$$

Demethylation of N,N-dimethylanilines is a reaction known to involve the intermediate formation of a radical cation, as has been demonstrated by anodic and chemical oxidation. Monodemethylation of N,N-dimethylaniline (without a substituent in the 4-position) has been observed by Penton and Zollinger [233], but only in dry acetonitrile and in strict absence of bases (moisture!). In the opinion of the present authors Colonna's experiments with 1-methyl-3-methylthio-2-phenylindole (*139*, $X = SCH_3$) provide a better evidence for radical intermediates than the demethylation of N,N-dimethyl-4-acetylaniline, since that demethylation may also be a heterolytic process.

Halogens as electrofugic leaving groups were studied by Fischer and Zollinger [121]. The rates of substitution of the group X in 1-X-2-naphthol-6-sulfonic acid by 4-chlorobenzenediazonium ions relative to the parent compound (X=H) are 0.0070:0.0089:0.149 for X=Cl, Br, and I, respectively. A remarkable observation is the catalysis by thiosulfate ions of the azo coupling of 1-bromo-2-naphthol-6-sulfonic acid. It is surprising because one might expect that the respective reactions with X=Cl or I would be catalyzed in the same way. In the analogous azo coupling reaction with 2-naphthol-1-sulfonic acid SO_3 is the electrofugic leaving group [122]. The mechanism was studied in detail. Due to its complexity reference has to be made to the original paper.

An interesting example of the influence of electrofugic leaving groups comes from a recent patent. Shuttleworth [234] reports that yields in azo coupling reactions of 5-nitrothiazol-2-diazonium ion (*142*) with N,N-dimethylaniline (*143*, X=H) and related coupling components are never more than 50%. The yield can be significantly increased if 4-N,N-dimethylaminobenzoic acid (*143*, X=COOH) is used instead of dimethylaniline. In this benzoic acid derivative, the electrofugic leaving group is CO_2, since the proton of the carboxylic group is already dissociated in a pre-equilibrium.

$$\text{142} + \text{143 } (X = H,\ = COO^-) \longrightarrow \text{O}_2\text{N-thiazolyl-N=N-C}_6\text{H}_4\text{-N(CH}_3)_2 + X^+ \quad (67)$$

4.4 Mechanisms of N-Coupling: Formation and Rearrangement of Diazoamino Compounds (Triazenes)

Diazonium ions react with all known nucleophiles. The addition of hydroxide, alkoxide, cyanide and sulfite ions does not belong to the subject of this review, but the formation of diazoamino compounds (1,3-triazenes) formed with ammonia, primary and secondary amines does. In analogy with the other reactions discussed in this paper, where carbon atoms are the site of attack by diazonium ions (C-coupling), these reactions are called N-coupling reactions.

It is surprising that the mechanism of N-coupling has been investigated only recently. Beránek and Večeřa investigated the kinetics of the reactions of substituted benzenediazonium ions with dimethylamine [235)] and aniline [236)]. The reactions are first order with respect to the diazonium ion and to the (nonprotonated) aniline. The Hammett ϱ-value is not significantly different from that obtained for C-coupling reactions (see also Sect. 4.6). There is a report by El'tsov et al. [237)] that the N-coupling of p-toluidine and diethanolamine with several substituted benzenediazonium ions in polar aprotic media shows general base catalysis. It is therefore likely that the reactions follow in principle the S_E2 mechanism (see Sect. 4.3) similar to the C-coupling.

The most significant difference from C-coupling is the facile reversibility of N-coupling. In acidic solutions the diazonium ion and the (protonated) amine are regenerated. This back reaction is important in the context of the diazoamino rearrangement discussed later in this Section.

Primary aromatic amines (e.g. aniline) and secondary aliphatic-aromatic amines (e.g. N-methylaniline) usually form diazoamino compounds in coupling reactions with benzenediazonium salts. If the nucleophilic reactivity of the aryl residue is raised by substituents or fused rings, as in m-toluidine and 1- and 2-naphthylamine, aminoazo formation takes place (C-coupling). However, the possibility has also been noted that in aminoazo formation, initial attack of the diazonium ion may still be at the amine N-atom, but the σ_N-complex might rearrange too rapidly to allow its identification [235)].

Penton and Zollinger [233,238] reported recently that this could indeed be the case. The coupling reaction of m-toluidine and N,N-dimethylaniline with 4-methoxybenzenediazonium tetrafluoroborate in dry acetonitrile showed a number of unusual characteristics, in particular an *increase* in kinetic deuterium isotope effect with temperature. Predominant C-coupling occurs ($\geqq 86\%$ for m-toluidine). but on addition of a tert.-alkyl ammonium chloride the rate became much faster, and predominantly diazoamino compounds (with loss of a methyl group from N,N-dimethylaniline) were formed. Therefore, initial attack of the diazonium ion is probably at the amine N-atom and aminoazo formation occurs via rearrangement.

A detailed investigation of these reactions led to the results which are summarized in Table 2: The kinetic isotope effect for N,N-d_2-m-toluidine, the kinetic order with respect to the amine and the activation parameters ΔH^* and ΔS^* change with amine concentration as well as with temperature.

Table 2. General characteristics of reactions of 4-methoxybenzenediazonium tetrafluoroborate with m-toluidine and (partly) with N,N-dimethylaniline, in acetonitrile

At 12.9 °C[a]				At intermediate [HA], increase in T from
[HA][b]	0.004	0.008–0.128	0.256	12.9° to 47.2 °C
k_H/k_D	2.06	decrease	1.04	increase
n_{HA}[c]	1.75	decrease	1.25	increase
ΔH^*[d]	25.3	increase	44.3	decrease
ΔS^*[e]	−173	increase	−121	decrease

a For ΔH^* and ΔS^* kinetic data at 12.9° and 31.1°.
b Concentration of amine in mol $\cdot$ l^{-1}.
c Kinetic order with respect to amine.
d kJ mol^{-1}. Estimated error ± 2.5 kJ mol^{-1}.
e J mol^{-1} per degree. Estimated error ± 10 J mol^{-1} at 12.9°.

An increased kinetic isotope effect at a higher temperature is not compatible with a simple reaction but only with a mechanism in which a pre-equilibrium is shifted at higher temperatures to reacting species, whose intrinsic isotope effect is higher. The inconstancy of the kinetic order n_A and of the thermodynamic parameters as functions of [HA] and T, leads to analogous conclusions. The evaluation of all experimental results is compatible with mechanism (68). Formation of the diazoamino compound (N-coupling) takes place through *two* intermediates, a 1:1 addition complex $(HAArN_2^+)_N$ and the N-σ-complex.

The aminoazo product, however, is formed via two pathways, namely through the 1:1 addition complex $(HAArN_2^+)_N$ as side-equilibrium and through the intermolecular rearrangement, i.e. redissociation of this complex into the reagents, formation of another 1:1 addition complex $(HAArN_2^+)_C$ and the classical C-σ-complex. The second pathway starts from the first mentioned 1:1 complex $(HAArN_2^+)_N$ to which a second molecule of amine is added. This complex forms the aminoazo product by proton transfer to a base. The base may be the second amine molecule of the 1:2 complex.

$$
\begin{array}{ccccc}
 & (HAArN_2^+)_N \underset{k_{-5}}{\overset{k_5}{\rightleftharpoons}} \sigma_N \underset{k_{-6}[BH^+]}{\overset{k_6[B]}{\rightleftharpoons}} \text{Diazoamino} + BH^+ & & & \\
K_N \nearrow\!\!\swarrow & k_3[HA] \searrow\!\!\nwarrow k_{-3} & & & \\
HA + ArN_2^+ & (HAArN_2^+HA) & & & (68) \\
K_C \searrow\!\!\nwarrow & \searrow k_4[B] & & & \\
 & (HAArN_2^+)_C \underset{k_{-1}}{\overset{k_1}{\rightleftharpoons}} \sigma_C \xrightarrow{k_2[B]} \text{Aminoazo} + BH^+ & & &
\end{array}
$$

Depending on the magnitude of $K_N[HA]$ and $K_C[HA]$ (>1 or <1), the dependence of k_H/k_D, $\Delta H^{\neq}$ and $\Delta S^{\neq}$ on [HA] is explainable in addition to the change in isotope effects which are dependent on $k_2[B]/k_{-1}$ and $k_4[B]/k_{-3}$ as well as on the relative concentrations of the two 1:1 complexes, respectively.

The intramolecular rearrangement of the conjugate acid of the diazoamino compound into the σ_C-complex without an additional molecule of amine corresponds to a thermal [1, 3] sigmatropic rearrangement. Such a mechanism can be ruled out, however, by the antarafacial pathway required from orbital symmetry considerations (Woodward-Hoffmann rules).

The conclusions of Penton and Zollinger [233] are supported further by an investigation by the same authors [239] on a system in which no detectable amount of aminoazo, but only diazoamino compound is formed, namely the reaction of 4-chlorobenzenediazonium tetrafluoroborate with 4-chloroaniline in acetonitrile. The reaction is catalyzed markedly by relatively strong bases in the order H_2O < dimethylformamide < pyridine < DMSO < hexamethylphosphoramide. The order with respect to the last four additives is 1; with water it is less than one. This is strong evidence for a steady-state σ_N-complex.

An evaluation of the observed (overall) rate constants as a function of the water concentration (5 to 25% in acetonitrile) yields no constant values for k_1 and k_2/k_{-1}, however. Tentatively this can be explained by changes of the water structure. Arnett et al. [240] have found that bulk water has about twice as large an H-bond acceptor capacity towards pyridinium ions as monomeric water and twice as strong an H-bond donor property towards pyridines. In the present case this should lead to an increase in N—H stretching frequency in the σ-complex (H-acceptor effect) and possibly to increased stabilization of the incipient diazoamino compound (H-donor effect). Water decreases the ion pairing of the diazonium salt and therefore increases its reactivity [127, 130, 241]. This results in an increase of the rate of formation of the σ-complex (k_1).

Evidence that the diazoamino rearrangement can be partly intramolecular was brought forward qualitatively by Ogata et al. [242]: 1,3-Bis(4-methylphenyl)triazene *144* rearranges in acidic ethanol, affording 2-amino-5,4'-dimethylazobenzene *145*. In the presence of N,N-dimethylaniline as a scavenger of diazonium ion, 4-dimethylamino-4′-methylazobenzene *146* was obtained in addition to the product of rearrangement *145*.

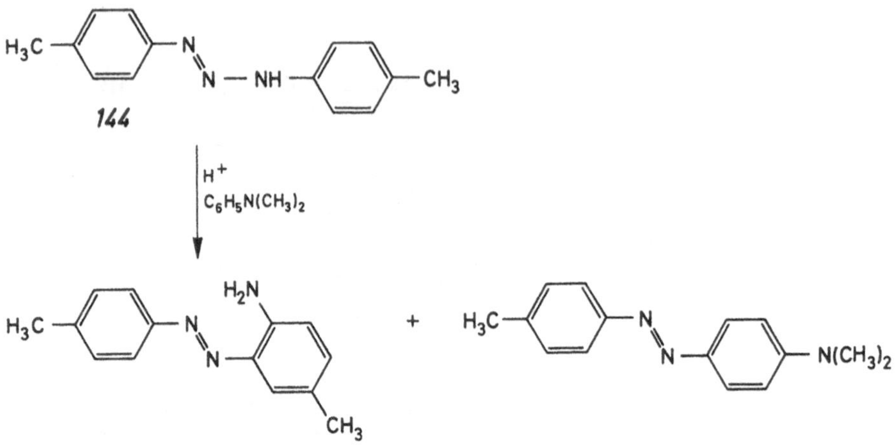

144

H^+
$C_6H_5N(CH_3)_2$

145 146

The ratio of the product yields [*145*]/[*146*] as a function of the reciprocal of the concentration of N,N-dimethylaniline (DMA) should be linear and, extrapolated to 1/[DMA] = 0, [*145*]/[*146*] should be zero. This ratio indeed decreases linearly with 1/[DMA] and gives the expected slope but only at low concentrations of DMA ([DMA] <0.27 mol $\times$ l^{-1}). At higher concentration (up to [DMA] = 2.0 mol $\times$ l^{-1}) [*145*]/[*146*] increases again significantly. The authors assume that for the intramolecular fraction of the rearrangement the nucleophilicity of the complex formed from the conjugate acid of the triazene *147* may be increased by hydrogen bonding with DMA *148*.

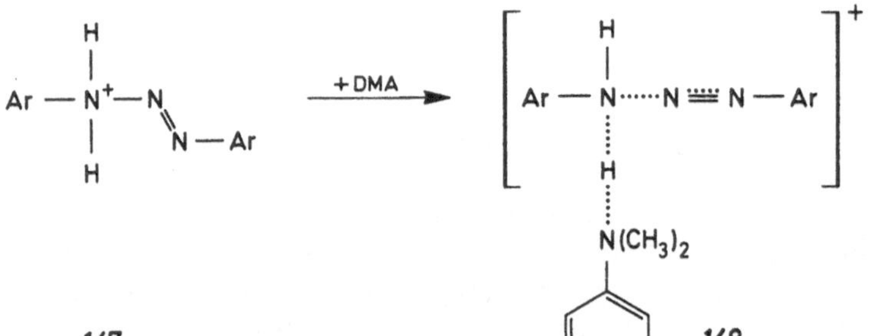

+DMA

147 148

This hypotheses is further supported by a viscosity effect found by the same authors [242] on the o/p-ratio of the rearrangement of 1,3-diphenyltriazene into 2- and 4-aminoazobenzene mixtures: There is more ortho migration in solvents of higher viscosity.

Both results of Ogata et al. [242] coincide with the conclusions which Penton and Zollinger [233] drew from their investigations when they postulated a second mechanism of rearrangement starting from the compound $(HAArN_2^+)_N$ in scheme (68).

The early work on diazoamino rearrangements has been well summarized by Shine [243,244]. Briefly, all the previous evidence supported an intermolecular, specifically acid-catalyzed process, the so-called Friswell-Green mechanism, postulated in 1884(!). Under certain conditions, particularly when not water, but the corresponding amine is used as solvent, a modification of this mechanism can occur. Thus Goldschmidt et al. [245] found in 1924 that the decomposition of the protonated diazoamino compound to amine and diazonium ion can be catalyzed by the anion of the acid when the latter is weak, for example, nitrobenzoic acid. With mineral acids the anion is a weak nucleophile and no evidence was found for such a pathway, but it was postulated [246] that here the amine itself can catalyze the fission of the protonated diazoamino compound. Neither of these processes has been observed in aqueous or partially aqueous solution, for example, in 95% aqueous ethanol [247].

Since aminoazo compounds are synthesized in many cases by rearrangement of diazoamino compounds, the nature of the mechanism has implications for the syntheses. Authors of preparative organic chemistry text books, e.g. [248], quote only a two-stage synthesis, i.e. starting from amine and diazonium salt with the respective diazoamino compound as (isolated) intermediate. Yet if under aqueous or partly aqueous conditions only the Friswell-Green mechanism holds, it should be possible, as predicted in 1961 [246], to prepare aminoazo compounds in a one-stage synthesis even from amines which predominantly undergo N-coupling. One-stage processes are described in patents, e.g. [249], but we know that there are industrial productions of aminoazo compounds which follow the classical two-stage method. A recent comparative investigation [250] of the one-stage and the two-stage process for the synthesis of 1-methylamino-4′-nitroazobenzene from N-methylaniline and 4-nitrobenzenediazonium salt showed that the one-stage method is simpler and gives at least as high yields. It is therefore no advantage to synthesize aromatic aminoazo compounds by a two-stage process.

A group of Egyptian chemists [251] published a paper in which they showed that arenediazophenyl sulfides ($Ar—N_2—S—C_6H_5$) react with 2-naphthol quantitatively yielding the corresponding azo compounds. This investigation is mentioned here because the authors claim that the reaction follows a mechanism analogous to the Friswell-Green mechanism of coupling of diazoamino compounds. In spite of relatively extensive kinetic data, it is difficult to review that paper critically as the kinetic order with respect to 2-naphthol is not reported.

4.5 Orientation Effects

The coupling reaction is probably that electrophilic aromatic substitution which is characterized to the highest degree by its sensitivity to orientation. In practically all cases the aromatic substrate reacts only if a strong electron donor (O^-, NH_2 etc.) is present. The reaction takes place exclusively at the o- and p-positions; m-substitution has never been observed nor a reaction at the 3-position of 2-naphthol and 2-naphthylamine (in contrast to, for example, sulfonation).

The o/p-ratio is of primordial industrial importance for the synthesis of technical dyes, particularly for coupling components like 1-naphthol and derivatives of 1-naphthol-3-sulfonic acid. The dyes with an arylazo group in the 2-position form an intramolecular hydrogen bond (*149*) which is, of course, impossible for the 4-isomer.

This hydrogen bond decreases the acidity of the OH group by two to three powers of ten (pK_a 11–12 compared to 8–9 for the 4-isomer). Such dyes have significantly different light absorption spectra if the OH group is dissociated. The change of colour is also present in dyeings on textile materials. If they are washed under slightly alkaline conditions (pH 8–9) 4-isomers will show a different shade, whereas the 2-isomers are unchanged[6] [253,254].

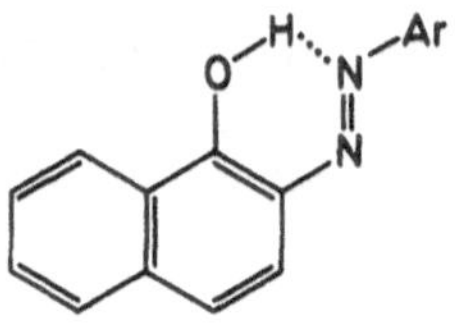

149

The basis for what may be called the 'intrinsic' o/p-ratio has not been studied in detail in the past. It is not known why the o/p ratio of azo coupling of phenol is in the range of 0.01 to 0.05, whereas in other electrophilic substitutions it is significantly higher.

The implications of the kinetics of the S_E2 mechanism on the o/p-ratio of the azo coupling reaction of 1-naphthol and its derivatives with a sulfonic group in the 3-position have already been discussed in detail in Section 4.3.

An interesting and industrially important case of selectivity in azo coupling reactions is the pH-dependence of the position of substitution by arenediazonium ions on aminonaphtholsulfonic acids. Since the turn of the century it was known in industry on an empirical basis that with such a coupling component, azo coupling takes place at that six-membered ring of the naphthalene system which contains the amino group if the reaction is run under acidic conditions, but on the nucleus which contains the hydroxy group if an alkaline solution was used. A theoretical explanation for this pH dependence of the reaction sites was given in 1952 [152], but was verified experimentally only after 1966 [255–259].

We will discuss it here for the model case of azo coupling of 2-amino-7-naphthol (the presence or absence of sulfonic groups is not relevant for the mechanistic problem of orientation). The experimental results with specific aminonaphtholsulfonic acids in this Section can be discussed on the basis of the following paragraphs, equation (70) and Figure 3 written for 2-amino-7-naphthol. 2-Amino-7-naphthol (*151*) can be substituted at the 1- or the 8-position. The substitution at the 1-position is determined only by the amino ⇄ ammonium equilibrium. As discussed in Section 4.1, an aromatic amine is reactive in azo coupling reactions only in its basic equilibrium form, e.g. for 2-amino-7-naphthol in form *151*, but not in form *150*.

6 Both isomers are in addition subject to an equilibrium between the hydroxy-azo and the quinone-hydrazone tautomer. This, however, does not belong to the subject of this paper. For a review see Ball and Nicholls [252].

The reaction at the 8-position depends on the naphthol ⇄ naphtholate equilibrium. In 2-amino-7-naphthol the naphtholate equilibrium form *152* is at least 8 powers of ten more reactive than the naphthol form *151*.

150 ⇄ *151* ⇄ *152* (70)

The pK_a values of the two equilibria of equation (70) are $pK_1 = 4.3$ and $pK_2 = 9.9$, respectively. Therefore, in solutions of $pH < 4.3$ the relative concentration of the amino form *151* will decrease by one power of ten per pH unit. As it is only the amino form which reacts with the diazonium ion, we expect that the logarithm of the rate of coupling at the 1-position will show a linear increase with pH (see Fig. 3, dotted line). An analogous situation exists for coupling at the 8-position. The rate will increase by one power of ten per pH unit up to pH 9.9 and then level off at higher pH values (Fig. 3, solid line). The decrease of both curves at $pH > 11$ is due to the diazonium ion ⇄ diazotate ion equlibrium (see Sect. 4.1).

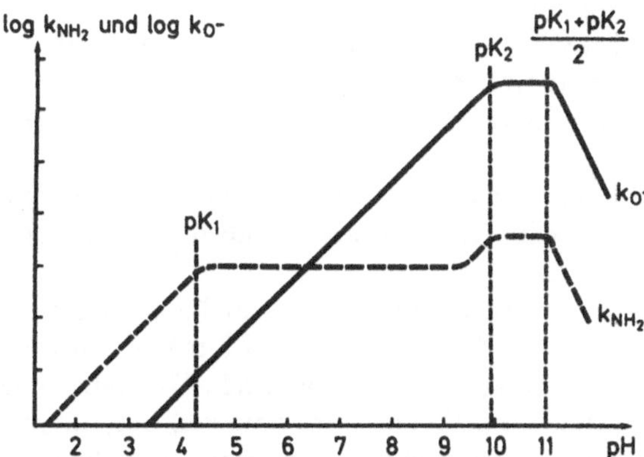

Fig. 3. Schematic representation of the dependence of the azo coupling rate of amino-naphthols on the acidity of the medium

Since 2-naphtholate is a stronger nucleophile than 2-naphthylamine, the intrinsic azo coupling rate constant k for naphtholate is larger. In the example shown in Fig. 3 the ratio of the rate constants k_{O^-} and k_{NH_2} is 10^3.

The measured rate constant (k_s, see Sect. 4.1) for the 1-position is, however, faster below pH 6.4, whereas at high pH values reaction at the 8-position dominates.

The ratio of products reacted at the 8- and the 1-positions, respectively, at a given pH-value can easily be determined from the graphical representation in this Fig., for example, at pH 4 this ratio is $1:10^2$, at pH 8, however, it is $10^3:1$.

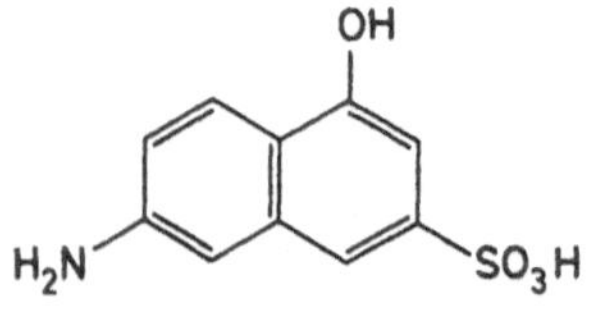

153 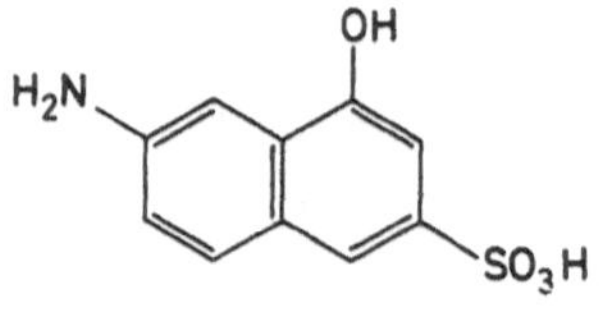 *154*

The small increase of the reaction rate for the 1-position (dotted line) at pH 9.5 to 10.0 reflects experimental results obtained by Ikeda et al. [255–257]. In their first paper [255] these authors determined the relative rate constants for azo coupling reactions of 6-amino-1-naphthol-3-sulfonic acid (J-acid, *153*) with 4-methyl- and 4-chlorobenzenediazonium ions in competitive experiments to the azo coupling reaction of 1-naphthol-4-sulfonic acid at pH 5.5–9.1. The rate at the 2-position of J-acid, i.e. the reaction directed by the O^--group, is 300–4000 times faster than that at the 5-position, which is directed by the amino group. These authors made two interesting observations:

1) The rate at the 5-position is 5–6 times larger if the OH-group in the 1-position is dissociated. In other words, extrapolated to equation (70) this means that the reactivity of the 1-position of the equilibrium form *152* is 5–6 times higher than that of *151*. This effect is taken into account in Fig. 3 by the slightly higher level of the dotted curve above pH 9.9. To our knowledge, this is the only clear case of an electrophilic substitution on naphthalene where an influence of the acid-base equilibrium of a substituent in one of the six-membered rings on the rate of substitution in the *other* ring has been measured quantitatively.

2) The ratio of rates of the 2- and the 5-positions is higher (3000–4000) with the 4-chlorobenzenediazonium ion than with the 4-methyl derivative (300–700). On the basis of the classical selectivity/reactivity rule one would expect the reverse order.

The azo coupling reactions of 7-amino-1-naphthol-3-sulfonic acid (γ-acid, *154*) and 8-amino-1-naphthol-3,6-disulfonic acid (H-acid) were studied by the same authors in a similar way [256,257].

As we found more recently [259], these orientation effects can be disguised by mixing effects of the type discussed in Section 4.2. In reactions of 3-trifluoromethylbenzenediazonium ions with 7-amino-1-naphthol-3-sulfonic acid *154*, rates and product ratios which would be predicted from Figure 3 are obtained only in very dilute solutions ($<10^{-2}$ mol/l). In systems with higher concentrations of the reactants the product with the 3′-trifluoromethylphenylazo group in the 8-position is not only dominant in acidic solution, as expected, but also up to pH 9. In this pH range also considerable amounts of bisazo product are obtained. These are observations which are typical for mixing effects observed in very fast reactions. With the unsubstituted benzenediazonium ions, a less reactive electrophile, however, rates and products are determined only by equilibria of the type of (70).

In older textbooks it was mentioned that *di*substitution of aminonaphtholmono- and disulfonic acids like *153* and *154* by diazonium ions is possible only if the first azo coupling is made under acidic and the second under alkaline conditions. Furthermore, it was said that with γ-acid (*154*) a second coupling after reaction in an acidic medium is impossible. Schneider [260] used more modern chromatographic techniques and showed that these claims are wrong. It must be said, however, that in order to get a high yield of the bisazo product, the acidic coupling should be run before the alkaline. The introduction of an arylazo residue on naphthalene reduces its reactivity in the second substitution. Therefore the reaction with the higher intrinsic rate, i.e. that azo coupling directed by the dissociated OH group should be carried out in the second step.

4.6 Comparative Studies of Reactivities

Azo coupling reaction rates of substituted benzenediazonium ions with a given nucleophile under standard conditions (solvent, temperature, pH, etc.) are an ideal case for the application of the Hammett relationship in its classical form (71): k_0 and k_σ refer to the rate constants of the benzenediazonium ion and its m- or p-substituted derivatives, respectively, with the given coupling component. ϱ and σ are Hammett's reaction and substituent constants, respectively.

$$\log k_\sigma - \log k_0 = \varrho\sigma \qquad (71)$$

Equation (71) was tested in azo coupling reactions of substituted benzenediazonium ions with 2-naphthol-6-sulfonic acid (as dianion), and 2-naphthylamine-6-sulfonic acid (monoanion) for the first time in 1953 [261] and later for a large number of nucleophiles (C- and N-coupling reactions) by the group of Czechoslovakian chemists in Pardubice and by others (references see Table 3). In general, good linear correlations between $\log k_\sigma$ and the substituent constant σ_m were obtained (fairly good with σ_p). Sometimes 4-methoxybenzenediazonium ion is less reactive by a factor up to 3 if the classical substituent constant $\sigma_{p\text{-}OCH_3} = 0.12$ is used. Although this observation can easily be explained on the basis of mesomeric structures *155a*, *155b*, a word of warning is necessary. As shown by Wepster [262], Hammett's σ-values are not independent of the reaction studied; they are not 'constants' in a literal sense.

$H_3CO-C_6H_4-N^+{\equiv}N$ ⟷ $H_3CO^+{=}C_6H_4{=}N^+{=}N^-$

155a *155b*

$H_3C-C(O^-){=}CH-N{=}N-Ar$

156 Ar = $p\text{-}CH_3-C_6H_4$
157 Ar = $p\text{-}NO_2-C_6H_4$

Table 3. Evaluation of azo coupling rates of substituted benzenediazonium ions with various nucleophiles

Coupling component	ϱ	log k_0 [a]	Remarks[b]	Ref.
1. C-coupling reactions				
Acetone (anion)	1.89	9.50		85)
Nitroethane (anion)	2.95	3.34		168)
Acetylacetonate ion	3.45	5.11		86)
Acetoacetanilide ion	3.06	6.80		84)
ω-Methylglyoxal-p-tolyl-hydrazone anion (*156*)	2.62	6.85		87)
ω-Methylglyoxal-p-nitro-phenylhydrazone anion (*157*)	2.73	5.74		87)
Resorcinol	3.45	−1,87		158)
Resorcinol, monoanion	3.76	5.57		158)
Resorcinol, dianion	3.12	9.87		158)
3-Methoxyphenolate ion	3.74	5.71		158)
1-Naphthol	~4.8	—	c	160)
1-Methoxynaphthalene	4.03	—	c	160)
1-Naphtholate ion	4.15	7.50		120)
1-Naphthol-4-sulfonic acid	3.94	4.56		120)
(dianion)	3.55	2.78	15.9 °C, I = 0.1	209)
2-Naphtholate ion	3.55	3.86		42)
2-Naphthol-6-sulfonic	3.46 (3.43)	4.07 (3.96)	0 °C	184, 263)
acid (dianion)	3.85	2.86	20 °C, I = 0.25	261)
	3.18	4.76	20 °C, I = 0.01	184)
2-Naphthol-3,6-disulfonic acid (trianion)	3.44	3.30	10 °C	59)
1-Phenyl-2,3-dimethyl-pyrazolin-5-one (anion)	3.60	3.87		160)
1-(4′-Sulfophenyl)-3-methyl-pyrazolin-5-one (dianion)	3.29	5.07		160)
N-Methylaniline	3.98	1.72	20% EtOH	264)
N-Dimethylaniline	1.98 to 4.38	−1.86 to 0.38	d	235, 265)
2-Naphthylamine-6-sulfonic acid (monoanion)	4.26	0.35	20 °C, I = 0.25	261)
4 derivatives of 1-naphthol-3-sulfonic acid (anions)	5.3 to 5.9	2.8 to 5.1		258)
2-Methylindole	3.3	0.96		114)
Pyrrole, 1- and 2-methyl-pyrrole	1.6–1.7	0.83 (1.40; 2.45)[e]		108)
1,2- and 2,4-dimethylpyrrole	1.1–1.2	2.96 (4.13)[f]		108)
Pyrrole	4.63[g]	2.51		161)
1-Methylpyrrole	4.28[g]	2.99		161)
1-Ethoxycarbonylpyrrole	4.64[g]	1.48		161)
2. N-coupling reactions				
Dimethylamine	3.47			266)
2,2′-Di(hydroxyethyl)amine	1.81	1.00		267)
	1.77	3.55	CH_3CN	267)
	1.44	3.48	$CH_3CON(CH_3)_2$	267)
	1.29	—	$[(CH_3)_2N]_3PO$	267)
Aniline	4.16	−0.92	0 °C, H_2O	236)
	3.90	—	20 °C, 20% EtOH	236)

Table 3. (continued)

Coupling component	ϱ	log k_0 [a]	Remarks[b]	Ref.
N-Methylaniline	3.94	2.82	20% EtOH	264,268)
6-Amino-1-naphthol-3-sulfonic acid (anion)	4.04	2.25		258)
6-Phenylamino-1-naphthol-3-sulfonic acid	4.15	0.96		258)

a k_0 in mol · l^{-1} · s^{-1}

b Aequeous solutions, if no solvent given. Temperature and ionic strength (I) are given only in cases where for these parameters more then one value was available

c Only determined with disubstituted benzenediazonium ion, assuming additivity of substituent effects

d In 20% EtOH [235] and in nitromethane, acetonitrile, acetone, methyl, ethyl ketone and dimethylacetamide [265]

e In brackets values for reactions with 1- and 2-methylpyrrole, respectively

f In brackets for reaction with 2,4-dimethylpyrrole

g Unreliable values as obtained with a narrow range of σ-constants

Although there is a certain trend of ϱ values decreasing with increasing rate constants (Table 3), this correlation is far from being linear. This can be seen in particular if ϱ values for the same coupling component, e.g. resorcinol in different acid-base equilibrium forms, are compared. The reactivity of the unionized resorcinol molecule is 7.44 and 11.74 powers of ten lower than its monoanion and dianion, respectively. The monoanion has a comparable reactivity to 3-methoxyphenolate ion as expected.

The comparison between the reactivities of these four coupling species is qualitatively easy to understand. Yet it is difficult to rationalize that the ϱ values for coupling reactions of these four compounds differ by less than one unit. On the basis of selectivity-reactivity correlations one expects larger differences between these ϱ values. Therefore, in our opinion they cannot be used for general reactivity evaluations in azo coupling reactions.

There are some investigations in which ϱ values of reactions other than azo coupling are taken as an indication of early or late transition states (low ϱ values = early and high ϱ values = late transition states in similar reactions). Among these investigations is a discussion of additions of nucleophiles to the β-nitrogen atom of substituted benzenediazonium ions (OH^-, CN^-, N_3^-, $ArSO_2^-$, amines, azo coupling components) by Zollinger [195]. It was claimed that in such additions to diazonium ions which resulted in ϱ values between 2.3 and 2.6 (OH^-, CN^-, N_3^- and $ArSO_2^-$), (N-coupling of amines, C-coupling of phenols and naphthols) had ϱ values above 3.0. an early transition state is involved. In contrast, additions with late transition states

This claim is well substantiated, as the rate relationships are supported by the respective equilibrium ϱ values. Those additions, where the *rates* have ϱ values between 2.3 and 2.6, have significantly higher ϱ values for the respective *equilibria* (3.5 to 5.2). Therefore, one can indeed, in our opinion, correlate ϱ values for rates with the position of the transition states along the reaction coordinates. If a ϱ value for a rate is similar to ϱ for the respective equilibrium, the structure of the transition state is similar to the addition product of the equilibrium. If ϱ (rate) is close to 1.0, the transition state is early, i.e. similar to the structure of the reagent (in these cases to the diazonium ion).

These conclusions on early and late transition states in additions to aromatic diazonium ions are also supported by (qualitative) hard/soft—acid/base considerations.

A more interesting problem than the influence of substituents in the electrophilic reagent of azo coupling is the extremely high selectivity of the C-coupling reactions, relative to other electrophilic aromatic substitutions. Unsubstituted benzene does not react with any arenediazonium ion, 1,3,5-trimethoxybenzene reacts very slowly with strongly electrophilic diazonium ions only; aromatic amines (e.g. N,N-dimethylaniline) or phenolate ions react very fast, in some cases close to diffusion control.

Unfortunately H. C. Brown's $\varrho\sigma^+$ treatment of electrophilic aromatic substitutions [269)] cannot be applied to azo coupling. One would have to compare the rate constants of phenolate ion, undissociated phenol, and N,N-dimethylaniline under the same reaction conditions and calculate the ϱ value with the help of the σ_p^+ values for O^-, OH and $N(CH_3)_2$. As mentioned before, Wepster's group [262,270)] and others showed that σ_p^+ values for these substituents are dependent on the specific reaction, but they are not known for azo coupling reactions.

Štěrba et al. [271–275)] published some work on this problem by investigating the kinetics of azo coupling reactions of 4-sulfobenzenediazonium ions with various 4-substituted phenolate ions. Therefore the reaction took place at the 2-position, i.e. ortho to O^- and meta to the other substituent. A $\varrho\sigma$ plot with the σ_m values of these substituents gave a fairly good linear correlation with 5 substituents but not at all with OCH_3. In addition, the resulting ϱ value (—3.77) cannot be meaningful. Electrophilic substitutions with a much lower selectivity than that of azo coupling reactions give more negative values (ϱ = —9.1 and —6.0 for acylation and nitration, respectively) [269)]. In comparison to those substitutions one expects for azo coupling reactions values for ϱ which are significantly more negative than —10.

Analogous comments can be made concerning the other investigations of that group as well as some work made by Golubkin et al. [276)] and by Hashida et al. [277)].

A good linear $\varrho\sigma$-relationship was obtained by Hashida et al. [164)] for azo coupling reactions with acetoacetanilides substituted in the p-position. The small influence of these substituents (ϱ = —0.68) is understandable since the substitution does not take place at the benzene ring. An analogous result was found by Tužálková et al. [278)] for azo coupling reactions of 3′- and 4′-substituted 1-phenyl-3-amino- and 1-phenyl-3-methylpyrazoline-5-ones.

Hashida et al. [164)] also found a good linear relationship for the azo coupling rates of 11 β-dicarbonyl compounds and the pK_a values of these C-acids. However, two dicarbonyl compounds with a trifluoromethyl group attached to one of the carbonyl functions were exceptions. Furthermore, no relationship between log k and pK_a could be observed for azo coupling reactions of active methylene compounds other than β-dicarbonyls.

In conclusion, we summarize the results of this Section by the comment that, in spite of the high sensitivity of the azo coupling reaction to structural parameters, only few results have been obtained which significantly improve our basic knowledge on the reactivity in these systems. However, for a better understanding of electrophilic aromatic substitution reactions, investigations of the azo coupling reaction, due to its high selectivity, still offer possibilities to improve our knowledge of the mechanism of the whole reaction class in the future.

5 References

1. Zollinger, H.: Azo and Diazo Chemistry, New York—London, Interscience 1961
2. Kirmse, W.: Angew. Chem. *88*, 273 (1976)
3. Kirmse, W., Baron, W. J., Seipp, U.: ibid. *85*, 994 (1973)
4. Kirmse, W., Seipp, U.: Chem. Ber. *107*, 745 (1974)
5. Kirmse, W., Schnurr, O., Jendralla, H.: ibid. *112*, 2120 (1979)
6. Curtin, D. Y., Klanderman, B. H., Tavares, D. F.: J. Org. Chem. *27*, 2709 (1962)
7. Lunt, R. S. III.: Dissert. Abstr. *29*, 3108-B (1968–9)
8. Scherer, K. V.: private communication
9. Scherer, K. V., Lunt, R. S. III.: J. Am. Chem. Soc. *88*, 2860 (1966)
10. Zwanenburg, B., Klunder, A. J. H.: Tetrahedron Lett. *1971*, 1717
11. Kirmse, W., Schnurr, O.: J. Am. Chem. Soc. *99*, 3539 (1977)
12. Reimlinger, H.: Angew. Chem. *75*, 788 (1963)
13. Bott, K.: ibid. *76*, 992 (1964)
14. Bott, K.: Chem. Ber. *108*, 402 (1975)
15. Saalfrank, R. W., Ackermann, E.: Liebigs Ann. Chem. *1981*, 7
16. Saalfrank, R. W., Ackermann, E.: Chem. Ber. *114*, 3456 (1981)
17. Szele, I., Tencer, M., Zollinger, H.: Helv. Chim. Acta *66*, in print (1983)
18. Collman, J. P., Yamada, M.: J. Org. Chem. *28*, 3017 (1963)
19. Helwig, R.: Ph. D. Thesis, University of Tübingen
20. Huisgen, H.: Angew. Chem. *67*, 439 (1955)
21. Staudinger, H., Meyer, J.: Helv. Chim. Acta *2*, 619 (1919)
22. Wittig, G., Haag, W.: Chem. Ber. *88*, 1654 (1955)
23. Coleman, G. H., Gilman, H., Adams, C. E., Pratt, P. E.: J. Org. Chem. *3*, 99 (1938)
24. Regitz, M., Heck, G.: Chem. Ber. *97*, 1482 (1964)
25. Kaiser, E. M., Warner, C. D.: J. Organomet. Chem. *31*, C17 (1971)
26. Wenkert, E., McPherson, C. A.: J. Am. Chem. Soc. *94*, 8084 (1972)
27. Müller, E., Huber-Emden, H.: Liebigs Ann. Chem. *660*, 54 (1962)
28. Severin, T.: Angew. Chem. *70*, 745 (1958)
29. Eisert, B., Greiber, B., Caspari, I.: Liebigs Ann. Chem. *659*, 72 (1962)
30. Eisert, B., Regitz, M., Heck, G., Schall, H. in: Methoden der organischen Chemie (Houben-Weyl), 4th ed., Vol. X/4, p. 873, Stuttgart, Georg Thieme Verlag 1968
31. Ciganek, E.: J. Org. Chem. *30*, 4198 (1965)
32. Schönberg, A., Junghans, K.: Chem. Ber. *98*, 820 (1965)
33. Nesnow, S., Shapiro, R.: J. Org. Chem. *34*, 2011 (1969)
34. Hartke, K., Uhde, W.: Tetrahedron Lett. *1969*, 1697
35. Schmiechen, R.: ibid. *1969*, 4995
36. Pyrek, J. S., Achmatowicz, Jr., O.: ibid. *1970*, 2651
37. Huisgen, R., Reissig, H.-U., Voss, S.: ibid. *1979*, 2987
38. Huisgen, R.: Angew. Chem. *75*, 742 (1963)
39. Huisgen, R.: J. Org. Chem. *33*, 2291 (1968)
40. Schöllkopf, U., Wiskott, E., Riedel, K.: Liebigs Ann. Chem. *1975*, 387
41. IUPAC Nomenclature of Organic Chemistry (eds.) Rigaudy, J., Klesney, S. P., Oxford, Pergamon Press 1979
42. Kazitsyna, L. A., Klyueva, M. D.: Dokl. Akad. Nauk SSSR *204*, 99 (1972)
43. Ershov, V. V., Nikiforov, G. A., de Jonge, C. R. H. I.: Quinonediazides, Studies in Organic Chemistry 7, Amsterdam—Oxford—New York, Elsevier 1981
44. Allan, Z., Podstata, J.: Tetrahedron Lett. *1967*, 1911
45. Allan, Z., Podstata, J.: Collect. Czech. Chem. Commun. *34*, 125 (1969)

46. Kazitsyna, L. A., Klyueva, M. D., Romanova, K. V.: Dokl. Akad. Nauk SSSR *183*, 105 (1968)
47. Nikiforov, G. A., Plekhanova, L. G., Efremenko, A. A., Pobedimski, D. G., Ershov, V. V.: Izv. Akad. Nauk SSSR, Ser. Khim. *1971*, 146
48. Huisgen, R., Fleischmann, R.: Liebigs Ann. Chem. *623*, 47 (1959)
49. Schulz, R., Schweig, A.: Angew. Chem. *91*, 737 (1979)
50. Reference 1, p. 41, 153
51. Ried, W., Dietrich, R.: Liebigs Ann. Chem. *666*, 113 (1963)
52. Ried, W., Wagner, K.: ibid. *681*, 45 (1965)
53. Ried, W., Kraemer, R.: ibid. *681*, 52 (1965)
54. Ried, W., Junker, P.: ibid. *709*, 85 (1967)
55. Reference 1, p. 215
56. Weaver, M. A., Shuttleworth, L.: Dyes and Pigments *3*, 81 (1982)
57. Butler, R. N.: Chem. Rev. *75*, 241 (1975)
58. Štěrba, V.: Diazonium-Diazo Equilibrium, in: The Chemistry of Diazonium and Diazo Groups (ed. Patai, S.), p. 84, Chichester, John Wiley & Sons 1978
59. Sawaguchi, H., Hashida, Y., Matsui, K.: Kogyo Kagaku Zasshi *74*, 1859 (1971)
60. Butler, R. N., Scott, F. L.: J. Org. Chem. *32*, 1224 (1967)
61. Hattori, K., Lieber, E., Horowitz, J. P.: J. Am. Chem. Soc. *78*, 411 (1956)
62. Horowitz, J. P., Grakauskas, V. A.: J. Am. Chem. Soc. *79*, 1249 (1957)
63. Gehlen, H., Dost, J.: Liebigs Ann. Chem. *655*, 144 (1963)
64. Tedder, J. M.: Adv. Heterocycl. Chem. *8*, 1 (1967)
65. Reimlinger, H., van Overstraeten, A.: Chem. Ber. *99*, 3350 (1966)
66. Reimlinger, H., Merenyi, R.: ibid. *103*, 3284 (1970)
67. Reimlinger, H., King, G. S. D., Peiren, M. A.: ibid. *103*, 2821 (1970)
68. Simonov, A. M., Andreichikov, Yu. P.: Zh. Org. Khim. *5*, 779 (1969)
69. Spiliadis, A., Bretcann, D., Eftimescu, C., Schip, R. T.: Romanian Pat. 50786 (1968); Chem. Abstr. *70*, 4114h (1969)
70. Pan'kov, A. K., Pevsner, M. S., Bagal, L. I.: Khim. Geterotsikl. Soedin. *1972*, 713
71. Wieland, H.: Liebigs Ann. Chem. *328*, 197 (1903)
72. Goerdeler, J., Haubrich, H.: Chem. Ber. *93*, 397 (1960)
73. Stevenson, H. B., Johnson, J. R.: J. Am. Chem. Soc. *59*, 2525 (1937)
74. Putokhin, N. I., Yakovlev, V. I.: Sb. Nauchn. Tr. Kuibyshev: Ind. Inst. *1953*, 175
75. Albert, A.: Heterocyclic Chemistry, pp. 80–85, London, Athlone Press, 1968
76. Kalatzis, E.: J. Chem. Soc. B, *1967*, 273
77. Henricksen, L., Antrup, H.: Acta Chem. Scand. *26*, 3342 (1972)
78. Kolodyazhnaya, S. N., Simonov, A. M., Zheltikova, N. N., Pozharskii, A. F.: Khim. Geterotsikl. Soedin. *1973*, 714
79. Temple, C., Kussner, C. L., Motgomery, J. A.: J. Org. Chem. *32*, 2241 (1967)
80. Tišler, M., Stanovnik, B.: Heterocycles *4*, 1115 (1976)
81. Sheppard, W. A., Webster, O. W.: J. Am. Chem. Soc. *95*, 2695 (1973)
82. Hünig, S., Boes, O.: Liebigs Ann. Chem. *579*, 28 (1953)
83. Macháček, V., Panchartek, J., Štěrba, V., Večeřa, M.: Collect. Czech. Chem. Commun. *33*, 3154 (1968)
84. Macháček, V., Panchartek, J., Štěrba, V., Večeřa, M.: ibid. *35*, 844 (1970)
85. Macháček, V., Macháčková, O., Štěrba, V.: ibid. *35*, 2954 (1970)
86. Macháček, V., Panchartek, J., Štěrba, V.: ibid. *35*, 3410 (1970)
87. Macháček, V., Macháčková, O., Štěrba, V.: ibid. *36*, 3187 (1971)
88. Reference 1, pp. 199–210
89. Wulfman, D. S.: Synthetic Applications of Diazonium Ions, in: The Chemistry of Diazonium and Diazo Groups (ed. Patai, S.), pp. 268–274, Chichester, John Wiley & Sons 1978
90. Hegarty, A. F.: Kinetics and Mechanisms of Reactions Involving Diazonium and Diazo Groups, in: The Chemistry of Diazonium and Diazo Groups (ed. Patai, S.), pp. 543–545, Chichester, John Wiley & Sons 1978
91. Enders, E., Pütter, R.: Methoden zur Herstellung und Umwandlung von Formazanen, in: Methoden der Organischen Chemie (Houben-Weyl), Vol. X/3, pp. 467, 627, Stuttgart, Georg Thieme Verlag 1965
92. Yao, H. C., Resnick, P.: J. Am. Chem. Soc. *84*, 3514 (1962)
93. Curtin, D. Y., Poutsma, M. L.: ibid. *84*, 4887 (1962)

94. Wahl, H., Lebris, M. T.: Bull. Soc. Chim. Fr. *1952*, 436
95. Wahl, H., Lebris, M. T.: ibid. *1954*, 1281
96. Lebris, M. T., Wahl, H.: ibid. *1954*, 248
97. Behr, L. C.: Indazoles and Condensed Types, in: The Chemistry of Heterocyclic Compounds *22* (ed. Wiley, R. H.), p. 295, New York—London—Sydney, Interscience 1967
98. Huisgen, R., Nakaten, H.: Liebigs Ann. Chem. *586*, 84 (1954)
99. Tröndlin, F., Werner, R., Rüchardt, C.: Chem. Ber. *111*, 367 (1978)
100. Reference 89, p. 273
101. Curtin, D. Y., Tveten, J. L.: J. Org. Chem. *26*, 1764 (1961)
102. Ainley, A. D., Robinson, R.: J. Chem. Soc. *1937*, 369
103. Dyatkin, B. L., Zhuravkova, L. G., Martynov, B. I., Sterlin, S. R., Knunyants, I. L.: J. Chem. Soc., Chem. Commun. *1972*, 618
104. Schwander, H. R.: Dyes and Pigments *3*, 133 (1982)
105. Hung, M.-H., Stock, L. M.: J. Org. Chem. *47*, 448 (1982)
106. Chin, A., Hung, M.-H., Stock, L. M.: ibid. *46*, 2203 (1981)
107. Park, K. K., Archer, M. C., Wishnok, J. S.: Chem. Biol. Interact. *29*, 139
107a. Hocking, J. D., Harris, J. I.: FEBS Lett. *34*, 280 (1973)
108. Butler, A. R., Pogorzelec, P., Shepherd, P. T.: J. Chem. Soc., Perkin II, *1977*, 1452
109. Butler, A. R., Shepherd, P. T.: J. Chem. Research *1978*, (S) 339 and (M) 4471
110. Butler, A. R., Shepherd, P. T.: J. Chem. Soc., Perkin II *1980*, 113
111. Schofield, K.: Heteroaromatic Nitrogen Compounds, p. 76, London, Butterworths 1967
112. Gossauer, A.: Die Chemie der Pyrrole, p. 105, Berlin, Springer 1974
113. Gore, S. T., Mackie, R. K., Tedder, J. M.: J. Chem. Soc., Perkin I *1976*, 1639
114. Challis, B. C., Rzepa, H. S.: J. Chem. Soc., Perkin II *1975*, 1209
115. Genkina, M. K., Eraksina, V. N., Shagalov, L. B., Suvorov, N. N.: Zh. Org. Khim. *15*, 2530 (1979)
116. Genkina, N. K., Mirzametova, R. M., Buyanov, V. N., Suvorov, N. N.: ibid. *17*, 667 (1981)
117. Sarma, P. K., Barooah, S. K.: Indian J. Chem. *15B*, 382 (1977)
118. Spande, T. F., Glenner, G. G.: J. Am. Chem. Soc. *95*, 3400 (1973)
119. Wittwer, Ch., Zollinger, H.: Helv. Chim. Acta *37*, 1954 (1954)
120. Kropáčova, H., Panchartek, J., Štěrba, V., Valter, K.: Collect. Czech. Chem. Commun. *35*, 3287 (1970)
121. Fischer, P. B., Zollinger, H.: Helv. Chim. Acta *55*, 2139 (1972)
122. Fischer, P. B., Zollinger, H.: ibid. *55*, 2146 (1972)
123. Reference 1, p. 178
124. Jaecklin, A. P., Skrabal, P., Zollinger, H.: ibid. *54*, 2870 (1971)
125. Bradley, W., Thompson, J. D.: Nature *178*, 1069 (1956)
126. Messmer, A., Szimán, O.: Angew. Chem. *79*, 237 (1967)
127. Penton, J. R., Zollinger, H.: Helv. Chim. Acta *54*, 573 (1971)
128. Gutman, V.: Chimia *31*, 1 (1977)
129. Bagal, I. L., Skvortsov, S. A., El'tsov, A. V.: Zh. Org. Khim. *14*, 361 (1978)
130. Hashida, Y., Ishida, H., Sekiguchi, S., Matsui, K.: Bull. Chem. Soc. Jpn. *47*, 1224 (1974)
131. Hashida, Y., Tanabe, F., Sekiguchi, S., Matsui, K.: Nippon Kagaku Kaishi *1975*, 1761
132. Hashida, Y., Fujinuma, K., Aikawa, K.: ibid. *1975*, 1369
133. Hashida, Y., Tanabe, F., Matsui, K.: ibid. *1980*, 865
134. Juri, P. N., Bartsch, R. A.: J. Org. Chem. *44*, 143 (1979)
135. Szimán, O., Messmer, A.: Tetrahedron Lett. *1968*, 1625
136. Poindexter, M., McKay, B.: J. Org. Chem. *37*, 1674 (1972)
137. Rufer, D. A.: Ph. D. Thesis, ETH Zurich 1981
138. Hashida, Y., Matsumura, K., Ohmori, Y., Matsui, K.: Nippon Kagaku Kaishi *1979*, 1745
139. Moss, R. A., Rav-Acha, C.: J. Am. Chem. Soc. *102*, 5045 (1980)
140. Zollinger, H.: Helv. Chim. Acta *36*, 1723 (1953)
141. Gokel, G. W., Cram, D. J.: J. Chem. Soc., Chem. Commun. *1973*, 481
142. Kyba, E. P., Helgeson, R. C., Madan, K., Gokel, G. W., Tarnowski, T. L., Moore, S. S., Cram, D. J.: J. Am. Chem. Soc. *99*, 2564 (1977)
143. Korzeniowski, S. H., Gokel, G. W.: Tetrahedron Lett. *1977*, 1637
144. Ahern, M. F., Gokel, G. W.: J. Chem. Soc., Chem. Commun. *1979*, 1019
145. Szele, I., Zollinger, H.: Helv. Chim. Acta *61*, 1721 (1978)

146. Ellwood, M., Griffiths, J.: J. Chem. Soc., Chem. Commun. *1980*, 181
147. Ellwood, M., Gregory, P., Griffiths, J.: Eur. Pat. Appl. 28, 464 (Oct. 8, 1980); Chem. Abstr. *95* P 134370r (1981)
148. Kobayashi, H., Sonoda, T., Iwamoto, H.: Chem. Lett. *1981*, 579
149. Conant, J. B., Peterson, W. D.: J. Am. Chem. Soc. *52*, 1220 (1930)
150. Wistar, R., Bartlett, P. D.: ibid. *63*, 413 (1941)
151. Pütter, R.: Angew. Chem. *63*, 188 (1951)
152. Zollinger, H., Wittwer, C.: Helv. Chim. Acta *35*, 1209 (1952)
153. Zollinger, H.: ibid. *39*, 1600 (1956)
154. Allan, Z. J.: Collect. Czech. Chem. Commun. *16*, 620 (1951)
155. Zollinger, H.: Helv. Chim. Acta *36*, 1070 (1953)
156. Kaválek, J., Panchartek, J., Štěrba, V., Valter, K.: Collect. Czech. Chem. Commun. *35*, 3470 (1970)
157. Štěrba, V., Valter, K.: ibid. *37*, 270 (1972)
158. Macháčková, O., Štěrba, V., Valter, K.: ibid. *37*, 1851 (1972)
159. Cox, A., Goodman, P. D., Kemp, T. J., de Moira, P., Pinot, P.: Int. J. Chem. Kinet. *13*, 97 (1981)
160. Dobáš, I., Štěrba, V., Večeřa, M.: Collect. Czech. Chem. Commun. *34*, 3905 (1969)
161. Mitsumura, K., Hashida, Y., Sekiguchi, S., Matsui, K.: Bull. Chem. Soc. Jpn. *46*, 1770 (1973)
162. Hashida, Y., Mitsumura, K., Sekiguchi, S., Matsui, K.: ibid. *46*, 3263 (1973)
163. Reference 1, p. 213
164. Hashida, Y., Kobayashi, M., Matsui, K.: Bull. Chem. Soc. Jpn. *44*, 2506 (1971)
165. Schwarzenbach, G., Felder, E.: Helv. Chim. Acta *27*, 1701 (1944)
166. Hashida, Y., Shimoda, I., Sekiguchi, S., Matsui, K.: Kogyo Kagaku Zasshi *74*, 73 (1971)
167. Dobáš, I., Štěrba, V., Večeřa, M.: Collect. Czech. Chem. Commun. *34*, 3895 (1969)
168. Macháček, V., Panchartek, J., Štěrba, V., Tunka, J.: ibid. *33*, 3579 (1968)
169. Noelting, E., Grandmougin, E.: Bull. Soc. Chim. Fr. *1891*, 873
170. Noelting, E., Grandmougin, E.: Ber. Dtsch. Chem. Ges. *24*, 1601 (1891)
171. Hanna, S. B., Hunziker, E., Saito, T., Zollinger, H.: Helv. Chim. Acta *52*, 1537 (1969)
172. Hunziker, E., Penton, J. R., Zollinger, H.: ibid. *54*, 2043 (1971)
173. Zollinger, H.: Chimia *27*, 186 (1973)
174. Ott, R. J., Rys, P.: Helv. Chim. Acta *58*, 2074 (1975)
175. Nabholz, F., Ott, R. J., Rys, P.: ibid. *60*, 2926 (1977)
176. Nabholz, F., Rys, P.: ibid. *60*, 2937 (1977)
177. Bourne, J. R., Crivelli, E., Rys, P.: ibid. *60*, 2944 (1977)
178. Belevi, H., Bourne, J. R., Rys, P.: ibid. *64*, 1599 (1981)
179. Belevi, H., Bourne, J. R., Rys, P.: ibid. *64*, 1618 (1981)
180. Belevi, H., Bourne, J. R., Rys, P.: ibid. *64*, 1630 (1981)
181. Besse, J., Schwarz, W., Zollinger, H.: ibid. *64*, 502 (1981)
182. Schwarz, W., Zollinger, H.: ibid. *64*, 513 (1981)
183. Besse, J., Zollinger, H.: ibid. *64*, 529 (1981)
184. Kaválek, J., Panchartek, J., Štěrba, V.: Collect. Czech. Chem. Commun. *35*, 3470 (1970)
185. Zollinger, H.: Helv. Chim. Acta *38*, 1597 (1955)
186. Zollinger, H.: ibid. *38*, 1617 (1955)
187. Zollinger, H.: ibid. *38*, 1623 (1955)
188. Melander, L.: Arkiv Kemi *2*, 213 (1950)
189. Berglund-Larsson, U., Melander, L.: ibid. *6*, 219 (1953)
190. Hammond, G. S.: J. Am. Chem. Soc. *77*, 2444 (1955)
191. Snyckers, F., Zollinger, H.: Helv. Chim. Acta *53*, 1294 (1970)
192. Binks, J. H., Ridd, J. H.: J. Chem. Soc. *1957*, 2398
193. Rys, P., Skrabal, P., Zollinger, H.: Angew. Chem. *84*, 921 (1972), Angew. Chem. Intern. Ed. *11*, 874 (1972)
194. Geidysh, L. S., Nikiforov, G. A., Ershov, V. V.: Izv. Akad. Nauk SSSR, Ser. Khim. *1969*, 2716
195. Zollinger, H.: Acc. Chem. Res. *6*, 335 (1973)
196. Jermini, C., Koller, S., Zollinger, H.: Helv. Chim. Acta *53*, 72 (1970)
197. Duthaler, R. O., Förster, H. G., Roberts, J. D.: J. Am. Chem. Soc. *100*, 4974 (1978)
198. Hanna, S. B., Jermini, C., Zollinger, H.: Tetrahedron Lett. *1969*, 4415
199. Hanna, S. B., Jermini, C., Loewenschuss, H., Zollinger, H.: J. Am. Chem. Soc. *96*, 7222 (1974)
200. Schowen, R. L.: Progr. Phys. Org. Chem. *9*, 275 (1972)

201. Stamm, O. A., Zollinger, H.: Helv. Chim. Acta *40*, 1955 (1957)
202. Eigen, M.: Discuss. Faraday Soc. *39*, 7 (1965)
203. Bell, R. P., Evans, P. G.: Proc. Roy. Soc. A *291*, 297 (1966)
204. Bell, R. P.: Adv. Phys. Org. Chem. *4*, 1 (1966)
205. Bell, R. P.: The Proton in Chemistry, pp. 122, 129, 186. London, Chapman and Hall 1973
206. Grunwald, E., Eustace, D.: Chapt. 4, in: Proton Transfer Reactions (ed. Caldin, E. F., Gold, V.), London, Chapman and Hall 1975
207. Williams, I. H., Spangler, D., Femec, D. A., Maggiora, G. M., Schowen, R. L.: J. Am. Chem. Soc. *102*, 6619 (1980)
208. Williams, I. H., Spangler, D., Femec, D. A., Maggiora, G. M., Schowen, R. L.: ibid. *105*, 31 (1983)
209. Demian, B.: Bull. Soc. Chim. Fr. *1973*, 769
210. Demian, B.: Tetrahedron Lett. *1972*, 3043
211. Hashida, Y., Katoka, S., Matsui, K.: Nippon Kagaku Kaishi *1975*, 1213
211a. Jencks, W. P.: Chem. Soc. Reviews *10*, 345 (1981) and previous papers quoted there
212. Kishimoto, S., Kitahara, S., Manabe, O., Hiyama, H.: Nippon Kagaku Kaishi *1973*, 1975
213. Kishimoto, S., Kitahara, S., Manabe, O., Hiyama, H.: ibid. *1974*, 1962
214. Kishimoto, S., Kitahara, S., Manabe, O., Hiyama, H.: ibid. *1981*, 428
215. Snyckers, F., Zollinger, H.: Tetrahedron Lett. *1970*, 2759
216. Gloor, B., Zollinger, H.: Helv. Chim. Acta *54*, 553 (1971)
217. Gloor, B., Zollinger, H.: ibid. *54*, 563 (1971)
218. Brunning, W., Holtzer, A.: J. Am. Chem. Soc. *82*, 4865 (1961)
219. Schick, M. J.: Phys. Chem. *68*, 3585 (1964)
220. Hashida, Y., Nakajima, K., Sekiguchi, S., Matsui, K.: Kogyo Kagaku Zasshi *72*, 1132 (1969)
221. Chmàtal, V., Allan, Z. J.: Collect. Czech. Chem. Commun. *30*, 1205 (1965); see also references to patent literature in Gloor et al. [217]
222. Clemson, A., Symons, C., Rezakhani, S., Whiting, M. C.: J. Chem. Res. (S) *1980*, 162
223. Koller, S., Zollinger, H.: Helv. Chim. Acta *53*, 78 (1970)
224. Christen, M., Koch, W., Simon, W., Zollinger, H.: ibid. *45*, 2077 (1962)
225. Bagal, I. L., Sire, L. E., El'tsov, A. V.: Zh. Org. Khim. *11*, 1263 (1975)
226. Bubnov, N. N., Bilevitch, K. A., Poljakova, L. A., Okhlobystin, O. Yu.: J. Chem. Soc., Chem. Commun. *1972*, 1058
227. Citterio, A., Minisci, F., Albinati, A., Bruckner, S.: Tetrahedron Lett. *1980*, 2909
228. Citterio, A., Minisci, F.: J. Org. Chem. *47*, 1759 (1982)
229. Citterio, A., Minisci, F., Vismara, E.: ibid., in press
230. Colonna, M., Greci, L., Poloni, M.: J. Chem. Soc., Perkin II *1982*, 455
231. Jackson, A. H., Shannon, P. V. R., Tinker, A. C.: J. Chem. Soc., Chem. Commun. *1976*, 796
232. Bisson, J. M., Hanson, P., Slocum, D.: J. Chem. Soc., Perkin II *1978*, 1331
233. Penton, J. R., Zollinger, H.: Helv. Chim. Acta *64*, 1728 (1981)
234. Shuttleworth, L.: Eastman Kodak, U.S. Pat. 4247458 (1981)
235. Beránek, U., Večeřa, M.: Collect. Czech. Chem. Commun. *34*, 2753 (1969)
236. Beránek, U., Večeřa, M.: ibid. *35*, 3402 (1970)
237. Bagal, I. L., Skvortsov, S. A., El'tsof, A. V.: Zh. Org. Khim. *14*, 2244 (1978)
238. Penton, J. R., Zollinger, H.: J. Chem. Soc., Chem. Commun. *1979*, 819
239. Penton, J. R., Zollinger, H.: Helv. Chim. Acta *64*, 1717 (1981)
240. Arnett, E. M., Chwala, B., Bell, L., Taagepera, M., Hehre, W. J., Taft, R. W.: J. Am. Chem. Soc. *99*, 5279 (1977)
241. Juri, P. N., Bartsch, R. A.: J. Org. Chem. *45*, 2028 (1980)
242. Ogata, Y., Nakagawa, Y., Inaishi, M.: Bull. Chem. Soc. Jpn. *54*, 2953 (1981)
243. Shine, H. J.: Aromatic Rearrangements, Chapter 3, New York, Elsevier 1967
244. Shine, H. J.: M.T.P. Reviews, Series One *3*, 89 (1973)
245. Goldschmidt, H., Johnson, S., Overwien, E.: Z. Phys. Chem. *110*, 251 (1924)
246. Reference 1, p. 185
247. Yamabe, T.: Bull. Chem. Soc. Jpn. *42*, 3565 (1969)
248. Vogel, A. I.: Practical Organic Chemistry, p. 626, London, Longmans 1956
249. American Cyanamid, U.S. Pat. 4018751 (1975)
250. Kelly, R. P., Penton, J. R., Zollinger, H.: Helv. Chim. Acta *65*, 122 (1982)
251. Sakla, A. B., Masoud, N. K., Sawiris, Z., Ebaid, W. S.: ibid. *57*, 481 (1974)

252. Ball, P., Nicholls, C. H.: Dyes and Pigments *3*, 5 (1982)
253. Zollinger, H.: Chemie der Azofarbstoffe, p. 217, Basel, Birkhäuser-Verlag 1958
254. Rys, P., Zollinger, H.: Fundamentals of the Chemistry and Applications of Dyes, p. 49, Chichester—New York, John Wiley 1972
255. Ikeda, T., Manabe, O., Hiyama, H.: Kogyo Kagaku Zasshi *70*, 319 (1967)
256. Ikeda, T., Manabe, O., Hiyama, H.: ibid. *70*, 323 (1967)
257. Ikeda, T., Manabe, O., Hiyama, H.: ibid. *70*, 327 (1967)
258. Panchartek, J., Štěrba, V.: Collect. Czech. Chem. Commun. *34*, 2971 (1969)
259. Kaminski, R., Lauk, U., Skrabal, P., Zollinger, H.: Helv. Chim. Acta *66*, in print (1983)
260. Schneider, L.: Helv. Chim. Acta *51*, 67 (1968)
261. Zollinger, H.: ibid. *36*, 1730 (1953)
262. Bekkum, H. V., Verkade, P. E., Wepster, B. M.: Recl. Trav. Chim. Pays-Bas *78*, 815 (1959)
263. Panchartek, J. Štěrba, V., Vorliček, J., Večeřa, M.: Collect. Czech. Chem. Commun. *33*, 894 (1968)
264. Beránek, V., Kořinková, H., Vetešnik, P., Večeřa, M.: ibid. *37*, 282 (1971)
265. Bagal, I. L., Skvortsov, S. A., El'tsov, A. V.: Zh. Org. Khim. *14*, 361 (1978)
266. Remeš, M., Diviš, J., Zvěrina, V., Matrka, M.: Collect. Czech. Chem. Commun. *38*, 1049 (1973)
267. Skvortsov, S. A., Barnakov, Ch. N., Bagal, I. L.: Zh. Org. Khim. *17*, 133 (1981)
268. Ritchie, C. D., Wright, D. J.: J. Am. Chem. Soc. *93*, 6574 (1971)
269. Brown, H. C., Stock, L. M.: ibid. *84*, 3298 (1962)
270. Hoefnagel, A. J., Wepster, B. M.: ibid. *95*, 5357 (1973)
271. Dobáš, I., Štěrba, V., Večeřa, M.: Chem. Ind. London *1968*, 1814
272. Dobáš, I., Štěrba, V., Večeřa, M.: Collect. Czech. Chem. Commun. *34*, 3746 (1969)
273. Dobáš, I., Panchartek, J., Štěrba, V., Večeřa, M.: ibid. *35*, 1288 (1970)
274. Štěrba, V., Valter, K.: ibid. *37*, 1327 (1972)
275. Kulič, J., Titz, M., Večeřa, M.: ibid. *40*, 405 (1975)

Some Aspects of Organic Synthesis Using Organoborates

Akira Suzuki

Department of Applied Chemistry, Faculty of Engineering, Hokkaido University, Sapporo 060, Japan

Table of Contents

I Introduction

Organoboranes readily obtainable by the hydroboration of alkenes and alkynes have become valuable reagents and intermediates for conversion and carbon—carbon bond formation reactions. Reviews [1-6] and books [7-10] have highlighted many uses of such compounds in organic synthesis. Among such synthetic applications of organoboron compounds, this article extends the perspective of organic synthesis using tetracoordinate organoborates by supplementing earlier important citations with reports of recent applications.

The facile addition of the boron—hydrogen bond to carbon—carbon multiple bonds of unsaturated organic derivatives makes the corresponding tricoordinate organoboranes. Thus trialkylboranes are readily prepared from 3 moles of alkenes and 1 mole of borane (BH_3). The hydroboration reaction has various characteristics. The reaction is extremely facile, usually being complete within a few minutes at temperatures below 25 °C. In general, hydroboration proceeds through the anti-Markovnikov addition of the B—H moiety to C=C bonds. Furthermore, the addition of B—H to a double bond is cis from the less hindered side of the unsaturated organic compound. The ready availability of organoboranes with a wide variety of structures by the hydroboration of alkenes and alkynes is one of most remarkable advantages, compared with other organometallic compounds.

In addition to that, organoboranes thus obtained have highly favorable characteristics for organic synthesis. For example, as shown in Eq. 1, the organoborane (*2*) prepared from the olefin (*1*) at room temperature gives the alcohol (*4*) by alkaline hydrogen peroxide oxidation. The organoborane (*2*) is easily isomerized by heating at 150 °C to *3*, which is oxidized with hydrogen peroxide under alkaline conditions to produce the alcohol (*5*). Consequently, two different alcohols can be selectively obtained in good yields from an alkene.

In spite of having such a merit for synthetic applications, organoboranes are chemically inert. For instance, trialkylboranes are markedly stable toward water, alcohols and phenols. Organoboranes do not undergo Grignard-type reactions with carbonyl compounds, and are not effective as a catalysts in the chain-growth reaction which is well-known in the case of alkylaluminum compounds. The reasons why organoboranes have properties so different from those of other organometallics could be elucidated by the following three characteristics of organoboranes: (A) A

```
            HB<                          Δ
C-C=C-C  -------->   C-C-C-C    -------->   C-C-C-C
                       |                          |
                       B<                         B<

   1                   2                      3              (1)

                       | [O]                  | [O]
                       v                      v

                     C-C-C-C    -------->   C-C-C-C
                       |                          |
                       OH                         OH

                       4                      5
```

small difference in the electronegativity between boron and carbon. In alkyllithium compounds, the C—Li bond has an electronegativity difference of 1.5. According to the Pauling's calculation, 43% of ionic character is expected in the carbon—lithium bond, whereas the carbon—boron bond is considered to have only 6% of ionic character. This is the major reason why organoboranes are inert to ionic reactions. (B) The second point is the vacant electron structure of the boron atom. Organoboranes are, therefore, readily attacked on the boron atom by bases and nucleophiles. This property provides a basis for a number of unique and selective synthetic reactions using organoboranes, as described later. (C) The third point is that the C—B bond length is almost the same length as that of a C—C bond. For example in trimethylboroxine and trimethylborane, the C—B bond lengths were reported to be 1.57 and 1.56 Å, respectively. Consequently, such steric features of organoboranes should be considered in their reactions.

II Preparation of Tetracoordinate Organoborates

Although there are sevaral methods of preparing organoboranes such as by transmetallation, by ligand exchange reaction, and by Friedel-Crafts type reaction, the best and most convenient is the procedure via hydroboration, as mentioned previously. No detailed discussion of the hydroboration is attempted here because many authoritative and comprehensive books [7,8,11] are available.

Organoboranes thus obtained usually exist as electron-deficient trigonal species with the empty p-orbital on the boron atoms and act as Lewis acids or electrophiles. Therefore, they have a strong tendency to react with bases or nucleophiles to form the corresponding tetracoordinate organoborates ("ate" complexes, *6*) (Eq. 2). As X^-s, both hetero-atom anions and carbanions are employed. Although the

$$R_3B \quad + \quad X^- \quad \longrightarrow \quad R_3\bar{B}X \qquad (2)$$

6

complexation reaction of trialkyboranes with organolithium compounds is known to be highly general, hindered alkyllithiums with β-hydrogen atoms tend to give tri-organoborohydrides in their reaction with triorganoboranes (Eqs. 3 [12] and 4 [13]).

$$\text{(thexyl-bicyclic borane)} + CH_3\text{-}\underset{CH_3}{\overset{CH_3}{C}}\text{-}Li \longrightarrow \left[\text{(thexyl-bicyclic }\bar{B}\text{-}H)\right] Li^+ + \underset{CH_3}{\overset{CH_3}{}}\!\!>C{=}CH_2 \qquad (3)$$

$$\text{thexyl} = \underset{CH_3}{\overset{CH_3}{CH}}\text{—}\underset{CH_3}{\overset{CH_3}{C}}\text{—}$$

$$\underset{H}{\overset{R}{}}\!\!>C{=}C<\!\!\underset{B(C_6H_{11})_2}{\overset{X}{}} + CH_3\text{-}\underset{CH_3}{\overset{CH_3}{C}}\text{-}Li \longrightarrow \left[\underset{H}{\overset{R}{}}\!\!>C{=}C<\!\!\underset{\bar{B}H(C_6H_{11})_2}{\overset{X}{}}\right]Li^+ + \underset{CH_3}{\overset{CH_3}{}}\!\!>C{=}CH_2 \qquad (4)$$

It was also reported that charge-delocalized organolithiums derived from carbon acids whose pK_a are lower than ca. 20, such as cyclopentadienyllithium, indenyllithium and α-lithioacetophenone, do not form the corresponding lithium organoborates to any detectable extent [14].

There is an alternative method for the preparation of organoborates which involves the reaction of triorganoborohydrides with alkenes or alkynes. The reaction with alkenes gives addition products [15] (Eq. 5), whereas that with alkynes provides substitution products [16] (Eq. 6). As 1-alkynyltrialkylborates are readily prepared by the reaction of trialkylboranes with lithium acetylides, the complexation reaction seems to be most convenient for producing tetracoordinate organoborates.

$$NaR_3BH \quad + \quad CH_2{=}CHR' \longrightarrow [R_3\bar{B}CH_2CH_2R']Na^+ \tag{5}$$

$$NaR_3BH \quad + \quad CH{\equiv}CR' \longrightarrow [R_3\bar{B}C{\equiv}CR']Na^+ \tag{6}$$

The structures of organoborates are not well established in the majority of cases. According to X-ray crystallographic studies of a few organoborates, it was reported that, in some cases such as lithium dimesitylborodihydride bis(dimethoxyethane) [17] and 2,2,5,5-tetra-n-hexyl-4,6-bis(diisopropylimino)-2,5-diborata-1,3-dioxacyclohexane [18], the borates exist as negatively charged tetrahedral boron anions. In some cases, however, the monomeric representation is inadequate. Thus, crystalline lithium tetramethylborate is in existence as a polymeric species [19]. Recently, many results on NMR spectral studies of organoborates have been reported, which offer valuable information of the structure of such borate complexes [14, 20], although these are not discussed here in detail.

In this article, we tentatively adopt the simple monomeric tetrahedral structure for organoborates.

III Synthesis via Organoborates (R_3B—X—Y) with Latent Leaving Groups (Y) in the α-Position

As depicted in Eq. 2, triorganoboranes are readily attacked with nucleophiles on the boron atom, accordingly the complexes (*7*) can be obtained as shown in Eq. 7. If the nucleophile (X—Y) has a latent leaving group (Y), the elimination of Y may bring about a concurrent 1,2-migration of R on boron to X to form the product (*8*). On the other hand, if the contribution of the structure (*9*) is considerable, there may be another

$$R_3B \quad + \quad :\bar{X}\text{-}\overset{+}{Y} \rightleftharpoons \underset{\substack{| \\ R \\ \mathbf{7}}}{R_2\bar{B}}\text{—}X\text{—}\overset{+}{Y} \longrightarrow \underset{\substack{| \\ R \\ \mathbf{8}}}{R_2B\text{-}X} \quad + \quad Y$$

$$\updownarrow$$

$$\underset{\substack{| \\ R \\ \mathbf{9}}}{R_2\bar{B}}\text{—}\overset{+}{X}\text{—}Y \longrightarrow \underset{\substack{| \\ R \\ \mathbf{10}}}{R_2B\text{-}X\text{-}Y} \tag{7}$$

possibility where the migration of R to X occurs without elimination of Y to give the product (*10*), as demonstrated in the reaction with carbon monoxide [7-10,21].

There have been many reports on the carbon—hetero bond formation via the intermediates (*7* and *8*). One of such representative examples is the oxidation of organoboranes with alkaline hydrogen peroxide [22]. The following mechanism has been proposed for the base—catalyzed reaction (Eq. 8).

$$H_2O_2 + OH^- \rightleftharpoons HO\text{-}O^- + H_2O \qquad (8\text{-}1)$$

$$HO\text{-}B(R)(OH) + {}^-O\text{-}OH \rightleftharpoons HO\text{-}\overset{-}{B}(R)(OH)\text{—}O\text{—}OH \xrightarrow{-OH^-} HO\text{-}B(OH)\text{-}O\text{-}R$$

$$\xrightarrow{H_2O} ROH + H_3BO_3 \qquad (8\text{-}2)$$

Organoboranes, though themselves relatively inert to halogens, undergo such reactions readily under basic conditions. For instance, a recent study indicates that the usually sluggish reactions of trialkylboranes with bromine or iodine are greatly accelerated by the presence of methanolic sodium methoxide [23], and the reaction was confirmed to proceed with inversion of configuration of the alkyl group (Eq. 9) [24]. The following S_H2 mechanism seems to be considered (Eq. 10). The reactions, however, have potential problems. One of them is the necessity of

$$\text{(norbornyl–B, H)} \xrightarrow[I_2]{NaOMe} \text{(norbornyl–I, H)} \qquad (9)$$

$$R\text{-}\overset{-}{B}\text{—}OMe \xrightarrow{I_2} \circ I \text{○○} I \text{○○} R \text{○○} B \text{—}OMe \qquad (10)$$

using a strong base which could either react with sensitive functional groups in complex molecules or initiate dehydroiodination reactions. Recently a mild and convenient procedure for conversion of organoboranes into alkyl iodides was reported via the reaction of iodine monochloride in the presence of sodium acetate in methanol (Eq. 11). This reaction proceeds via an inversion of configuration at the carbon attached to boron (Eq. 12) [25]. Organic iodides are also produced by the reaction of organoboranes with sodium iodide in the presence of mild oxidizing agents such as chloramine-T [26]. Organoboranes also react with either bromine or bromine chloride in aqueous media to yield corresponding bromides [27].

$$R_3B \xrightarrow[NaOAc/MeOH]{\text{excess ICl}} \underset{97\text{-}82\%}{2RI} \qquad (11)$$

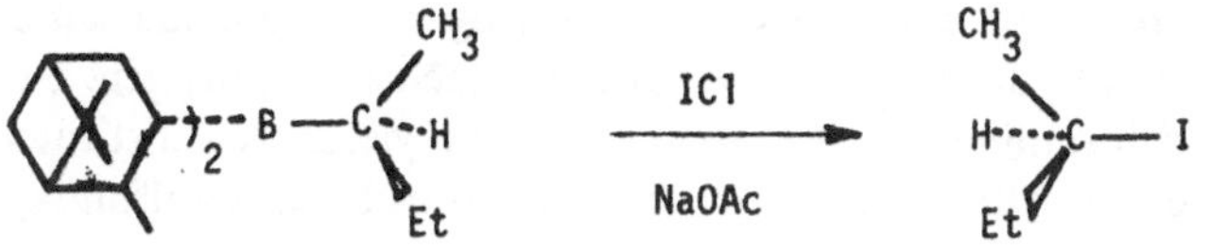

$$\left(\text{Ipc}\right)_2\text{B}-\text{C}(\text{CH}_3)(\text{H})(\text{Et}) \xrightarrow[\text{NaOAc}]{\text{ICl}} \text{H}\cdots\text{C}(\text{CH}_3)(\text{Et})-\text{I} \quad (12)$$

(92% optical purity) (87% optical purity)

It is well-known that organoboranes can be utilized to yield primary amines by treating them with chloramine or hydroxylamine-0-sulfonic acid [7-10]. Recently Kabalka et al. reported a convenient synthesis of primary alkyl amines via the reaction of organoboranes with ammonium hydroxide in the presence of sodium hypochlorite (Eq. 13) [28]. The reaction presumably proceeds through the in situ formation of chloramine.

$$R_3B \quad + \quad 2NH_4OH \xrightarrow{\text{NaOCl}} 2RNH_2 \quad (13)$$

96-72%

The reaction of organic azides with triethylborane in refluxing xylene, followed by treatment with methanol, gives secondary amines in good yields (Eq. 14) [29]. The observed reaction rate indicates that the reaction proceeds via initial coordination

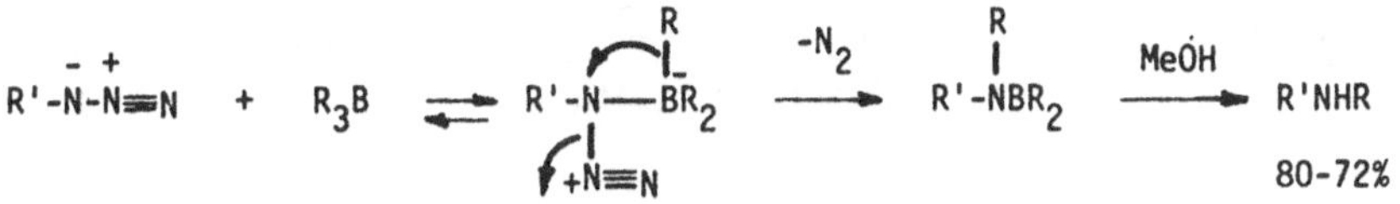

$$R'\text{-}\overset{-}{N}\text{-}\overset{+}{N}{\equiv}N \quad + \quad R_3B \rightleftharpoons R'\text{-}N(\overset{+}{N}{\equiv}N)\text{-}\overset{-}{B}R_2(R) \xrightarrow{-N_2} R'\text{-}N(R)BR_2 \xrightarrow{\text{MeOH}} R'NHR \quad (14)$$

80-72%

between the azide and triethylborane, followed by transfer of an ethyl group from boron to nitrogen. The reaction is slow with sterically hindered azides and fails when both sterically hindered azides and organoboranes are used. Use of dialkylchloroboranes [30] and alkyldichloroboranes [31] in place of trialkylboranes results in fairly rapid reactions even in the case of relatively hindered organoboranes and azides.

When vinylic azides, such as α-azidostyrene and 2-azido-1-alkenes, are used instead of alkyl azides, a different type of reaction occurs, in which alkyl group migration takes place from boron to the vinylic carbon, followed by hydrolysis to give the corresponding ketones (Eq. 15) [32].

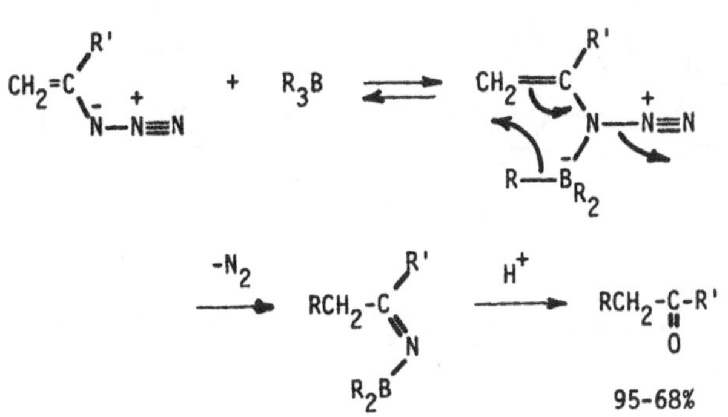

$$CH_2{=}C(R')\text{-}\overset{-}{N}\text{-}\overset{+}{N}{\equiv}N \quad + \quad R_3B \rightleftharpoons CH_2{=}C(R')\text{-}N(\overset{+}{N}{\equiv}N)\text{-}\overset{-}{B}R_2(R)$$

$$\xrightarrow{-N_2} RCH_2\text{-}C(R'){=}N\text{-}BR_2 \xrightarrow{H^+} RCH_2\text{-}\underset{\|}{C}\text{-}R' \;(O)$$

95-68% (15)

There are examples of similar reactions leading to nitrogen compounds which possess substituents other than hydrogen or alkyl groups attached to nitrogen. For example, the reaction of chloramine-T with organoboranes yields N-substituted toluenesulfonamides (Eq. 16) [33]. Although the reaction of p-nitrobenzenesulfonoxy-

$$R_3B \; + \; {}^{-}NClTs\;Na^{+} \longrightarrow R_2\overset{-}{B}(R)\text{—}NTs(Cl)\;Na^{+} \longrightarrow R_2B(R)NTs \xrightarrow{NaOH} \underset{84\text{-}69\%}{RNHTs} \tag{16}$$

$(Ts = p\text{-}MeC_6H_4SO_2)$

carbamate with organoboranes under basic conditions was unsatisfactory for the synthesis of N-alkylcarbamates, the reaction was found to occur smoothly in an organic-aqueous two-phase system in the presence of phase-transfer catalysts, such as quaternary ammonium halides, to give N-alkylcarbamates (Eq. 17) [34]. Organoboranes react with iron(III) azide in the presence of hydrogen peroxide to yield the corresponding azidoalkanes. The reaction seems to proceed by a free-radical chain reaction path in which iron(II) ions, contained inherently in a small amount even after sufficient purification, participate (Eq. 18) [35].

$$O_2N\text{-}C_6H_4\text{-}SO_3NHCO_2Et \; + \; R_3B \xrightarrow[PhCH_2\overset{+}{N}Et_3\cdot Cl^{-}\;(catalyst)]{NaHCO_3\text{-}H_2O/CH_2Cl_2} \underset{97\text{-}71\%}{RNHCO_2Et} \tag{17}$$

$$R_3B \; + \; N_3^{-} \; + \; H_2O_2 \xrightarrow[Fe^{2+}\;(catalyst)]{Fe^{3+}} \underset{100\text{-}56\%}{RN_3} \tag{18}$$

Many carbon—carbon bond formation reactions, where migration of alkyl groups from boron to carbon with the concurrent elimination of leaving groups occur, have been reported. Some of such earlier representative syntheses are shown in Eqs. 19 [36], 20 [37], and 21 [38].

$$Et_3B \; + \; \text{2-bromocyclohexanone} \xrightarrow{t\text{-}BuOK} \text{[2-Br-2-}(\overset{-}{B}Et_3)\text{cyclohexanone]} \longrightarrow \text{2-Et-2-}BEt_2\text{-cyclohexanone} \xrightarrow{t\text{-}BuOH} \underset{68\%}{\text{2-ethylcyclohexanone}} \tag{19}$$

$$R_3B \; + \; :\overset{+}{N}{\equiv}\overset{-}{N}\text{-}CHCO_2Et \longrightarrow R_2\overset{-}{B}(R)\text{—}CHCO_2Et(\overset{+}{N}{\equiv}N) \longrightarrow R_2B(R)CHCO_2Et \xrightarrow{HX} \underset{83\text{-}40\%}{RCH_2CO_2Et} \tag{20}$$

$$R_3B + Me_2\overset{+}{S}\text{-}\overset{-}{C}HCO_2Et \longrightarrow R_2\overset{R}{\overset{|}{B^-}}\text{—}\underset{+SMe_2}{\underset{|}{C}}HCO_2Et \longrightarrow R_2B\overset{R}{\overset{|}{C}}HCO_2Et \xrightarrow{HX} \underset{52\text{-}31\%}{RCH_2CO_2Et} \quad (21)$$

Recently, the following synthesis of secondary alcohols by the reaction of trialkylboranes with bis(phenylthio)methyllithium has been found (Eq. 22)[39]. The

$$R_3B + LiCH(SPh)_2 \longrightarrow [R_3\overset{-}{B}CH(SPh)_2]\overset{+}{Li} \xrightarrow{-LiSPh} R_2B\overset{R}{\overset{|}{C}}HSPh$$

$$\xrightarrow{HgCl} \underset{Cl\ \ R}{RB\text{—}\overset{R}{CH}} \xrightarrow{H_2O_2,\ OH^-} \underset{88\text{-}75\%}{R_2CHOH} \quad (22)$$

reaction of trialkylboranes with thiomethoxymethyllithium or 2-lithiothiomethoxy-*1,3-thiazoline*, followed by treatment of the resultant α-thioorganoborate complexes with methyl iodide, produces the one-carbon homologation in high yields[40]. Using a stereochemically defined α-bromo- or α-iodo-organoboranes, it has been found that the base-induced migration of an alkyl group from boron to the α-halo carbon in the absence of solvent is stereospecific and occurs with essentially complete inversion at the migration terminus. Remarkably, in the presence of the normal hydroboration solvents such as THF or dimethyl sulfide, the α-halo-carbon suffers complete loss of stereochemistry. The THF appears to cause the epimerization of the organoborane without causing migration[41].

Cyanotrialkylborates (*11*), which can be readily prepared and are stable, are of interest due to their synthetic applicability in organic synthesis. Although no detailed introduction of the reactions[9] is given here, a brief account is presented. Cyanotrialkylborates react smoothly with 1 equivalent of trifluoroacetic anhydride (TFAA) at −78 °C and then at room temperature. Oxidation of the intermediates with hydrogen peroxide under basic conditions gives the corresponding symmetrical ketones in excellent yields[42]. The reaction is considered to proceed via the reaction path, as indicated in Eq. 23. Unsymmetrical ketones can also be obtained

$$R_3B + CN^- \longrightarrow \underset{11}{R_3\overset{-}{B}CN} \xrightarrow{TFAA} R_2\overset{R}{\overset{|}{B^-}}\text{—}C\equiv\overset{+}{N}\text{-}COCF_3 \longrightarrow RB(R)\text{—}C(R)(=N\text{—}C(CF_3)=O)$$

$$\longrightarrow \underset{12}{RB\text{—}CR_2\text{—}N{=}C(CF_3)\text{—}O} \xrightarrow{H_2O_2,\ OH^-} \underset{100\text{-}84\%}{R_2C{=}O} \quad (23)$$

by using mixed thexyldialkylboranes (*14*), prepared by a modified hydroboration with thexylborane (*13*) (Eq. 24) [42]. An exceptional cyanidation reaction for the ketone

$$(CH_3)_2C{=}C(CH_3)_2 + BH_3 \longrightarrow \underset{13}{\text{Thx-}BH_2} \xrightarrow{\text{Olefin A}} \text{Thx-}B(R)H \xrightarrow{\text{Olefin B}} \underset{14}{\text{Thx-}B(R)R'} \xrightarrow{CN^-} \xrightarrow[-78 \to 0^\circ]{\text{TFAA}} \underset{85\text{–}76\%}{R\text{–}\underset{\|}{\overset{}{C}}\text{–}R'\ (=O)} \quad (24)$$

synthesis was reported. The seven-membered cyclic borane obtained from linalyl acetate and thexylborane, loses 2,3-dimethyl-2-butene, and the resulting dialkylborane gives an alcohol on treatment with sodium cyanide, followed by trifluoroacetic anhydride and oxidation, as revealed in Eq. 25 [43]. Interestingly, when an excess of TFAA is used as an electrophile, the third migration of an alkyl group from boron to carbon occurs to give trialkylmethanol (Eq. 26) [44]. The reaction is much affected by the reaction temperatures and solvents used. Protonation of cyanotrialkylborates in diglyme with strong acids such as methanesulfonic acid in place of TFAA, followed by a period of heating at 100 °C, leads to the unexpected intermediates (*15*). Oxidation with alkaline hydrogen peroxide produces ketones but the yield is low (45–40%) (Eq. 27) [45].

OAc; $\text{Thx-}BH_2$; OAc; CN^-; OAc; B–CN; $-$ 2,3-dimethyl-2-butene; OAc; B; $-$ 2,3-dimethyl-2-butene; OAc; CN^-; OAc; B–CN, H; B–H; TFAA; OAc; H_2O_2, OH^-; OAc; OH; B—$OCOCF_3$, $OCOCF_3$ (25)

12 $\xrightarrow[CF_3COO]{TFAA}$ [cyclic intermediate: R, R, R–$\bar{B}$–C–N^+–$COCF_3$; CF_3COO; O–C–CF_3] → CF_3COO–B(CR_3)–O–C(CF_3)=N–$COCF_3$ ($CF_3CON{=}C$) $\xrightarrow{H_2O_2,\ OH^-}$ R_3COH 96–70% (26)

$4R_3BCN \xrightarrow{HX} 4R_3BC{\equiv}\overset{+}{N}H \rightarrow$ 15 + $2R_3B$

15 $\xrightarrow{H_2O_2,\ OH^-}$ $2R_2CO$ 45–40% (27)

All three possible migrations from boron to carbon in the cyanoborate process involve retention of configuration at the migrating center. For each of the three migrations, the relative migratory aptitudes of alkyl groups are in the order n- > s- > t-, a result rationalized in terms of the charge density at carbon in the transition state [46]. Recently the synthesis of natural products has been reported, in which the cyanoborate reaction highligths the migratory aptitude of alkyl groups. As a consequence of the low migratory tendency of the thexyl group (2,3-dimethyl-2-butyl), thexyldialkylboranes play an important role in many organoboron-mediated carbon-carbon bond-forming reactions. A variety of thexyldialkylboranes are available via sequential hydroboration of the appropriate alkenes with thexylborane, as depicted in Eq. 24. Zweifel and Pearson [47] reported an alternative preparation of mixed thexyldialkylboranes with thexylchloroborane. This method enhances

Thexyl–BH_2 $\xrightarrow{HCl}$ Thexyl–B(H)(Cl) $\xrightarrow{CH_2{=}CHC_8H_{17}}$ Thexyl–B($C_{10}H_{21}$)(Cl) $\xrightarrow{H_{11}C_5CH{=}CH(CH_2)_3MgCl}$ Thexyl–B($C_{10}H_{21}$)–$(CH_2)_3CH{=}CHC_5H_{11}$ $\xrightarrow{1.\ NaCN;\ 2.\ C_6H_5COCl;\ 3.\ H_2O_2,\ OH^-}$ O=C($C_{10}H_{21}$)–$(CH_2)_3CH{=}CHC_5H_{11}$ (16) (28)

the synthetic utility of thexyl-substituted organoboranes in the synthesis of ketones via the cyanidation reaction, which is exemplified by the preparation of (Z)-6-heneicosen-11-one (*16*), the sex pheromone of the Douglas fir tussoch moth (Eq. 28).

Garst and Bonfiglio reported a facile synthesis of δ-coniceine via the cyanoborate reaction [47a)]. Recently, Brown et al. found a new synthetic procedure of thexylchloroborane by the hydroboration of 2,4-dimethyl-2-butene with chloroborane-dimethyl sulfide [48)], and also reported that thexylalkylchloroboranes are readily reduced by the reagent potassium triisopropoxyborohydride (KIPBH). The resulting thexylalkylborane then readily hydroborates a second mole of olefin to give the mixed thexyldialkylborane in high purity (Eq. 29) [49)].

$$\text{ThxB(Cl)H} + CH_2{=}CHR \longrightarrow \text{ThxB(Cl)}CH_2CH_2R \xrightarrow{\text{KIPBH}} \text{ThxB(H)}CH_2CH_2R \xrightarrow{CH_2=CHR'} \text{ThxB}(CH_2CH_2R')(CH_2CH_2R) \qquad (29)$$

The reaction of organoboranes with carbon monoxide is one of the most versatile reactions among synthetic applications of organoboranes. Triorganoborane-CO adducts, although not stable enough to be isolated, are isoelectronic with cyanotrialkylborates. A wide variety of primary, secondary and tertiary alcohols, aldehydes, and ketones have been synthesized by this reaction. Since such reactions have been extensively reviewed [7-10,21)], only recent progress is presented here.

Lithium trialkylborohydrides react readily with carbon monoxide at —25 °C to presumably give highly valuable synthetic intermediates (*17*), whereas the corresponding potassium compounds do not. A recent paper reports that in the case of undissociated alkali metal trialkylborohydrides which fail to react at 0–25 °C with carbon monoxide, rapid reaction is induced by the presence of a small quantity of free trialkylboranes (Eq. 30) [50)]. A possible mechanism for the carbonylation is given in Eq. 31. Reversible coordination of the trialkylborane with carbon monoxide is followed by reduction of the carbonyl adduct by the trialkylborohydride. Subsequent migration of an alkyl group from boron to carbon forms *17* and trialkylborane. The latter is recycled until all the trialkylborohydride is consumed. Thus the trialkylborane serves as a catalyst in this reaction. Reduction by lithium aluminum hydride of the intermediate formed in the hydride-induced carbonylation of B—R-9-BBN provides a valuable new method for the stereospecific homologation of B—R-9-BBN → B—RCH_2-9-BBN [51)].

$$MR_3BH + CO \xrightarrow[\text{THF, -25 °C}]{R_3B\text{ (catalyst)}} R_2B\underset{\displaystyle 17}{CH(OM)R} \qquad (30)$$

$$R_3B + CO \rightleftharpoons R_3\bar{B}{:}C{\equiv}\overset{+}{O}{:} \longrightarrow R_3\bar{B}{:}\overset{+}{C}{=}\ddot{O}{:} \qquad (31\text{-}1)$$

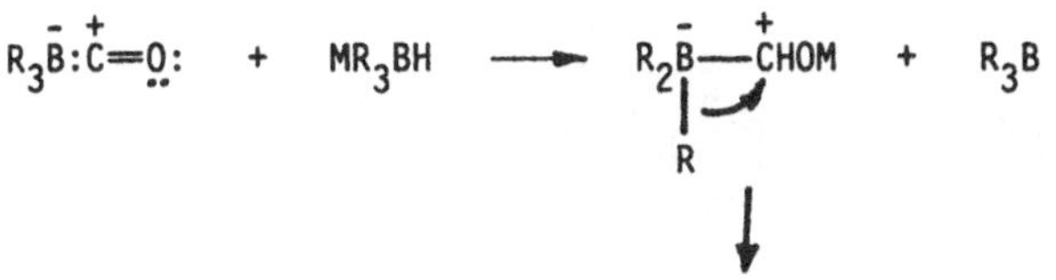

$$R_3\bar{B}{:}\overset{+}{C}{=}\ddot{O}{:} \quad + \quad MR_3BH \longrightarrow R_2\overset{-}{B}(R){-}\overset{+}{C}HOM \quad + \quad R_3B \qquad (31\text{-}2)$$

$\downarrow$

17

One of applications of organoborane-carbonylation reactions for steroid synthesis is revealed in Eq. 32 [52].

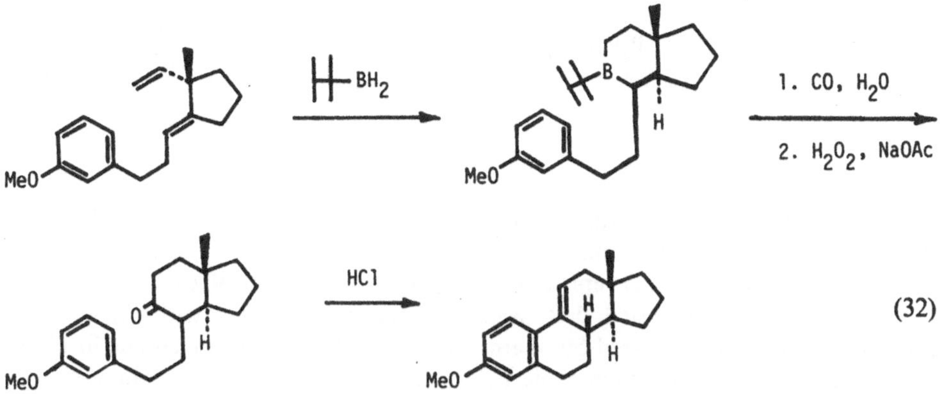

MeO — $\mathrm{BH_2}$ — MeO — 1. CO, H_2O / 2. H_2O_2, NaOAc

MeO — O — H — HCl — MeO — H — H (32)

The carbonylation reaction of organoboranes can be utilized to produce various kinds of organic compounds, which are also prepared virtually by the cyanoborate process. For example, both reactions give corresponding trialkylmethanols by the three-migration of alkyl groups on boron. Until most recently, it has been assumed that both of reactions would lead to the same tertiary alcohol and that in any specific case the choice of reaction would be governed solely by convenience and the tolerance of the organoborane to the reaction conditions. Recently Pelter et al. showed that this assumption is invalid since different tertiary alcohols can be obtained from the same organoborane depending on the method employed, as shown in Eq. 33 [53].

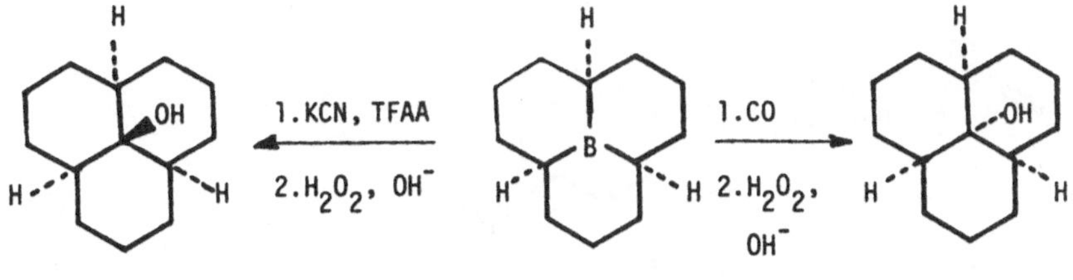

H — OH — H — H — 1. KCN, TFAA / 2. H_2O_2, OH^- — H — B — H — H — 1. CO / 2. H_2O_2, OH^- — H — OH — H — H (33)

The reaction of organoboranes with catechol dichloromethylene ether in the presence of methyllithium provides a synthesis of symmetrical ketones (Eq. 34) [54]. From the result of a crossover experiment by using a mixture of two different trialkylboranes, the reaction was confirmed to proceed through intramolecular reaction pathways.

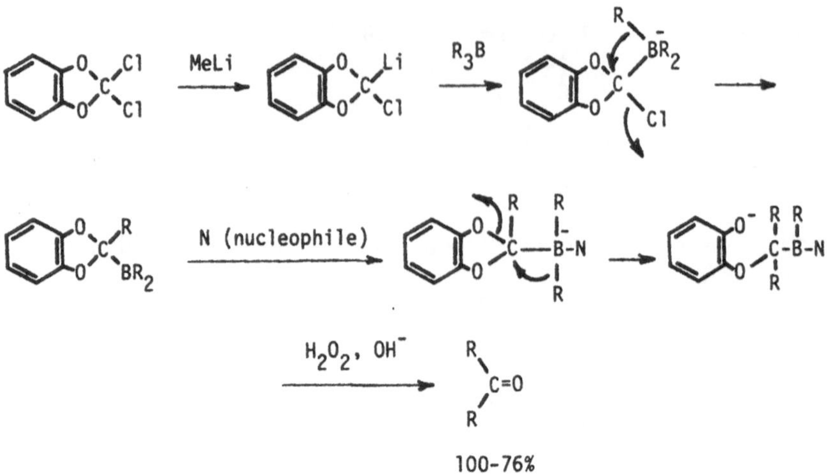

$$\text{catechol-C}(Cl)_2 \xrightarrow{\text{MeLi}} \text{catechol-C}(Li)(Cl) \xrightarrow{R_3B} \text{catechol-C}(Cl)(\overset{-}{B}R_3) \longrightarrow$$

$$\text{catechol-C}(R)(BR_2) \xrightarrow{\text{N (nucleophile)}} \text{catechol-C}(R)-\overset{-}{B}(R)_2-N \longrightarrow \text{catechol(O}^-)\text{-C}(R)_2-B(R)-N$$

$$\xrightarrow{H_2O_2,\ OH^-} R_2C{=}O \qquad (34)$$

100–76%

In recent years, many reports have been published dealing with new syntheses of many types of organic compounds from organoboranes. However, there was, until most recently, no successful report on the direct preparation of carboxylic acids. Recently it was demonstrated that organoboranes readily react with the dianion of phenoxyacetic acid, prepared by treatment with lithium diisopropylamide in THF, to give the corresponding carboxylic acids, although introduction of bulky alkyl groups is sluggish and required modified conditions (Eq. 35) [55]. This reaction is applicable for the synthesis of olefinic carboxylic acids by using alkenylboranes formed from monohydroboration of alkadienes with 9-BBN. Thus 6-octenoic acid is prepared from 1,4-hexadiene in 93% yield, as depicted in Eq. 36.

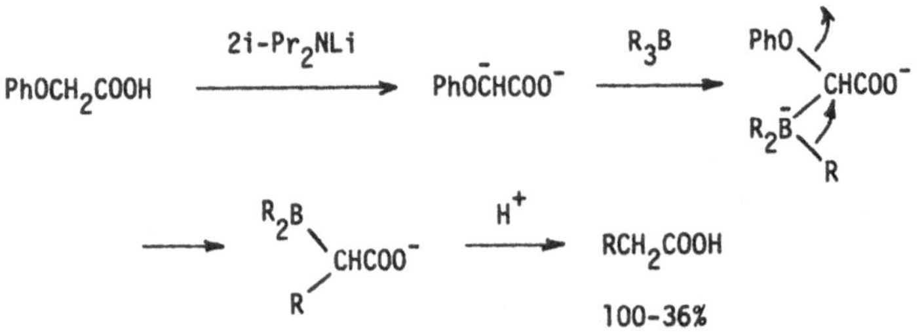

$$PhOCH_2COOH \xrightarrow{2\,i\text{-}Pr_2NLi} Ph O\overset{-}{C}HCOO^- \xrightarrow{R_3B} PhO\text{-}CHCOO^-(\overset{-}{B}R_3)$$

$$\longrightarrow R_2B(R)CHCOO^- \xrightarrow{H^+} RCH_2COOH \qquad (35)$$

100–36%

$$\text{1,4-hexadiene} + \text{HB(9-BBN)} \longrightarrow \text{CH}_3\text{CH=CH(CH}_2)_3\text{-B(9-BBN)}$$

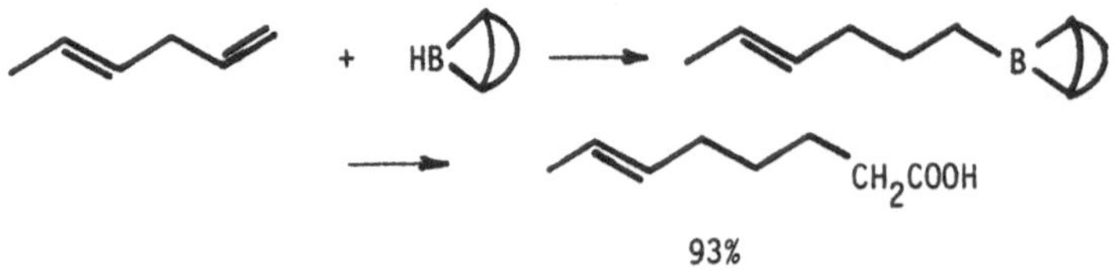

$$\longrightarrow \text{CH}_3\text{CH=CH(CH}_2)_3CH_2COOH \qquad (36)$$

93%

IV Synthesis via Alkynylborates and Alkenylborates

As shown in Eq. 7, the organoborates formed from organoboranes and basic or nucleophilic species have possibility to make new R—X bonds by the intramolecular

1,2-migration reactions. The following modified patterns of reaction are also expected (Eqs. 37 and 38).

$$R_3B \;+\; {}^{-}X{\equiv}Y \longrightarrow R_2\overset{R}{\overset{|}{B^{-}}}{-}X{\equiv}Y \xrightarrow{E^+} R_2B{-}\overset{R}{\overset{|}{X}}{=}Y{-}E \qquad (37)$$

$$R_3B \;+\; {}^{-}X{=}Y \longrightarrow R_2\overset{R}{\overset{|}{B^{-}}}{-}X{=}Y \xrightarrow{E^+} R_2B{-}\overset{R}{\overset{|}{X}}{-}Y{-}E \qquad (38)$$

A typical example of such reactions of 1-alkynyltrialkylborates is demonstrated in the synthesis of internal acetylenes. When 1-alkynyltrialkylborates (*18*), readily prepared from lithium acetylides and trialkylboranes, are treated with iodine, the corresponding internal alkynes are obtained in almost quantitative yields (Eq. 39) [56].

$$R_3B + LiC{\equiv}CR' \longrightarrow \underset{18}{R_3\bar{B}C{\equiv}CR'} \xrightarrow{I_2} R_2BC(R){=}C(I)R' \xrightarrow{-R_2BI} \underset{100\text{-}91\%}{RC{\equiv}CR'} \qquad (39)$$

The reaction appears to proceed through β-attack of iodine cation and migration of an alkyl group followed by elimination of the R_2BI. Probably, the most general method previously available for the synthesis of internal alkynes involves the reaction of alkali metal acetylides with organic halides or sulfates. However, that synthesis is only satisfactory for primary alkyl halides or sulfates, which readily undergo nucleophilic substitution reactions. In the case of secondary alkyl halides, elimination reactions also take place competitively. On the other hand, the borate-iodine reaction easily permits the introduction, not only of both primary and secondary alkyl groups, but also of aryl groups in excellent yields. One may wonder if the reaction is applicable for the preparation of terminal alkynes from monolithium acetylide and trialkylboranes. This is rather difficult, since ethynyltrialkylborates thus formed are subject to disproportionation leading to a mixture of the diboronyl derivatives and acetylene. However, Midland and Brown [57] overcame this difficulty by using monolithium acetylide-ethylenediamine complex, in place of monolithium acetylide itself. Most recently, relative migratory aptitudes of various alkyl groups have been determined in the iodine-induced rearrangement of lithium ethynyltrialkylborates [58]. The overall migratory aptitude order is bicyclooctyl > n-butyl > cyclohexyl, isobutyl, sec-butyl > thexyl. Except for the anomalous position of the bicyclo group, the order is generally primary > secondary > tertiary, a ranking which is compatible with both carbanion character and a favorable steric factor in the migrating group. The greater migratory aptitudes of alkenyl and alkynyl groups which are described later, compared to secondary groups, support this interpretation. Although the results reported by Slayden [58] provide some insight into the mechanism of organoborate rearrangements, further systematic studies should be necessary in order to elucidate the reaction mechanism.

The borate-iodine reaction is also useful for the syntheses of symmetrical (Eq. 40) [59] and unsymmetrical (Eq. 41) [60] conjugated alkadiynes, and the preparation of conjugated alkenynes (Eq. 42)[61]. The versatility of these reactions is exemplified in the synthesis of many natural products[61,62,63]. In Eq. 43, four-step synthesis of a natural sex pheromone of the European grape vine moth, *Lobesia botrana*, (7E, 9Z)-7,9-dodecadien-1-yl acetate, is illustrated[63].

$$Sia_2BX + 2LiC{\equiv}CR \xrightarrow{-LiX} [Sia_2\overset{-}{B}(C{\equiv}CR)_2]Li^+ \xrightarrow{I_2} RC{\equiv}C\text{-}C{\equiv}CR \quad \text{96-72\%} \tag{40}$$

Sia = 3-Methyl-2-butyl

$$Sia_2BOMe + LiC{\equiv}CR \longrightarrow [Sia_2\overset{-}{B}(OMe){-}C{\equiv}CR]Li^+ \xrightarrow{BF_3:OEt_2} Sia_2B\text{-}C{\equiv}CR$$

$$\xrightarrow{LiC{\equiv}CR'} [Sia_2\overset{-}{B}(C{\equiv}CR'){-}C{\equiv}CR]Li^+ \xrightarrow{I_2} RC{\equiv}C\text{-}C{\equiv}CR' \quad \text{73-61\%} \tag{41}$$

$$Sia_2BH + RC{\equiv}CH \longrightarrow (R)(H)C{=}C(H)(BSia_2) \xrightarrow{LiC{\equiv}CR'} \left[(R)(H)C{=}C(H)\overset{-}{B}Sia_2(C{\equiv}CR')\right]Li^+$$

$$\xrightarrow{I_2} (R)(H)C{=}C(H)(C{\equiv}CR') \quad \text{74-60\%} \tag{42}$$

$$C_5H_{11}C{\equiv}CCH_2OH \xrightarrow[\text{2.}H_2O \;\; \text{3.}Ac_2O]{\text{1.}KNH(CH_2)_3NH_2} HC{\equiv}C(CH_2)_6OAc \xrightarrow[\text{2.}LiC{\equiv}CC_2H_5]{\text{1.}Sia_2BH} Sia_2\overset{-}{B}(C{\equiv}CC_2H_5){-}CH{=}CH(CH_2)_6OAc$$

$$\xrightarrow[\text{3.}H_2O_2,\ NaOAc]{\text{1.}I_2 \;\; \text{2.}NaOAc} H_5C_2C{\equiv}C{-}CH{=}CH(CH_2)_6OAc \xrightarrow[\text{3.}H_2O_2,\ NaOAc]{\text{1.}Sia_2BH \;\; \text{2.}AcOH} H_5C_2{-}CH{=}CH{-}CH{=}CH(CH_2)_6OAc \tag{43}$$

In the reaction of 1-alkynyltrialkylborates (*18*), in addition to the iodine cation, many other chemical species can also be used as electrophiles. β-Alkylation of *18* occurs with alkyl migration from boron to the α-carbon atom under mild conditions to give dialkylvinylboranes. Oxidation of these intermediates provides a versatile route to regiospecifically α-substituted ketones, and hydrolysis yields alkenes in high yields as E—Z mixtures. The reaction was reported to proceed through a two-stage mechanism, as shown in Eq. 44[64]. Usual alkylating agents such as alkyl, allyl and benzyl halides, sulfates, α-bromoketones, ethyl α-bromoacetate, iodoacetonitrile, and prop-2-ynyl bromide, give major products which have the alkylating agent and the migrating group on the same side of the double bond. On the other hand,

$$\mathrm{Li^+}\ \ R_3\bar{B}C{\equiv}CR'\ (18)\ \xrightarrow[\text{slow}]{R''X}\ \mathrm{Li^+}\ R_3\bar{B}(\overset{+}{C}{=}C(R')R'')\ X^- \xrightarrow{\text{rapid}} \begin{cases} R_2B(R)C{=}C(R')R''\ \text{(major)} \xrightarrow{H^+} H(R)C{=}C(R')R'' \\ R(R_2B)C{=}C(R')R''\ \text{(minor)} \xrightarrow{H^+} R(H)C{=}C(R')R'' \end{cases} \xrightarrow{[O]} RCOCH(R')R''\ \ 86\text{-}78\% \tag{44}$$

the reaction of *18* with tributyltin chloride specifically produces the isomer in which the migrating group and entering electrophile are trans [65]. The reaction with chlorodiphenylphosphine also gives the latter type of product specifically [66].

Lithium 1-alkynyltrialkylborates (*18*) react in a stereoselective fashion with benzo-1,3-dithiolium fluoroborate to give vinylboranes which on oxidation yield protected 3-oxo-aldehydes, and on protonolysis give protected α,β-unsaturated aldehydes; the hydrolysis with mercuric oxide and boron trifluoride is highly selective giving rise eventually to stereoselectively defined α,β-unsaturated aldehydes (Eq. 45) [67].

$$R_3\bar{B}C{\equiv}CR'\ (18) + \text{benzo-1,3-dithiolium } BF_4^- \rightarrow \begin{cases} R(R_2B)C{=}C(R')\text{-CH(benzodithiole)} \xrightarrow[1\ h,\ 25^\circ]{i\text{-PrCOOH}} R(H)C{=}C(R')\text{-CH(benzodithiole)}\ (19) \xrightarrow[BF_3]{HgO} R(H)C{=}C(R')CHO \\ R_2B(R)C{=}C(R')\text{-CH(benzodithiole)} \xrightarrow[48h,\ 50^\circ]{i\text{-PrCOOH}} H(R)C{=}C(R')\text{-CH(benzodithiole)}\ (20) \xrightarrow[BF_3]{HgO} H(R)C{=}C(R')CHO \end{cases} \tag{45}$$

yield of *19*, 87-71%

ratio of *19*:*20*, 100-83:0-17

Ethynylbis(trialkylborates) prepared from dilithium acetylide and two moles of trialkylboranes are interesting intermediates for the synthesis of substituted ethylene derivatives by the reaction with cyanogen bromide. By controlling the amounts of cyanogen bromide, di-, tri- and tetraalkyl substituted ethylenes are produced respectively in reasonable yields with relatively high distribution ratios of the expected products (Eq. 46)[68]. The reaction of iodine with the ate-complexes, obtained from trialkylboranes and lithium 2-chloroethyne, gives the corresponding symmetrical alkynes (Eq. 47)[69]. In the case of the borate-iodine reaction of ethyl propynoate or ethynyl aryl ketones, a modified procedure is required, as an addition of butyllithium to the carbon-oxygen double bonds occurs competitively in the lithium acetylide forming step. Thus by using lithium diisopropylamide as a base, the expected ethyl 2-alkynoates or 1-alkynyl aryl ketones are prepared in good yields (Eq. 48)[70].

$$2R_3B + LiC{\equiv}CLi \longrightarrow R_3\bar{B}C{\equiv}C\bar{B}R_3 \xrightarrow{BrCN} \underset{88\text{-}46\%}{RCH{=}CHR},\ \underset{76\text{-}50\%}{RCH{=}CR_2},\ \underset{30\%}{R_2C{=}CR_2} \tag{46}$$

$$R_3B + LiC{\equiv}CCl \longrightarrow R_3\bar{B}C{\equiv}CCl \xrightarrow{I_2} \underset{86\text{-}55\%}{RC{\equiv}CR} \tag{47}$$

$$HC{\equiv}CCOOEt \xrightarrow{i\text{-}Pr_2NLi} LiC{\equiv}CCOOEt \xrightarrow{R_3B} \xrightarrow{I_2} \underset{81\text{-}73\%}{RC{\equiv}CCOOEt} \tag{48}$$

The reaction of lithium 1-propynyltributylborate with oxirane in dichloromethane affords, after protolysis, (Z)-3-methyl-3-octen-1-ol (*21*) as the major product, whereas the predominant product is the (E)-isomer (*22*) when THF is used as a solvent (Eq. 49)[71]. Consequently, great care should be taken in the alkylation of such 1-alkynyltrialkylborates, although they are versatile as synthetic reactions.

$$Bu_3\bar{B}C{\equiv}C\text{-}CH_3 + \overset{O}{\triangle} \xrightarrow{CH_2Cl_2} (Bu)(Bu_2B)C{=}C(CH_2CH_2OLi)(CH_3) \xrightarrow{H^+} (Bu)(H)C{=}C(CH_2CH_2OH)(CH_3)\ \ \mathbf{21}\ \ 62\%\ (21{:}22=92{:}\ 8) \tag{49}$$

$$\xrightarrow{THF} (Bu_2B)(Bu)C{=}C(CH_2CH_2OLi)(CH_3) \xrightarrow{H^+} (H)(Bu)C{=}C(CH_2CH_2OH)(CH_3)\ \ \mathbf{22}\ \ 72\%\ (21{:}22=40{:}60)$$

As previously mentioned, 1-alkynyltrialkylborates (*18*) have become increasingly important in the formation of carbon-carbon bonds via attack of electrophiles. However, such complexes cannot react with simple α,β-unsaturated carbonyl compounds such as methyl vinyl ketone, because of their weak electrophilicity. Recently it was ascertained that α,β-unsaturated carbonyl compounds react with *18* via a Michael-type reaction in the presence of titanium tetrachloride, and the usual alkaline hydrogen peroxide oxidation leads to the synthesis of δ-dicarbonyl compounds

$$R'C{\equiv}C\bar{B}R_3 \;(18) + CH_3COCH{=}CH_2 \xrightarrow{TiCl_4} CH_3C(OTiCl_3){=}CHCH_2C(R'){=}C(BR_2)R \xrightarrow{[O]} CH_3COCH_2CH_2CH(R')COR \quad 83\text{-}52\% \qquad (50)$$

(Eq. 50) [72]. Uncatalyzed Michael reactions of lithium 1-alkynyltrialkylborates with α,β-unsaturated nitro-compounds [73], alkylidenemalonates and alkylideneacetoacetates [74] proceed rather smoothly.

Sequential treatment of an ethynylalkanol acetate with butyllithium and a trialkylborane produces an allenic borane (*23*), which when protonated with acetic acid forms an allene, while upon treatment with water affords an acetylene derivative, presumably via a cyclic process (Eq. 51) [75]. A novel synthesis of homopropargylic

$$HC{\equiv}CCR'_2(OAc) \xrightarrow{BuLi} LiC{\equiv}CCR'_2(OAc) \xrightarrow{R_3B} R_2\bar{B}(R){-}C{\equiv}C{-}CR'_2(OAc) \rightarrow \underset{23}{(R_2B)RC{=}C{=}CR'_2} \xrightarrow{AcOH} \underset{92\text{-}70\%}{HRC{=}C{=}CR'_2}$$

$$23 \xrightarrow{H_2O} \underset{91\text{-}84\%}{RC{\equiv}CCHR'_2} \qquad (51)$$

and α-allenic alcohols has been realized from lithium chloropropargylide, trialkylboranes and aldehydes by the same type of reaction. The interesting feature of these synthetic transformations to the specifically formed alcohols is the temperature at which the organoborane precursor is maintained prior to its reaction with the aldehyde, as illustrated in Eq. 52 [76]. By a similar method, 1,3-enynols or 1,2,4-trienols are prepared [77].

The ate complexes prepared from 4-tosyloxy-1-butyne and trialkylboranes also provides a selective synthesis of cyclopropyl ketones and homopropargylic alcohols simply by changing the reaction temperature (Eq. 53) [78].

The protonation of 1-alkynyltrialkylborates (*18*) reported by Köster and Binger [79] in 1965, may be the first report concerning the reaction of such borate complexes. According to their report, 1-alkynylborates were found to give a mixture of the corresponding (Z)- and (E)-alkenes on treatment with hydrogen chloride. The lack

$$R_2\overset{R}{\underset{|}{B}}-C\equiv C-\overset{Cl}{\underset{|}{CH_2}} \xrightarrow{-90\,°C} \begin{matrix}R \\ R_2B\end{matrix}\!\!>C=C=CH_2 \underset{}{\overset{25\,°C}{\rightleftarrows}} RC\equiv CCH_2BR_2$$

$\begin{matrix}R \\ R_2B\end{matrix}>C=C=CH_2$ → 1. R'CHO, −78 °C; 2. H_2O_2, OH^- → $RC\equiv CCH_2CHR'$(OH), 89–74% (purity, >97%)

$RC\equiv CCH_2BR_2$ → 1. R'CHO, −78 °C; 2. H_2O_2, OH^- → R(R'CH(OH))C=C=CH₂, 86–73% (purity, >96%) (52)

$$HC\equiv CCH_2CH_2OTs \xrightarrow[2.\,R_3B,\,-78\,°C]{1.\,BuLi,\,-78\,°C} R_3\overset{-}{B}C\equiv CCH_2CH_2OTs$$

$$\xrightarrow{-78\,°C} \begin{matrix}R \\ R_2B\end{matrix}\!\!>C=\!\triangleleft \xrightarrow[28\,h]{25\,°C} RC\equiv CCH_2CH_2BR_2 \qquad (53)$$

$\begin{matrix}R \\ R_2B\end{matrix}>C=\triangleleft$ → H_2O_2, OH^- → RCO–◁, 75–55%

$RC\equiv CCH_2CH_2BR_2$ → H_2O_2, OH^- → $RC\equiv CCH_2CH_2OH$, 70%

of stereoselectivity had not recognized the protonation as a useful methodology in organic synthesis, although the alkaline hydrogen peroxide oxidation provides a convenient synthesis of ketones due to disappearance of the geometrical integrity. Later the protonation reaction received much attention by many chemists. For details of the discussions the readers should refer to other articles [1,9,10].

The reaction of ate complexes (*24*), formed from trialkylboranes and trimethylsilylpropargyl phenyl ether, with a mixture of acetic acid and hexamethylphosphoric triamide (HMPT) gives trimethylsilylacetylenes (*25*) selectively (Eq. 54) [80], whereas the corresponding trimethylsilylallenes (*26*) are selectively prepared by the reaction with sodium methoxide instead of acetic acid and HMPT (Eq. 54) [81]. In the latter, when primary alkylboranes are used, the corresponding allene derivatives are obtained in high purity, but secondary alkylboranes give reverse ratios of the isomer distribution.

Next, the reactions of 1-alkenyltrialkylborates will be discussed. One of the most interesting and versatile reactions of such 1-alkenylboranes readily obtainable

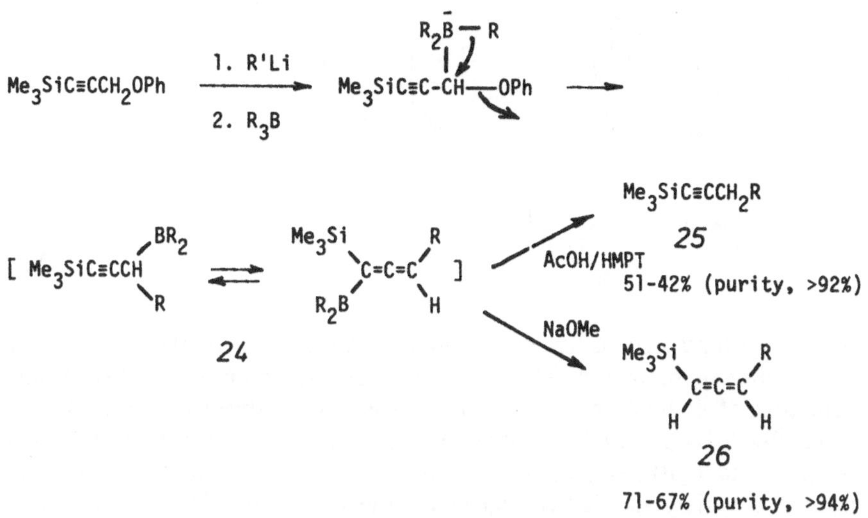

(54)

by the monohydroboration of 1-alkynes with dialkylboranes was discovered at the initial stage of these reactions by Zweifel et al. 1-Alkenyldialkylboranes upon treatment with iodine in the presence of a base give the corresponding (Z)-alkenes stereospecifically, the reaction mechanism of which may be considered via the alkenylborate intermediate (Eq. 55)[82]. On the other hand, treatment of (1-bromo-1-alkenyl)dialkylboranes (formed by the monohydroboration of 1-bromo-1-alkynes with dialkylboranes) with sodium methoxide, followed by protonolysis with a carboxylic acid yields the isomeric (E)-alkenes (Eq. 56)[83], and also (E)-isomers are obtained by the cyanobromination of vinylboranes in the absence of base (Eq. 57)[83a].

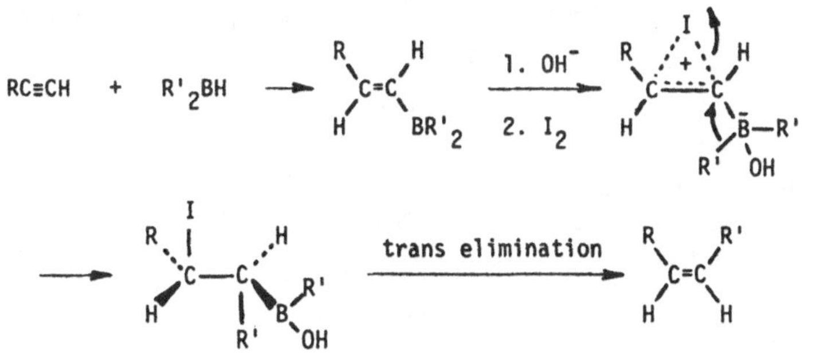

(55)

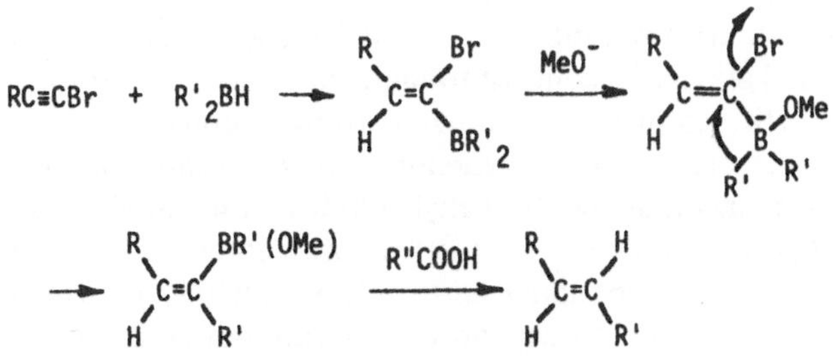

(56)

$$\text{RHC=C(H)BR'}_2 \xrightarrow{\text{BrCN}} [\text{bromonium intermediate, } \overset{-}{B}(R')(R')CN] \rightarrow \text{R(Br)HC–CH(R')B(R')CN} \xrightarrow{\text{cis elimination}} \text{RHC=CHR'} \tag{57}$$

However, the utility of this Zweifel synthesis was limited in the past by the limited availability of dialkylboranes, because direct hydroboration leads cleanly to the formation of dialkylboranes only in the case of relatively hindered alkenes such as 2-methyl-2-butene and cyclohexene. More generally, the hydroboration fails to stop at the R_2BH stages. Recent developments have provided a general preparation of a variety of dialkylboranes via the hydridation [84] of dialkylhaloboranes [85]. Thus, dialkylvinylboranes prepared via the hydridation of dialkylhaloboranes in the presence of an alkyne, react with iodine under basic conditions to produce disubstituted alkenes (Eq. 58) [86] and trisubstituted alkenes (Eq. 59) [87] of established stereochemistry. These results indicate a mechanism analogous to that

$$\text{1-methylcyclopentene} \xrightarrow{BH_2Br} (\text{2-methylcyclopentyl})_2\text{BBr} \xrightarrow[\text{2. LiAlH}_4]{\text{1. HC≡C–C}_6\text{H}_{11}} (\text{2-methylcyclopentyl})_2\text{B–CH=CH–C}_6\text{H}_{11} \xrightarrow[-25\ ^\circ\text{C}]{\text{NaOMe, I}_2} \text{(cis) 2-methylcyclopentyl–CH=CH–C}_6\text{H}_{11}\ \ 67\% \tag{58}$$

$$R_2BX \xrightarrow[\text{2. LiAlH}_4]{\text{1. R'C≡CR'}} R_2B(H)C{=}C(R')R' \xrightarrow{\text{NaOMe, I}_2} R'(H)C{=}C(R)R' \quad 76\text{–}69\% \tag{59}$$

depicted in Eq. 55. Apparently the reaction proceeds via trans addition, followed by trans elimination (dehaloboration), resulting in the trans stereochemistry of the two alkyl groups from the internal alkyne. In contrast to that, another method for the stereospecific synthesis of trisubstituted olefins via 1-alkenyltrialkylborates has been reported by Levy and his co-workers [88]. In this reaction, 1-alkenyltrialkylborates were prepared by sequential treatment of the 1-alkenyl iodides with butyllithium and trialkylborane. The borates thus obtained on reaction with iodine without a base at low temperature gave the corresponding trisubstituted olefins, in which the introduction of alkyl group on boron to carbon occurs through retention of configuration with respect to the boronyl group (Eq. 60)[88]. Presumably, this reaction involves a mechanism similar to that of halogenation reaction of 1-alkenylborates at the initial stage. Initial complexation of the vinyllithium with the trialkylborane gives an ate-complex. Subsequent formation of the iodonium ion followed by migration of an alkyl group from boron to carbon with concurrent anti opening of the iodonium ion occurs by the same manner as in Eq. 55, but the last step, dehaloboration, proceeds through a thermal cis elimination in the absence of base to produce the trisubstituted alkene.

$$(Et)(I)C{=}C(H)(C_6H_{13}) \xrightarrow[-65\ °C]{BuLi} \xrightarrow[-78\ °C]{Et_3B} (Et)(\bar{B}Et_3)C{=}C(H)(C_6H_{13}) \xrightarrow[-78\ °C]{I_2} (Et)(Et)C{=}C(H)(C_6H_{13})\quad 75\% \qquad (60)$$

In connection with such a reaction of 1-alkenylborates, isopropenylation of C=C bonds is carried out by the complexation of organoboranes with isopropenyllithium followed by iodination[89]. A convenient procedure of the conversion of methyl ketones to 1,1-dialkylethylenes via the corresponding 2,4,6-triisopropylbenzenesulfonylhydrazones (trisylhydrazones) has been recently reported. Trialkylboranes react with the carbanion derived from a methyl ketone trisylhydrazone to give vinyltrialkylborates, the treatment of which with iodine results in an alkyl migration from boron to carbon, followed by spontaneous deiodoboration to provide a highly convenient route to 1,1-dialkylethylenes from methyl ketones (Eq. 61)[90].

$$O{=}C(R')(CH_3) \longrightarrow 2,4,6\text{-}(i\text{-}Pr)_3C_6H_2\text{-}SO_2NHN{=}C(R')(CH_3) \xrightarrow[TMEDA]{BuLi} CH_2{=}\bar{C}\text{-}R'$$

$$\xrightarrow{R_3B} CH_2{=}C(\bar{B}R_3)\text{—}R' \xrightarrow{I_2} \xrightarrow{-IBR_2} CH_2{=}C(R')(R)\quad 97\text{-}63\% \qquad (61)$$

At the first glance, one may consider that both 1-alkynylborates and 1-alkenylborates provide the same type of reactions with various reagents. There are, however, some obvious differences between them. For example, 1-alkynylborates do not react with aldehydes and ketones, while the reaction of 1-ethenyltrialkylborates with aldehydes has been reported to proceed smoothly, giving ca. 1:1 mixture of diastereomeric 1,3-alkanediols (Eq. 62)[91]. Lithium ethenyltrialkylborates react with oxirane and methyloxirane affording 1,4- and 2,5-alkanediols respectively[92]. The same style of reaction is observed in the case of 1-alkynyltrialkylborates, as shown in Eq. 49.

Arylborates are expected to be more reluctant to undergo 1,2-migration reactions than 1-alkynyl- or 1-alkenylborates, as such their reactions would require the

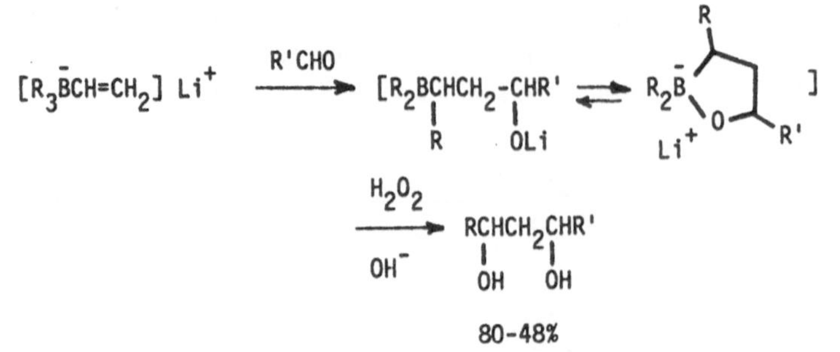

$$[R_3\bar{B}CH{=}CH_2]\,Li^+ \xrightarrow{R'CHO} [R_2BCH(R)CH_2\text{-}CHR'(OLi)] \rightleftharpoons [R_2\bar{B}(\text{cyclic, }R, R', O)\,Li^+] \xrightarrow[OH^-]{H_2O_2} RCH(OH)CH_2CH(OH)R' \quad 80\text{-}48\% \tag{62}$$

destruction of the aromatic systems. About a decade ago, the reaction of arenyltrialkylborates was carried out with the expectation of finding a new synthetic method of γ-ketoaldehydes from the ate-complexes (*27*), prepared from trialkylboranes and 2-lithiofuran. However, unexpectedly, this reaction was found to produce the (Z)-alkendiols, probably via the reaction pathway as shown in Eq. 63 [93]. Recently it has been disclosed that the ate complexes (*27*) give the corresponding 2-alkylfurans upon treatment with iodine (Eq. 64) [94]. The reaction has significant applicability for the introduction of both primary and secondary alkyl groups to the furan nucleus. On the other hand, it is known that there is a large difference of

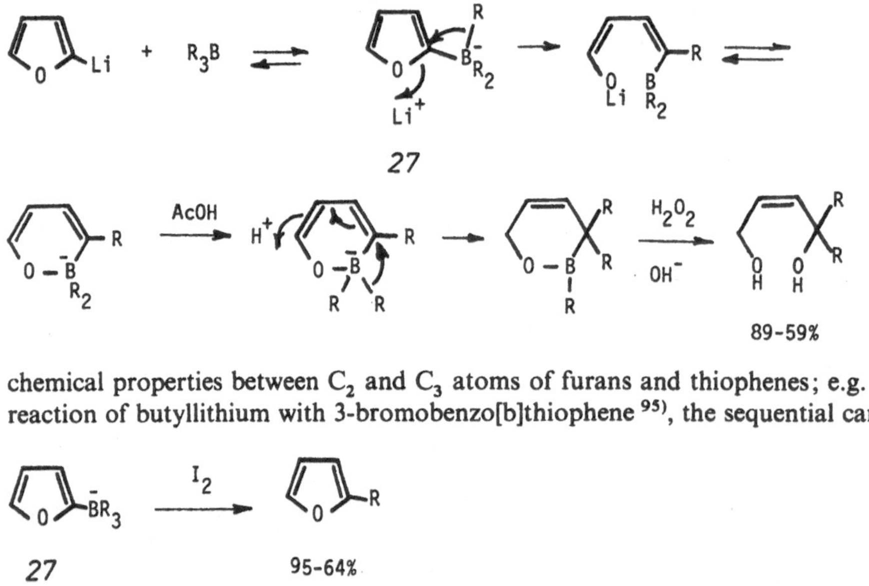

(63)

27

AcOH

H_2O_2

OH^-

89-59%

chemical properties between C_2 and C_3 atoms of furans and thiophenes; e.g. in the reaction of butyllithium with 3-bromobenzo[b]thiophene [95], the sequential carboxy-

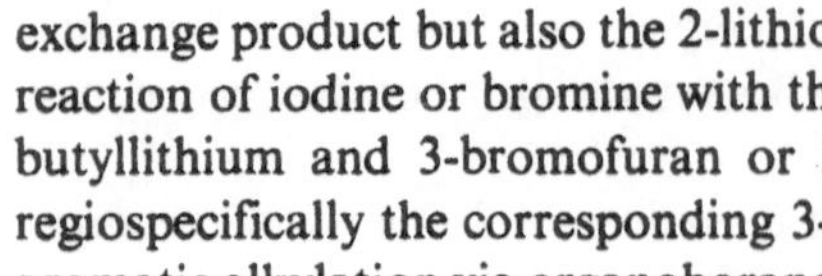

(64)

27 $\xrightarrow{I_2}$ 95-64%

lation even at —70 °C gives a mixture of the corresponding 3-carboxylic acid and 2-acid, indicating that the reaction gives not only the straightforward metal-halogen exchange product but also the 2-lithio derivative as the intermediate. Fortunately, the reaction of iodine or bromine with the ate-complexes prepared from trialkylboranes, butyllithium and 3-bromofuran or 3-bromothiophene, proceeds smoothly to give regiospecifically the corresponding 3-alkylfurans or 3-alkylthiophenes [96]. This novel aromatic alkylation via organoboranes is applicable for various heterocyclic nuclei [97].

A convenient synthesis of unsaturated nitriles by a stereospecific alkylative cleavage of pyridine ring via borate process has been reported. Namely, the reaction of 2-bromo-6-lithiopyridine with trialkylboranes affords 5-alkyl-5-dialkylboryl-2-(Z)-4-(E)-pentadienenitriles (*28*), which are versatile intermediates for the preparation of 5-alkyl-2-(Z)-4-(E)-pentadienenitriles (*29*), 5,5-dialkyl-4-pentenenitriles (*30*), and 5,5-dialkyl-2,4-pentadienenitriles (*31*), as depicted in Eq. 65 [98].

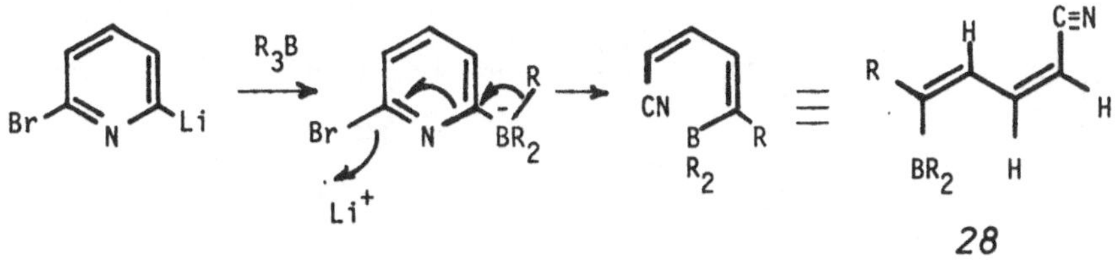

(65-1)

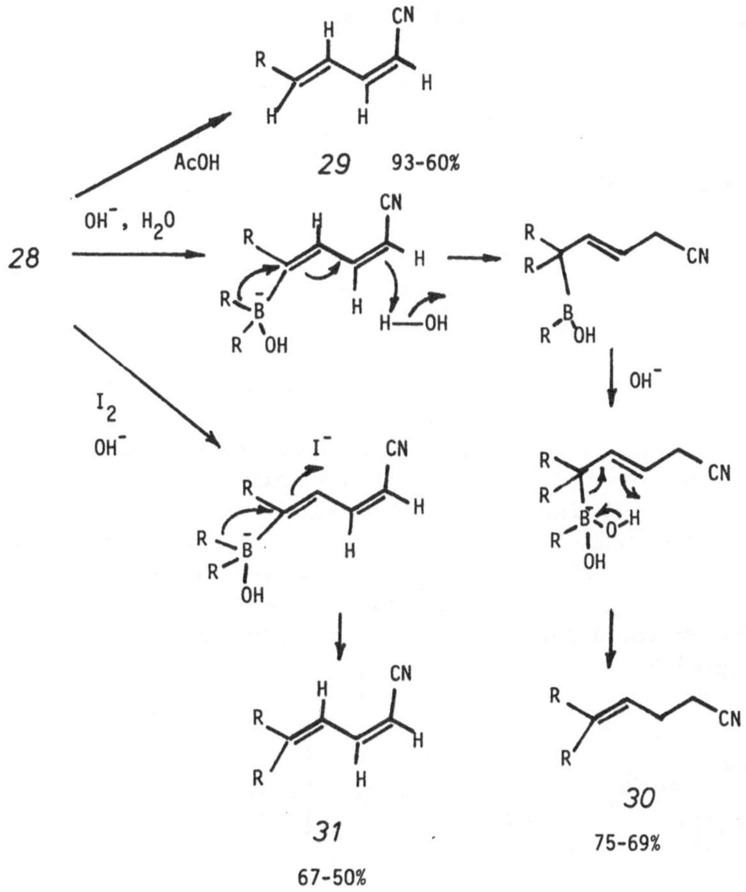

(65-2)

In addition to the synthetic applications of hetero-arylborates, the applicability of real arylborates has also been reported. Since no detailed discussion of such reactions is attempted in this article because of the page limit, the readers are recommended to consult other references [1,9,10].

Recently, the reactions of 1-alkenyltrialkylborates possessing a functionality such as a methoxy group at the C=C bond have been explored. Thus Levy et al. reported that alkenyltrialkylborate salts (*32*), prepared from α-methoxyvinyllithium and trialkylboranes, which are stable at −80 °C, upon warming to room temperature, affords alkenyldialkylmethoxyborates (*33*) via an alkyl group migration. These ate complexes (*33*) have been proved to be versatile intermediates, as illustrated in Eq. 66 [99].

$$R_3B + LiC(OMe)=CH_2 \xrightarrow{-78\ ^\circ C} R_3\bar{B}C(OMe)=CH_2\ (32) \xrightarrow{-78\ ^\circ C \rightarrow r.t.} [R_2\bar{B}(OMe)-C(R)=CH_2]\ Li^+\ (33) \quad (66\text{-}1)$$

$$33 \xrightarrow{H_2O_2,\ NaOH/EtOH} R-C(=O)-CH_3 \quad 97\text{-}75\%$$

$$33 \xrightarrow[2.\ H_2O_2,\ NaOH/EtOH]{1.\ HCl} R_2C(OH)-CH_3 \quad 100\text{-}87\% \quad (66\text{-}2)$$

$$33 \xrightarrow[2.\ H_2O_2,\ NaOH/EtOH]{1.\ R'X} R_2C(OH)-CH_2R' \quad 98\text{-}66\%$$

The reaction of acetic acid with the ate complexes obtained from organoboranes and 1-lithio-1-methoxyallene provides a synthesis of methoxycyclopropane derivatives (Eq. 67) [100]. The formation of 2,2-d_2-1-alkyl-1-methoxycyclopropane by protonolysis with deutero acetic acid, reveals that the reaction proceeds via the mechanism depicted in Eq. 67. The versatility of these ate complexes is further demonstrated by the preparation of 1,1-dialkylethylenes from 1,2-dimethoxyethenyltrialkylborates by the reaction with trichloroacetic acid, followed by treatment with a mixture of sodium acetate-acetic anhydride and $TiCl_4/Ti(OPr^i)_4$ [101], while upon treatment with alkyl fluorosulfonates, 1,2-dimethoxyethenyltrialkylborates give the corresponding ketones [102].

$$CH_2=C=C(OMe)H \xrightarrow[2.\ R-B\langle]{1.\ BuLi} CH_2=C=C(OMe)-\bar{B}(R)\langle \xrightarrow{AcOD} CH_2=CD-C(OMe)(R)-\bar{B}(OAc)\langle$$

$$\rightleftharpoons \langle\bar{B}(OAc)-CH_2CD=C(OMe)R \xrightarrow{AcOD} \text{2,2-}d_2\text{-1-R-1-methoxycyclopropane} \quad 81\text{-}76\% \quad (67)$$

The utility of allylic moieties in the construction of complex molecules has been numerously demonstrated. Recently, Yamamoto et al. have reported an interesting application of allylic borate complexes in the cross-coupling with organic halides [103], though the organic parts of organoboranes, prepared by hydroboration are not used in the reaction. Thus lithium allylic borate complexes, prepared by the addition of trialkylboranes in an ether solution of allylic lithium compounds, regioselectively react with allylic halides to produce head-to-tail 1,5-dienes (Eq. 68). Lithium crotylborates undergo a rapid reaction with aldehydes with high threoselectivity (Eq. 69). The selectivity is affected by the steric hindrance, as elucidated by steric requirement of the six-membered transition state. The reaction of lithium (alkylthioallyl)trialkylborates with allylic halides also gives head-to-tail 1,5-alkadienes in good yields [104].

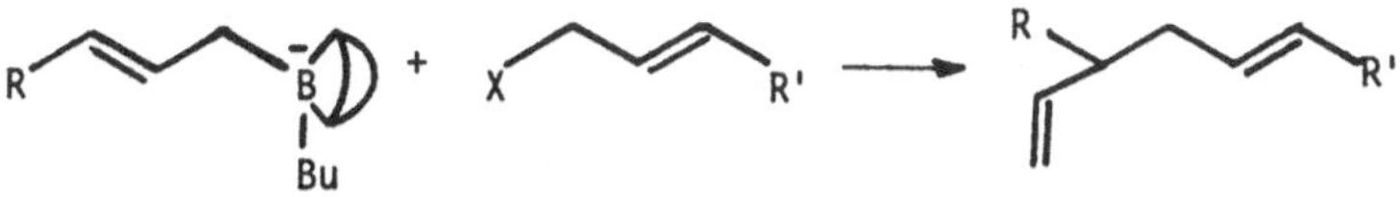

(68)

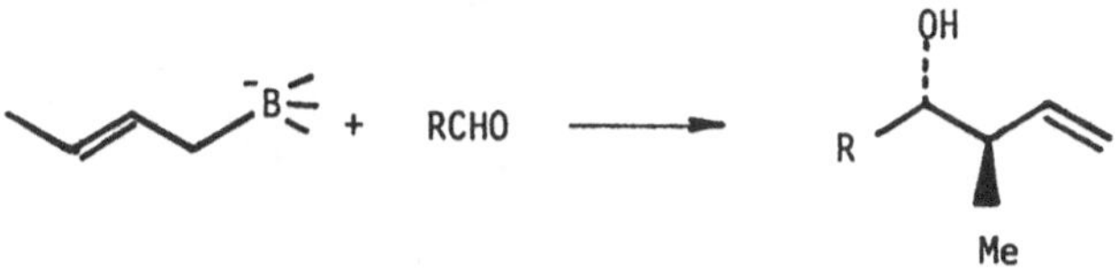

(69)

Recently, a lot of attention has been focused on enantioselective aldol condensations using organoboron compounds. But in this review, only related literature [105] has been cited, as the major purpose of the review is to present synthetic applications of organoboranes obtained by the hydroboration reaction of carbon-carbon multiple bonds.

The base-induced migration of an alkyl group from boron to carbon in a γ-acetoxyvinylborane occurs predominantly in an anti-fashion with respect to the leaving group to give a trans allylic alcohol upon oxidation (Eq. 70) [106].

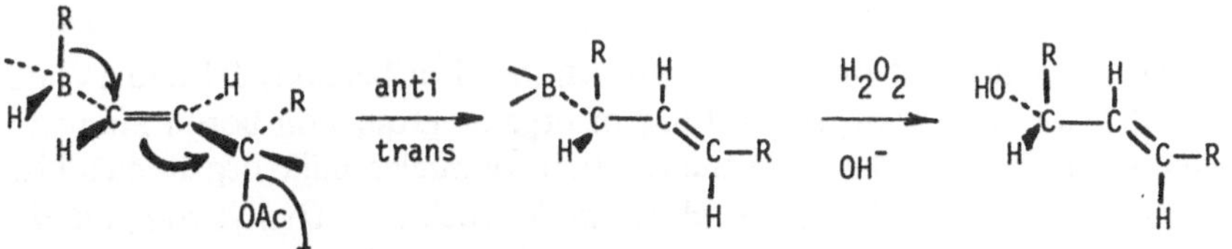

(70)

Although the following examples are not concerned with the synthetic applications of organic groups of organoboranes, these are included being synthetically important. Conversion of organoboranes into organic products is most commonly achieved by either oxidation or protonolysis. Whereas the alkaline hydrogen peroxide oxidation is a highly general reaction, protonolysis with carboxylic acids suffers from a few difficulties, such as the incompatibility with various acid-sensitive functional groups and the frequent need for high temperatures more than 100 °C. It has been recently discovered that 1-alkenyltrialkylborates undergo a selective protonolysis reaction of the alkenyl groups with aqueous sodium hydroxide to produce the corresponding alkenes (Eq. 71)[107]. Potassium enoxytrialkylborates, readily obtainable by treating potassium enolates with trialkylboranes, undergo a remarkably selective reaction with organic halides, the reaction of which provides a highly selective method for α-alkylation of ketones (Eq. 72)[108].

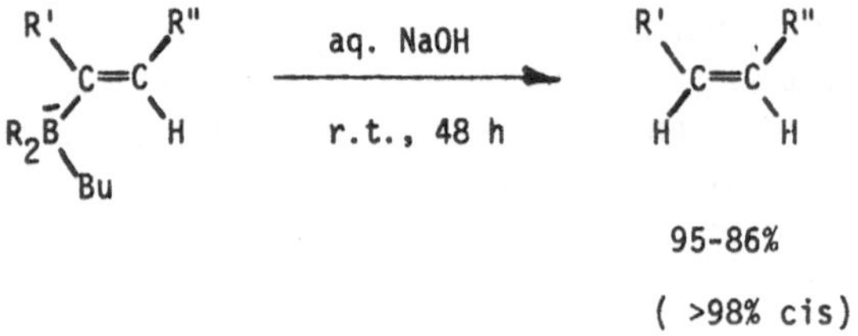

$$R'(R_2\bar{B}Bu)C{=}C(H)R'' \xrightarrow[\text{r.t., 48 h}]{\text{aq. NaOH}} R'(H)C{=}C(H)R'' \qquad (71)$$

95–86%

(>98% cis)

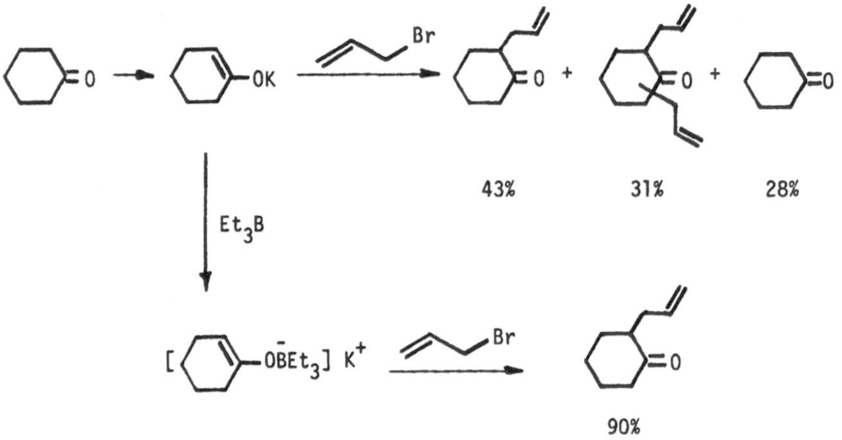

(72) Cyclohexanone → potassium cyclohexenolate (OK); with allyl bromide → 2-allylcyclohexanone 43% + 2,6-diallylcyclohexanone 31% + cyclohexanone 28%

Et$_3$B: [cyclohexenyl–O$\bar{B}$Et$_3$] K$^+$ + allyl bromide → 2-allylcyclohexanone 90%

V Synthesis via Copper(I) Tetraorganoborates and Related Compounds

Most of the new reactions of organoborates mentioned in Sections III and IV are regarded as intramolecular transfer reactions of organic groups on boron atoms to positive centers. On the other hand, in the reactions of nucleophilic organometallics such as Grignard reagents with organic electrophiles such as carbonyl compounds, the carbon-carbon bond formation occurs intermolecularly between the two reactants. Such intermolecular transfer reactions of organoborates are also known in some cases, one of which is the reaction of copper(I) methyltrialkylborates.

Though the detail of the reaction is not described here, it was discovered that organoboranes undergo spontaneous 1,4-addition in the presence of catalytic amounts

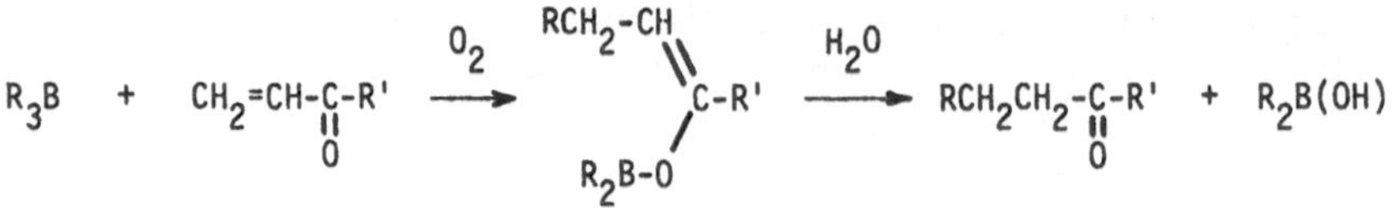

$$R_3B + CH_2{=}CH{-}\underset{\underset{O}{\|}}{C}{-}R' \xrightarrow{O_2} RCH_2{-}CH{=}C(R'){-}O{-}BR_2 \xrightarrow{H_2O} RCH_2CH_2{-}\underset{\underset{O}{\|}}{C}{-}R' + R_2B(OH) \qquad (73)$$

of oxygen to numerous α,β-unsaturated carbonyl and related compounds such as methyl vinyl ketone (Eq. 73)[109], acrolein[110], acetylacetylene[111], 1,3-butadiene monoxide[112], and 3,4-epoxy-1-butyne[113]. However, ethyl acrylate and acrylonitrile were found to fail to react with organoboranes under such reaction conditions, while vinyl phenyl ketone and 1-acyl-2-vinylcyclopropane[114] give the corresponding products in poor yields. This may be due to the polymerization of substrates, or that the unstable intermediates initially formed are subject to other reactions. In order to surmount the obstacles, many attempts were made, one of which is the reaction of copper(I) methyltrialkylborates, the alkyl groups of which are considered to be more anionic than those of the corresponding trialkylboranes. The reaction of substrates such as acrylonitrile and 1-acyl-2-vinylcyclopropane with lithium methyltrialkylborates, readily prepared from trialkylboranes and methyllithium, was found to fail to take place. However, it has become apparent that the reaction of the substrates with copper(I) methyltrialkylborates (*34*) obtainable without any difficulty via the metathesis between lithium methyltrialkylborates and cuprous halides, proceeds smoothly to provide satisfactory results. Thus the reaction of *34* with acrylonitrile gives the corresponding 1,4-addition products which upon hydrolysis with water afford alkyl cyanides in about 90% yields[115]. Such cuprous borates also readily react with 1-acyl-2-vinylcyclopropane to give γ,δ-unsaturated ketones in good yields[115]. As one of probable reaction routes, the mechanism involving a redox process of copper ions may be considered (Eq. 74).

$$R_3B + MeLi \longrightarrow [R_3\bar{B}Me]\,Li^+ \xrightarrow{CuX} \underset{34}{[R_3\bar{B}Me]\,Cu^+} \qquad (74\text{-}1)$$

$$34 + CH_2{=}CH{-}CH{-}CH{-}C(=O){-}R' \ (\text{cyclopropane ring via } CH_2) \longrightarrow [R_3\bar{B}Me]\,Cu^{2+} + \text{(allyl radical)}{-}CH_2CH{=}C(O^-)R'$$

$$\longrightarrow R_2BMe + Cu^+ + R{-}CH_2CH{=}CHCH_2CH{=}C(O^-)R' \xrightarrow{H_2O} R{-}CH_2CH{=}CHCH_2CH_2C(=O)R' \quad 84\text{-}75\% \qquad (74\text{-}2)$$

The ate-complexes (*34*) also react with benzylic halides[116], aroyl chlorides[117], allyl halides[118], and propargyl halides[118] to afford the corresponding products, respectively (Eq. 75). (E)-α,β-Unsaturated carboxylic esters can also be obtained by the reaction of such borate complexes with ethyl propiolate[119]. The same type of reaction between copper(I) methyltrialkylborates and 1-(1-pyrrolidinyl)-6-chloro-1-cyclohexene gives the corresponding alkylation products which are readily hydrolyzed to 2-alkylcyclohexanone[120].

Copper(I) 1-alkenyltrimethylborates (*35*), readily prepared by the procedure

$$\text{34: } [R_3\bar{B}Me]\,Cu^+ \xrightarrow[\text{115)}]{CH_2{=}CHCN} RCH_2CH_2CN\ (93\text{–}84\%);\quad \xrightarrow[\text{116)}]{PhCH_2Br} PhCH_2R\ (70\text{–}50\%);\quad \xrightarrow[\text{117)}]{PhCOCl} PhCOR\ (90\text{–}75\%);\quad \xrightarrow[\text{118)}]{CH_2{=}CHCH_2Br} RCH_2CH{=}CH_2\ (90\text{–}50\%);$$
$$\xrightarrow[\text{118)}]{HC{\equiv}CCH_2Br} RCH{=}C{=}CH_2\ (59\text{–}53\%);\quad \xrightarrow[\text{119)}]{HC{\equiv}CCOOEt} (E)\text{-}RCH{=}CHCOOEt\ (63\text{–}40\%);\quad \xrightarrow[\text{120)}]{\text{2-chloro-1-pyrrolidinocyclohexene}} \xrightarrow[OH^-]{H_2O_2} \text{2-R-cyclohexanone}\ (83\text{–}60\%) \tag{75}$$

depicted in Eq. 76, react easily with allyl bromide to give 1,4-alkadienes in excellent yields [121].

$$\text{catecholborane} + HC{\equiv}CR \longrightarrow \text{(E)-catechol-B–CH=CHR} \xrightarrow{3MeLi} [\,Me_3\bar{B}\text{–CH=CHR}\,]\,Li^+ \xrightarrow{CuI} [\,Me_3\bar{B}\text{–CH=CHR}\,]\,Cu^+\ (\mathbf{35}) \tag{76-1}$$

$$\mathbf{35} + CH_2{=}CHCH_2Br \longrightarrow CH_2{=}CHCH_2\text{–CH=CHR (E)}\quad 97\text{–}81\% \tag{76-2}$$

The reaction of cuprous methyltrialkylborates with β-bromoacrylates is also interesting. The reaction with (E)-β-bromoacrylate occurs smoothly to afford the corresponding α,β-unsaturated esters which are specifically (E)-isomers. On the other hand, the (Z)-β-alkylacrylates are produced from the (Z)-β-bromoacrylate (Eq. 77). This evidence, although the mechanism is not clear, seems to suggest that the reaction proceeds through the retention of configuration by the cis-addition and trans-elimination process [122].

Most recently, Brown et al. and Yamamoto et al. have independently explored organic boron-copper mediated reactions. For example, dialkenylchloroboranes (*36*), readily available through the reaction of alkynes with chloroborane-ethyl

$$[R_3\bar{B}Me]\,Cu^+ + \underset{H}{\overset{Br}{>}}C{=}C\underset{COOEt}{\overset{H}{<}} \longrightarrow \underset{H}{\overset{R}{>}}C{=}C\underset{COOEt}{\overset{H}{<}} \quad (98\text{-}65\%) \tag{77-1}$$

$$[R_3\bar{B}Me]\,Cu^+ + \underset{Br}{\overset{H}{>}}C{=}C\underset{COOEt}{\overset{H}{<}} \longrightarrow \underset{R}{\overset{H}{>}}C{=}C\underset{COOEt}{\overset{H}{<}} \quad (95\%) \tag{77-2}$$

etherate, react with 3 molar equivalents of methylcopper at 0 °C to produce symmetrical (E,E)-1,3-dienes in high yields (Eq. 78) [123]. In this reaction, it is essential

$$RC{\equiv}CR' \xrightarrow{BH_2Cl:OEt_2} \underset{H}{\overset{R}{>}}C{=}C\overset{R'}{<}\!\!\rightarrow_2 BCl\ \ (\mathit{36}) \xrightarrow[0\ °C]{3MeCu} \underset{H}{\overset{R}{>}}C{=}C(R')\text{-}C(R'){=}C\underset{R}{\overset{H}{<}} \quad (99\text{-}45\%) \tag{78}$$

to utilize 3 molar equivalents of methylcopper in order to achieve the effective conversion of the organoboranes into the dienes. The proposed reaction mechanism is described in Eq. 79, where R means $\underset{H}{\overset{R}{>}}C{=}C\overset{R'}{<}$

$$\begin{aligned} &R_2BCl + MeCu \longrightarrow R_2BMe + CuCl \\ &R_2BMe + MeCu \rightleftharpoons [R_2\bar{B}(Me)_2]\,Cu^+ \rightleftharpoons RCu + RB(Me)_2 \\ &RB(Me)_2 + MeCu \rightleftharpoons [R\bar{B}(Me)_3]\,Cu^+ \rightleftharpoons RCu + (Me)_3B \\ &RCu \longrightarrow 1/2R{-}R + Cu \end{aligned} \tag{79}$$

The above methylcopper-induced reaction is also applicable for cross-coupling of alkenylboranes with organic halides. The treatment of dialkenylchloroboranes (*36*) with 3 equivalents of methylcopper at −30 to −40 °C followed by addition of allylic halides gives 1,4-dienes (Eq. 80) [124]. In the case of simple alkyl halides, the cross-coupling products are obtained with the aid of PhSLi or $P(OPh)_3$ (Eq. 80) [124]. A similar cross-coupling reaction of alkenyl-9-BBN with allylic halides is realized by the use of 1 equivalent of methylcopper, whereas cross-coupling with alkyl halides fails even in the presence of ligands. Systems such as dialkenylchloroborane-methylcopper and alkenyl-9-BBN-methylcopper react with substituted allylic halides predominantly at a γ-position. The regioselectivity of the former system is quite similar to that of the free alkenylcopper, while the regioselectivity of the latter system is greater than that of the free alkenylcopper. From these results, the reactive intermediate of the alkenyl-9-BBN-methylcopper reaction appears to be an ate-complex such as *37* (Eq. 81).

$$\mathrm{(RR'C{=}C)_2BCl}\ \mathbf{36}\ \xrightarrow[-30\ °C]{3MeCu}\ \begin{cases}\xrightarrow{CH_2{=}CHCH_2X}\ \mathrm{RHC{=}CR'(CH_2CH{=}CH_2)}\quad 95\text{-}58\% \\ \xrightarrow[P(OPh)_3]{R''X}\ \mathrm{RHC{=}CR'R''}\quad 53\text{-}25\%\end{cases} \tag{80}$$

$$\mathrm{RHC{=}CH{-}B} \xrightarrow{MeCu} [\mathrm{RHC{=}CH{-}\bar{B}(Me)}]\,Cu^+\ \mathbf{37} \xrightarrow{CH_2{=}CHCH_2X} \mathrm{RHC{=}CH(CH_2CH{=}CH_2)}\quad 61\text{-}58\% \tag{81}$$

While exploring the organic boron–copper mediated reaction, interesting reactions of a new type of alkylating reagent, $RCu \cdot BF_3$ have been disclosed. By using Lewis acid mediated reaction of the organocopper reagent, regioselective γ-attack of allylic halides is realized irrespective of the degree of substitution at the two ends of the allylic systems and of the structural factors (cyclic or acyclic) involved. Among the Lewis acids examined, $BF_3 \cdot OEt_2$ has been found to be the most effective with respect to the selectivity and total yield. Direct displacement of the OH group of allylic alcohols is also achieved via $RCu \cdot BF_3$ (Eqs. 82 and 83). The examination of the stereochemistry by employing cyclohexenyl derivatives, gave complicated results. Thus, it was shown that the substitution proceeds through a formal anti S_N2' pathway in the case of cis-5-methyl-2-cyclohexenyl acetate and syn S_N2' in trans-5-methyl-2-cyclohexanol (Eqs. 84 and 85). The stereochemical integrity disappears in the reaction of cis- and trans-5-methyl-2-cyclohexenyl chloride and cis-5-methyl-2-cyclohexen-1-ol (Eq. 86) [125]. As a reactive intermediate of the reaction, the ate complex between RCu and BF_3 seems to be considered (Eq. 87).

$$\mathrm{CH_3CH{=}CHCH_2X} \xrightarrow{RCu\cdot BF_3} \mathrm{CH_3CH(R)CH{=}CH_2} \tag{82}$$

$$\mathrm{CH_3CH(X)CH{=}CH_2} \xrightarrow{RCu\cdot BF_3} \mathrm{CH_3CH{=}CHCH_2R} \tag{83}$$

X = halides, OH

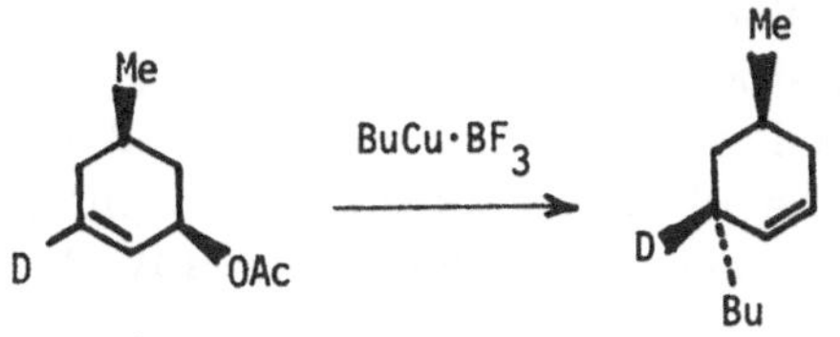

(84)

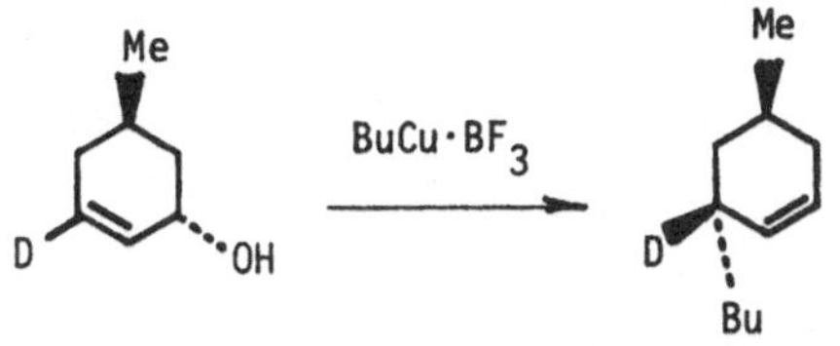

(85)

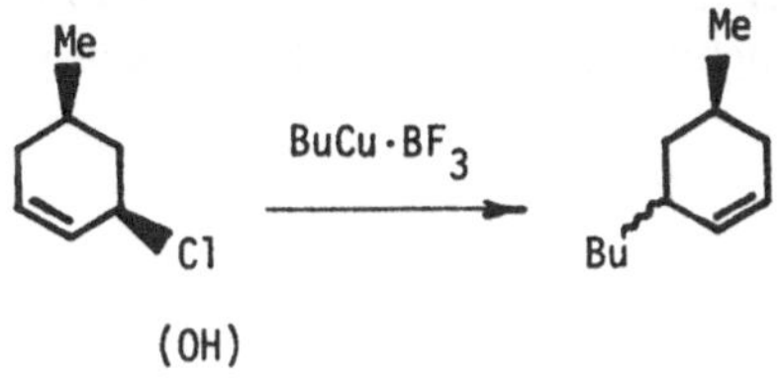

(86)

$$RCu + BF_3 \longrightarrow [R{-}\overset{F}{\underset{F}{B^-}}{-}F]\,Cu^+ \qquad (87)$$

$RCu \cdot BF_3$ is also useful for the conjugate addition to the α,β-unsaturated ketones and esters, whose double bonds are sterically crowded. Such conjugated additions occur selectively by a 1,4 manner. Certain α,β-unsaturated carboxylic acids also undergo a 1,4-addition through this reagent (Eq. 88) [126]. Methyl sorbate reacts with $BuCu \cdot BF_3$ to give predominantly the 1,4-adduct (Eq. 89). Two mechanistic

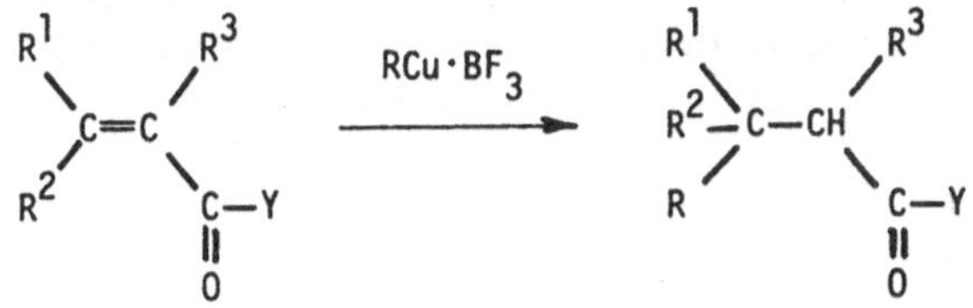

(88)

R^1, R^2, R^3 = H, aryl and/or alkyl

Y = H, R, OR, OH

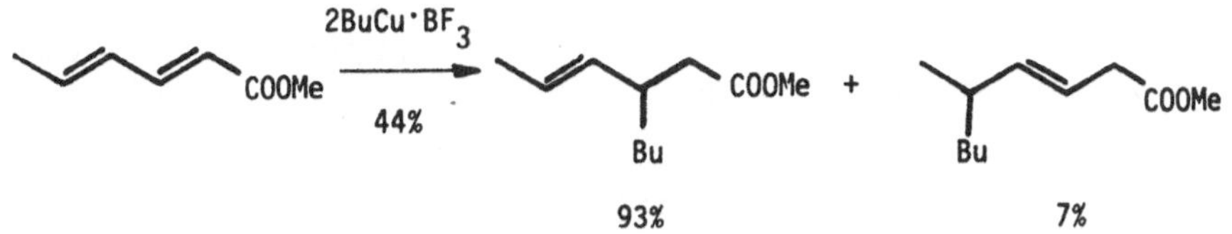

(89)

rationales are proposed for the conjugate addition. One involves the activation of reactivity by the coordination of BF_3 with a carbonyl group. Another explanation assumes a transition state schematically depicted in Eq. 90. Although the fact that the 1,4-addition proceeds predominantly in the case of methyl sorbate, may support the second explanation, much detailed study will be necessary to clarify the mechanism.

$$[R-\overset{F}{\underset{F}{B^-}}-F]\,Cu^+ + >C=C-\underset{O}{\overset{\|}{C}}- \longrightarrow [\text{cyclic transition state, } BF_3] \longrightarrow >\underset{R}{C}-C=\underset{OBF_3^-}{C}- \qquad (90)$$

Organocopper-organoborane complexes ($RCu \cdot BR'_3$) have been recently discovered to add to α,β-acetylenic carbonyl compounds with a high stereospecificity at low temperatures (—70 to —20 °C), which cannot be achieved with reagents previously available (Eq. 91) [127].

$$-C{\equiv}C-\underset{O}{\overset{\|}{C}}-Y + RCu \cdot BR'_3 \longrightarrow \xrightarrow{H_2O} \underset{R}{}C{=}C\overset{C(=O)Y}{\underset{H}{}} \qquad (91)$$

Y = OR", OH, R", H   98-30%

Sodium methoxyalkenyldialkylborates, obtained in the simple treatment of alkenyldialkylboranes with sodium methoxide, react readily with cuprous bromide-methyl sulfide at 0 °C to afford symmetrical conjugated dienes. The dienes are formed with retention of configuration predetermined from the stereochemistry of the initial alkenylborane intermediate [128]. Thus B-(1E)-1-hexen-1-yl-9-borabicyclo[3.3.1]nonane (*38*), gives a 95% yield of (5E,7E)-5,7-dodecadiene (Eq. 92). By a similar procedure, the (Z,Z)-diene is also prepared, as shown in Eq. 93. The exact nature of the reactive species generated in the reaction in uncertain. Although the intermediacy of a copper(I) borate complex formed by cation exchange with sodium appears likely, whether this intermediate decomposes directly to give a diene or dissociates to yield an alkenylcopper compound remains to be established.

$$\mathbf{38}\ (\text{9-BBN})\text{-}CH{=}CH\text{-}Bu \xrightarrow{NaOMe} [(\text{9-BBN})^-(OMe)\text{-}CH{=}CH\text{-}Bu]\,Na^+ \xrightarrow[0\ °C]{CuBr \cdot SMe_2} Bu\text{-}CH{=}CH\text{-}CH{=}CH\text{-}Bu + \text{B-OMe-9-BBN} \qquad (92)$$

95%

$$\mathrm{IC{\equiv}CBu} \xrightarrow{(C_6H_{11})_2BH} (C_6H_{11})_2B(I)C{=}C(H)Bu \xrightarrow[-78\ ^\circ C]{t\text{-BuLi}} (C_6H_{11})_2B(H)C{=}C(H)Bu \xrightarrow[0\ ^\circ C]{1.\ NaOMe\quad 2.\ CuBr\cdot SMe_2} BuCH{=}CH{-}CH{=}CHBu \quad 85\% \qquad (93)$$

On the other hand, sodium methoxyalkenyldialkylborates derived from 9-BBN or dicyclohexylborane, undergo stereospecific cross-coupling at —15 °C with allylic halides in the presence of cuprous bromide-methyl sulfide to yield stereodefined 1,4-alkadienes (Eq. 94)[129].

$$RCH{=}CH{-}B(9\text{-BBN}) \xrightarrow{NaOMe} [RCH{=}CH{-}\overset{-}{B}(OMe)(9\text{-BBN})]\,Na^+ \xrightarrow[2.\ CH_2{=}CHCH_2Br,\ -15\ \text{to}\ 25\ ^\circ C]{1.\ CuBr\cdot SMe_2,\ -15\ ^\circ C} RCH{=}CH{-}CH_2CH{=}CH_2 \quad 95\text{-}76\% \qquad (94)$$

The same type of coupling reaction with 1-halo-1-alkynes occurs readily at —40 °C to provide stereodefined conjugated enynes of high isomeric purity and in yields approaching quantitative (Eq. 95)[130]. Previously, Normant and his co-workers had reported that alkenylcopper intermediates, obtained from the corresponding Grignard reagents, could be coupled to 1-halo-1-alkynes in the presence of 1–2 equivalents of TMEDA to provide excellent yields of conjugated enynes (Eq. 96)[131], and pointed that TMEDA was essential to promote the clean coupling. In contrast, Brown et al.[130] found that TMEDA was not only necessary, but was actually detrimental to the desired process. Large amounts of conjugated diynes were formed instead. Such evidence may indicate that the cuprous halide-mediated reactions

$$RCH{=}CH{-}B(9\text{-BBN}) \xrightarrow[2.\ CuI,\ -40\ ^\circ C]{1.\ NaOMe} \xrightarrow[-40\ \text{to}\ 25\ ^\circ C]{BrC{\equiv}CBu} \xrightarrow{H^+} RCH{=}CH{-}C{\equiv}CBu \quad 98\text{-}90\% \qquad (95)$$

$$[R(Et)C{=}CH(Cu)]\,MgX_2 + R'C{\equiv}CX \xrightarrow[1\text{-}2\ \text{equiv of TMEDA};\ 2.\ H_3O^+]{1.\ Et_2O/THF,\ -15\ ^\circ C} R(Et)C{=}CH{-}C{\equiv}CR' \quad 84\text{-}77\% \qquad (96)$$

of alkenylborates proceed through different reaction routes with those of alkenylcopper compounds.

VI Synthesis via 1-Alkenylboranes in the Presence of Palladium Catalysts and Bases

Stereoselective syntheses of conjugate (E,E)-, (E,Z)-, and (Z,Z)-alkadienes are of considerable importance in organic chemistry in themselves, as well as in their utilization in other reactions such as Diels-Alder reaction. Recently a number of new methods for the preparation of conjugate dienes have appeared utilizing organoaluminum [132,133], copper [134], magnesium[135], mercury [136], silver [137], and zirconium [138] reagents. However, the scope of many of these reactions is limited by the nature of the organometallic compound involved or the procedure employed, e.g. they require stoichiometric amounts of metal compounds, and some of them can be best utilized only for the synthesis of symmetrical dienes or unfunctionalized dienes because of the reducing property of organometallic derivatives.

On the other hand, it is well known that stereodefined 1-alkyenyldialkylboranes are readily prepared by the monohydroboration of alkynes, i.e. dialkylboranes such as disiamylborane and catecholborane permit the monohydroboration of terminal alkynes, thus making readily available the corresponding (E)-1-alkenyldialkylboranes with high stereoselectivity, more than 99% (Eq. 97) [7,8]. Highly pure (Z)-1-alkenyldialkylboranes (purity, more than 98%) are prepared without any difficulty via the monohydroboration of 1-halo-1-alkynes with disiamylborane or dicyclohexylborane, followed by treatment with t-butyllithium (Eq. 98) [13]. Consequently, if such 1-alkenyldialkylboranes react with 1-alkenyl halides or 1-alkynyl

$$RC{\equiv}CH \quad + \quad HBY_2 \longrightarrow \underset{H}{\overset{R}{>}}C{=}C\underset{BY_2}{\overset{H}{<}} \tag{97}$$

$$RC{\equiv}CX \quad + \quad HBY_2 \longrightarrow \underset{H}{\overset{R}{>}}C{=}C\underset{BY_2}{\overset{X}{<}} \xrightarrow{t\text{-BuLi}} \underset{H}{\overset{R}{>}}C{=}C\underset{H}{\overset{BY_2}{<}} \tag{98}$$

halides, these reactions provide direct and convenient synthetic procedures for conjugate alkadienes or alkenynes. Many of the efforts made to find such cross-coupling reactions have met with no success [138a], but it was recently disclosed that the coupling can in fact take place smoothly in the presence of bases. The speculative thought at the initial stage of the exploration was based on the following consideration. The common mechanism of transition metal-catalyzed cross-coupling reactions [139] between organometallic compounds and organic halides involves sequential (a) oxidative addition, (b) transmetallation, and (c) reductive elimination. One of the reasons why 1-alkenylboranes cannot react with 1-alkenyl or 1-alkynyl halides seems to be the step (b), namely because the transmetallation process between R'MX and organoboranes does not occur readily, owing to the weak carbanion character of organic groups in the organoboranes. Therefore, if the organoborates formed from. 1-alkenylboranes and basic species such as alkoxide, acetate and hydroxide ions are used instead of alkenylboranes themselves, there may be a possibility that the transmetallation takes place more readily.

Actually, (E)-1-alkenyl-1,3,2-benzodioxaboroles (*39*) and (E)-1-alkenyldisiamylboranes (*40*) react with (E)-1-alkenyl halides or 1-alkynyl halides in the presence of a catalytic amount of tetrakis(triphenylphosphine)palladium and bases to give the corresponding conjugate (E,E)-dienes or (E)-enynes respectively in excellent yields with high regio- and stereospecificity (Eqs. 99 and 100) [140]. By employing the same catalyst and sodium ethoxide as a base, the reactions of (Z)-1-alkenyldisiamylboranes with (Z)- or (E)-alkenyl bromides afford the corresponding (Z,Z)-

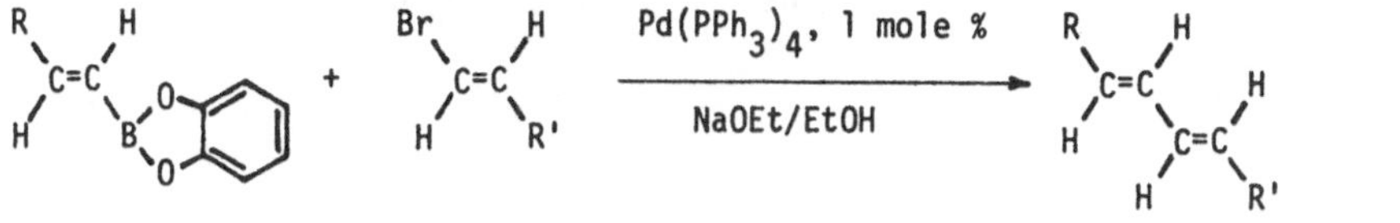

$$\text{(E)-RCH=CH-B(O}_2\text{C}_6\text{H}_4\text{)}\ (\mathbf{39}) + \text{(E)-BrCH=CHR'} \xrightarrow[\text{NaOEt/EtOH}]{\text{Pd(PPh}_3)_4\text{, 1 mole \%}} \text{(E,E)-RCH=CH-CH=CHR'} \quad (99)$$

86-80% (>96% purity)

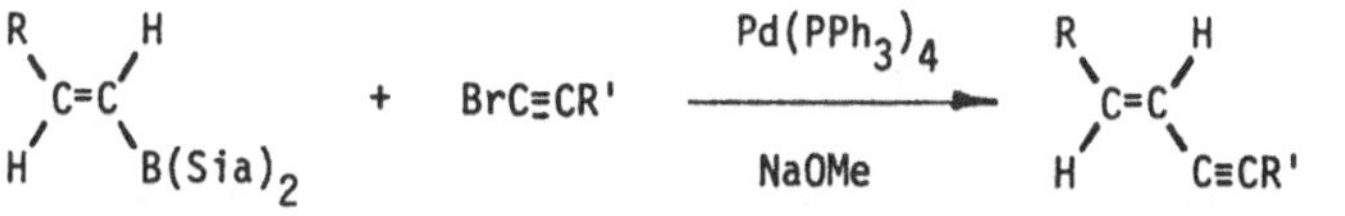

$$\text{(E)-RCH=CH-B(Sia)}_2\ (\mathbf{40}) + \text{BrC}\equiv\text{CR'} \xrightarrow[\text{NaOMe}]{\text{Pd(PPh}_3)_4} \text{(E)-RCH=CH-C}\equiv\text{CR'} \quad (100)$$

100-72% (>99% purity)

or (Z,E)-alkadienes with high stereospecificity (Eqs. 101 and 102) [141], though the yields are comparatively lower. The reaction of aryl bromides or iodides with (E)-1-alkenyl-1,3,2-benzodioxaboroles (*39*) provides a convenient method for stereoselective synthesis of arylated (E)-alkenes in high yields. These reactions are also

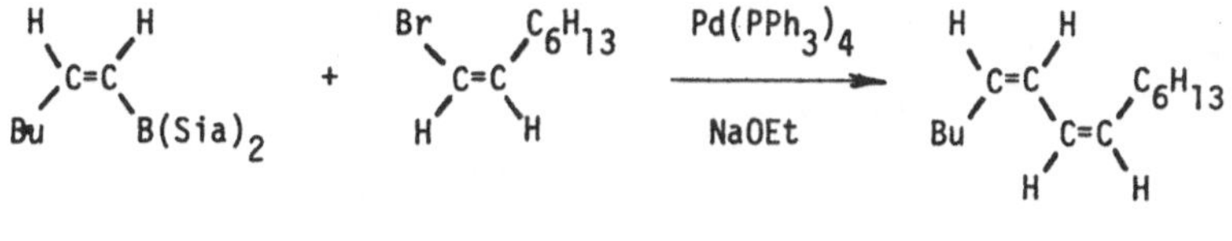

$$\text{(Z)-BuCH=CH-B(Sia)}_2 + \text{(Z)-BrCH=CHC}_6\text{H}_{13} \xrightarrow[\text{NaOEt}]{\text{Pd(PPh}_3)_4} \text{(Z,Z)-BuCH=CH-CH=CHC}_6\text{H}_{13} \quad (101)$$

49% (98% purity)

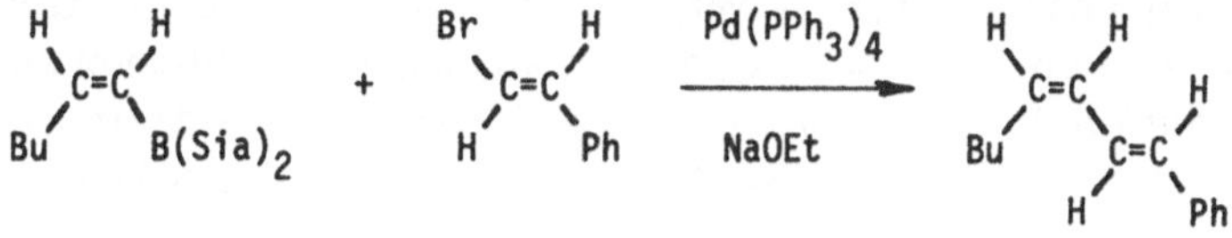

$$\text{(Z)-BuCH=CH-B(Sia)}_2 + \text{(E)-BrCH=CHPh} \xrightarrow[\text{NaOEt}]{\text{Pd(PPh}_3)_4} \text{(Z,E)-BuCH=CH-CH=CHPh} \quad (102)$$

49% (99% purity)

effectively catalyzed by tetrakis(triphenylphosphine)palladium and sodium ethoxide (Eq. 103) [142]. Allylic or benzylic halides also undergo the cross-coupling reaction with 1-alkenylboranes under the same conditions to give the corresponding 1,4-alkadienes or allylbenzenes (Eqs. 104 and 105) [143]. In the reaction with 1-bromo--2-butene, the carbon-carbon bond formation occurs at two positions, yielding

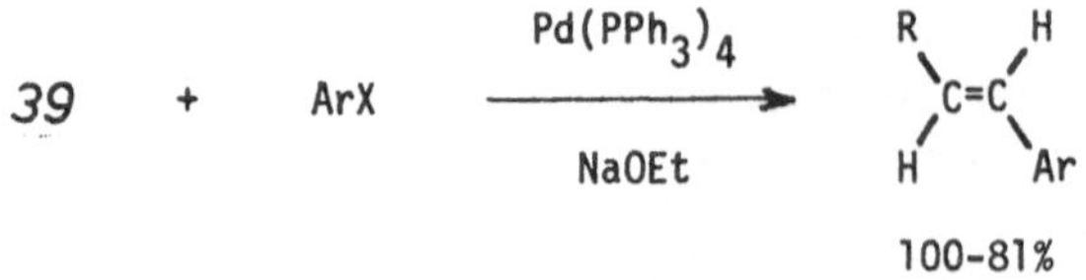

$$\mathbf{39} + ArX \xrightarrow[NaOEt]{Pd(PPh_3)_4} \underset{H}{\overset{R}{}}\!\!>C{=}C<\!\!\underset{Ar}{\overset{H}{}} \qquad (103)$$

100–81%

straight and branched chain dienes in the ratio of 72:28. The formation of these dienes suggests that the mechanism involves π-allylpalladium intermediates.

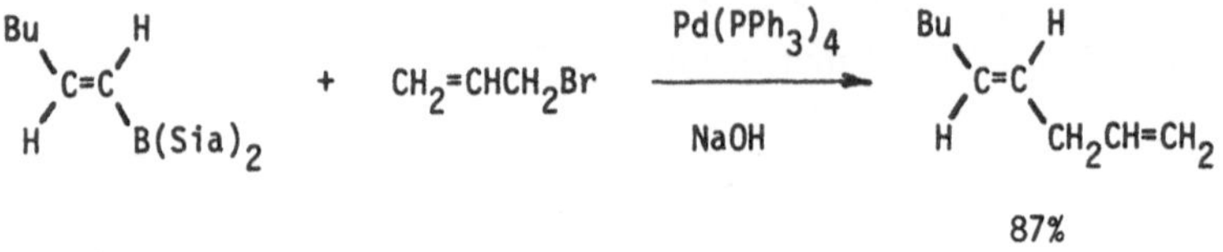

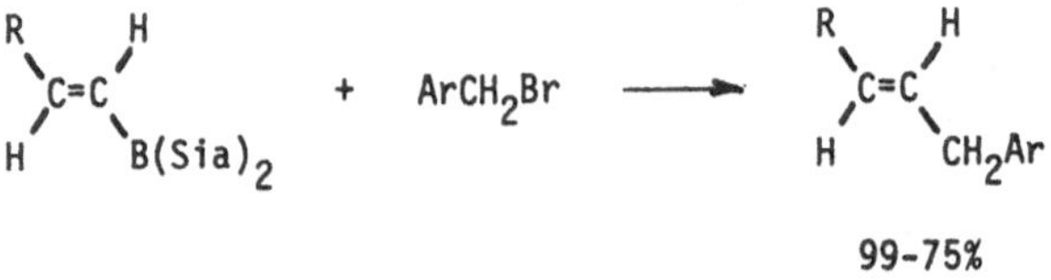

$$\underset{H}{\overset{Bu}{}}\!\!>C{=}C<\!\!\underset{B(Sia)_2}{\overset{H}{}} + CH_2{=}CHCH_2Br \xrightarrow[NaOH]{Pd(PPh_3)_4} \underset{H}{\overset{Bu}{}}\!\!>C{=}C<\!\!\underset{CH_2CH{=}CH_2}{\overset{H}{}} \qquad (104)$$

87%

$$\underset{H}{\overset{R}{}}\!\!>C{=}C<\!\!\underset{B(Sia)_2}{\overset{H}{}} + ArCH_2Br \longrightarrow \underset{H}{\overset{R}{}}\!\!>C{=}C<\!\!\underset{CH_2Ar}{\overset{H}{}} \qquad (105)$$

99–75%

Previously Davidson and Triggs [144] have reported that arylboronic acids react with sodium palladate to give the dimeric biaryls. The synthetic utility of this dimerization reaction is, however, limited owing to stoichiometric requirement of the palladium compound. On the other hand, the palladium-catalyzed cross-coupling reaction between arylboronic acids and haloarenes in the presence of bases provides a clean synthesis of biaryls (Eq. 106) [145]. In this case, sodium carbonate has been proven to be most effective base.

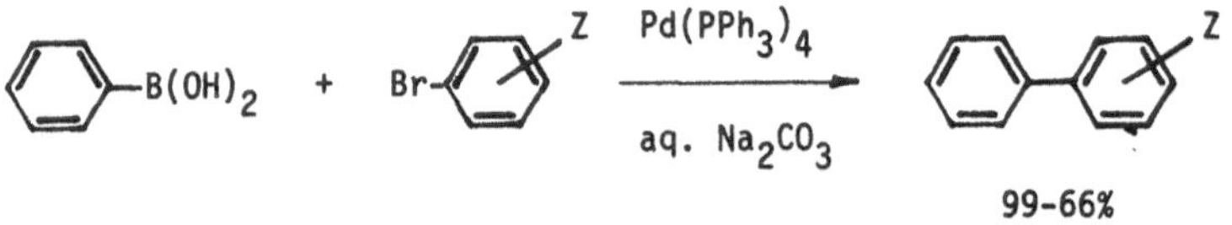

$$C_6H_5{-}B(OH)_2 + Br{-}C_6H_4{-}Z \xrightarrow[aq.\ Na_2CO_3]{Pd(PPh_3)_4} C_6H_5{-}C_6H_4{-}Z \qquad (106)$$

99–66%

The mechanism of the palladium-catalyzed "head-to-head" cross-coupling reactions described above are poorly understood. However, our recent investigation suggests that the "head-to-head" reaction can be accomodated by an oxidative addition of an organic halide R'X to Pd(0) followed by an exchange with sodium alkoxide, transmetallation with 1-alkenylborates and reductive elimination, as depicted in Scheme 1 [146].

In addition to the "head-to-head" cross-coupling reactions, it has been discovered that the reaction of phenyl or 1-alkenyl iodides with (E)-1-alkenyl-1,3,2-benzodioxaboroles produces the "head-to-tail" cross-coupling products, 2-phenyl-1-alkenes or 2-alkyl-1,3-alkadienes, respectively (Eq. 107) [147]. The reaction is profoundly affected by catalytic quantities of palladium compounds (Pd black prepared by reduction of $Pd(OAc)_2$ with diborane is especially effective) in the presence of triethylamine.

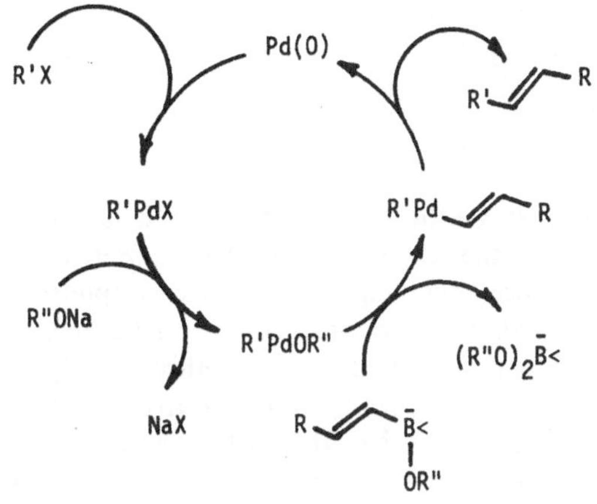

$$R'X \xrightarrow{Pd(0)} R'PdX \xrightarrow{R''ONa,\ -NaX} R'PdOR'' \xrightarrow{R\text{-}CH{=}CH\text{-}\bar{B}<(OR''),\ -(R''O)_2\bar{B}<} R'Pd\text{-}CH{=}CH\text{-}R \longrightarrow R'\text{-}CH{=}CH\text{-}R + Pd(0)$$

Scheme 1

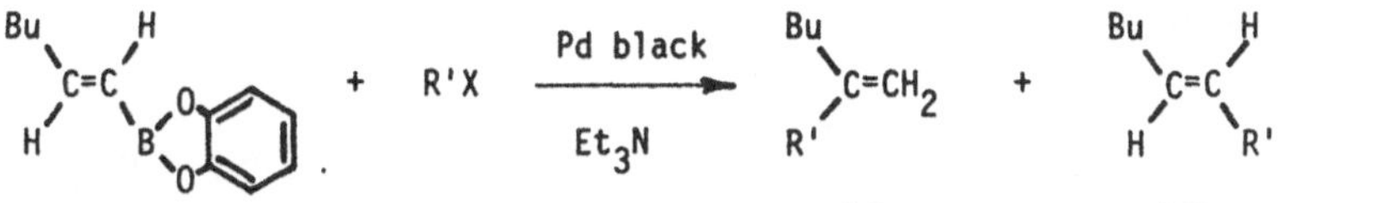

$$\text{Bu(H)C=CH-B(O}_2\text{C}_6\text{H}_4) + R'X \xrightarrow[Et_3N]{\text{Pd black}} \underset{\mathbf{41}}{\text{Bu(R')C=CH}_2} + \underset{\mathbf{42}}{\text{Bu(H)C=C(H)R'}} \qquad (107)$$

R' = phenyl, 94% (*41*:*42* = 96: 4)

R' = (E)-1-octenyl, 92% (*41*:*42* = 95: 5)

Vinylic ethers (*44*) can be synthesized in high yields by the cross-coupling of aryl or benzyl halides with tris(2-ethoxyvinyl)borane (*43*) in the presence of 1 mole % of a palladium compound such as tetrakis(triphenylphosphine)palladium and a base (Eq. 108) [148]. Since vinylic ethers (*44*) thus obtained can readily be hydrolyzed to aldehydes, this reaction provides a convenient procedure for converting aryl or benzylic halides into the corresponding aldehydes with two more carbon atoms.

$$3\text{EtO-C}\equiv\text{CH} + BH_3 \longrightarrow \underset{\mathbf{43}}{(\text{EtO-CH=CH})_3B} \qquad (108\text{-}1)$$

$$\mathbf{43} + RX \xrightarrow[NaOH]{Pd(PPh_3)_4} \underset{\substack{\mathbf{44} \\ 98\text{-}86\%}}{\text{RCH=CHOEt}} \xrightarrow{H_3O^+} RCH_2CHO \qquad (108\text{-}2)$$

R = Ar and $ArCH_2$

1-Alkenylboranes also react with carbon monoxide in the presence of palladium chloride and sodium acetate in methanol to yield α,β-unsaturated carboxylic acid esters while retaining the original configuration of the double bonds of 1-alkenylboranes (Eq. 109) [149].

Most recently, Rossi et al. have applied the cross-coupling of 1-alkenylboranes

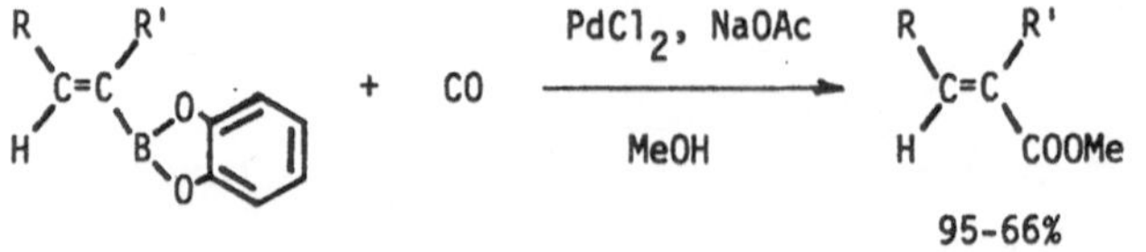

$$\text{RHC=C(R')B(O_2C_6H_4)} + CO \xrightarrow[\text{MeOH}]{\text{PdCl}_2,\ \text{NaOAc}} \text{RHC=C(R')COOMe} \quad (95\text{-}66\%) \tag{109}$$

with 1-alkenyl or 1-alkynyl halides for their synthesis of natural products. Thus, pure (7E,9Z)-7,9-dodecadien-1-yl acetate (*45*), the sex pheromone of *Lobesia botrana*, has been prepared by the following sequence involving: (i) the cross-coupling of (E)-8-(2-tetrahydropyranyloxy)-1-octenyldisiamylborane with 1-bromo-1-butyne, in the presence of tetrakis(triphenylphosphine)palladium and sodium methoxide; (ii) the acetylation of the crude product of the reaction; (iii) the (Z)-stereoselective reduction of the conjugate (E)-enyn-1-yl acetate (Eq. 110)[150]. (E)-9,10-Dodecadien-1-yl acetate (*46*), a sex pheromone component of *Diparopsis castanea*, has been analogously obtained (Eq. 111)[150].

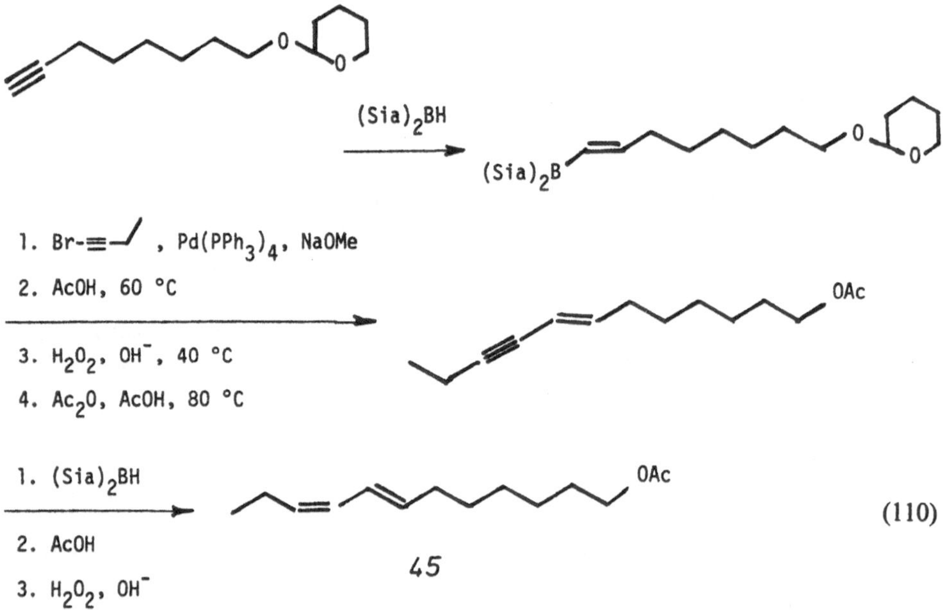

(Sia)$_2$BH

1. Br-≡-⁄, Pd(PPh$_3$)$_4$, NaOMe
2. AcOH, 60 °C
3. H_2O_2, OH^-, 40 °C
4. Ac_2O, AcOH, 80 °C

OAc

1. (Sia)$_2$BH
2. AcOH
3. H_2O_2, OH^-

OAc

45 (110)

(Sia)$_2$BH

1. ⁄Br, Pd(PPh$_3$)$_4$, NaOH
2. AcOH, 60 °C
3. H_2O_2, OH^-, 40 °C
4. Ac_2O, AcOH, 80 °C

OAc

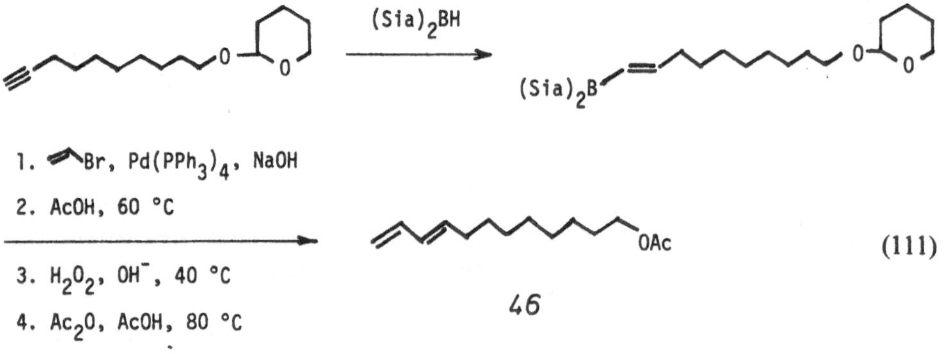

46 (111)

Stereospecific synthesis of (2Z,4E,6E)-3,7,11-trimethyl-2,4,6,10-dodecatetraene, trans (C_{10})-allofarnesene (*49*), isolated from *Perilla frutscens* Makino, was realized by the palladium-catalyzed cross-coupling of 4,8-dimethyl-1,3,7-nonatrienyl-1,3,2-benzodioxaborole (*48*) with (Z)-2-bromo-2-butene. The benzodioxaborole derivative (*48*) was prepared by hydroboration of 4,8-dimethylnona-3,7-dien-1-yne (*47*), obtained via two steps from geranial, with 1,3,2-benzodioxaborole (Eq. 112)[150a)]. Bombykol and its geometrical isomers were also synthesized selectively[150b)].

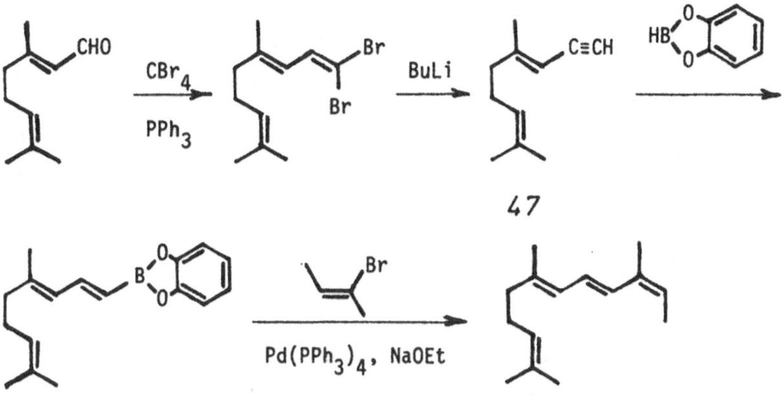

CHO $\xrightarrow[PPh_3]{CBr_4}$ Br, Br $\xrightarrow{BuLi}$ C≡CH (47) $\xrightarrow{HB(O_2C_6H_4)}$

B(O₂C₆H₄) (48) $\xrightarrow[Pd(PPh_3)_4,\ NaOEt]{(Z)\text{-}CH_3CH=C(Br)CH_3}$ 49 (112)

Before the above-mentioned palladium-catalyzed cross-coupling reactions of alkenylboranes with organic halides in the presence of bases had been brought to light, Heck and Dieck reported cross-coupling between (Z)-alkenylboronic acid and methyl acrylate in the presence of palladium acetate[151)]. Although this was probably the first report of these types of reactions, stoichiometric amounts of the palladium compound were required. Thereafter, Yatagai et al. discovered that alkenyldialkylboranes, derived from terminal alkynes, undergo an intramolecular migration reaction in the presence of stoichiometric quantities of palladium acetate and triethylamine to give (E)-alkenes (*50*) (Eq. 113)[152)]. On the other hand, under

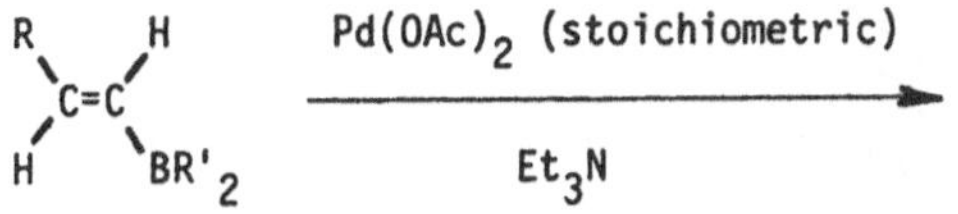

$$\underset{H}{\overset{R}{>}}C=C\underset{BR'_2}{\overset{H}{<}} \xrightarrow[Et_3N]{Pd(OAc)_2\ (stoichiometric)} \underset{H}{\overset{R}{>}}C=C\underset{R'}{\overset{H}{<}} \qquad (113)$$

50

94–58%

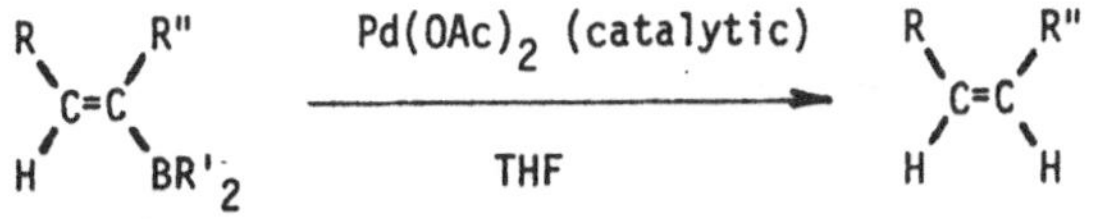

$$\underset{H}{\overset{R}{>}}C=C\underset{BR'_2}{\overset{R''}{<}} \xrightarrow[THF]{Pd(OAc)_2\ (catalytic)} \underset{H}{\overset{R}{>}}C=C\underset{H}{\overset{R''}{<}} \qquad (114)$$

51

95–69%

the aforementioned conditions or even in the presence of catalytic quantities of palladium acetate, in the absence of triethylamine, alkenylboranes derived from internal alkynes undergo a protonolysis reaction to produce (Z)-alkenes (*51*) (Eq. 114) [152], in which the proton source is considered to be the solvent.

Most recently, it has been disclosed that 1-alkenylboranes react with 3,4-epoxy-1-butene in the presence of a palladium or nickel complex to form the two corresponding coupling products, 2-ethenyl-3-alken-1-ol (*52*) and 2,5-alkadien-1-ol (*53*) (Eq. 115) [153].

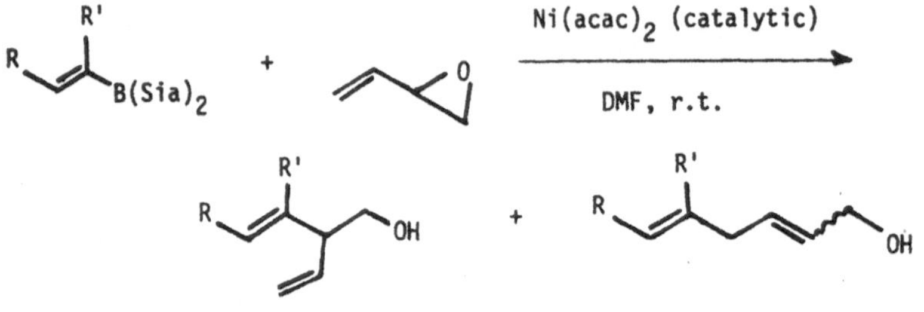

Ni(acac)$_2$ (catalytic)

DMF, r.t.

(115)

52 *53*

Total yield, 85-60%

Ratio of *52*:*53*, 0-40:100-60

VII Miscellaneous Reactions

Although in most of the tetracoordinate organoborate reactions aforementioned, the organic groups on the boron atom were utilized for synthetic applications, the tetraalkylborate (*54*) derived from B-butyl-9-borabicyclo[3.3.1]nonane and butyllithium was demonstrated to be a useful reducing agent for organic compounds. Thus, *54* reacts readily with certain alkyl halides to give the corresponding hydrocarbons (Eq. 116) [154]. The reagent is highly selective in reducing tertiary, benzyl and allyl halides (chlorides and bromides) to the corresponding hydrocarbons in good

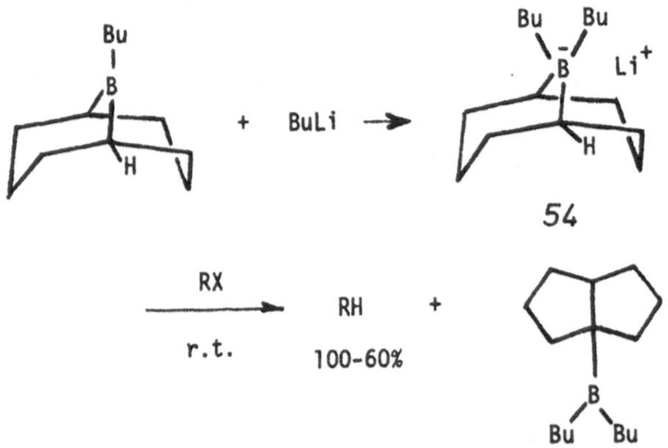

+ BuLi →

54

RX, r.t. → RH 100-60% +

(116)

55

yields, while primary and secondary alkyl and aryl halides remain unaffected. It appears certain that one of the bridge-head hydrogens acts as a hydride, since the borane byproduct formed has been identified as *55*.

The lithium dibutyl ate complex of 9-BBN (*54*) also exhibits high stereo-, chemo-, and regioselectivities in the reduction of carbonyl groups. Thus both cis- and trans-4-methylcyclohexanols with reasonably high isomeric purity are independently obtained from 4-methylcyclohexanone with a mere change of additives (Eq. 117) [155]. Furthermore, aldehydes can be chemoselectively reduced in the presence of ketones (Eq. 118), and the reagent even discriminates between the regioisomers of ketones (Eq. 119).

trans 90% ← 54, MeLi — 4-methylcyclohexanone — 54, MeOH → cis 84% (117)

54, MeOH; conv. 57%; 95%; 5% (118)

54, MeOH; conv. 56%; 91%; 9% (119)

There has been a report on the organic synthesis utilizing cis-bicyclo[3.3.0]oct-1-yldialkylboranes (*57*), which are easily prepared by the reaction of *56* with acetyl chloride. These organoboranes are valuable intermediates for the preparation of a variety of 1-substituted cis-bicyclo[3.3.0]octanes, which otherwise are difficult to produce (Eq. 120) [156].

The chemistry of transition-metal tetrahydroborate complexes has been the subject of numerous investigations over the last few years, which have focused primarily upon the unusual bonding and fluxional characteristics such compounds possess. The synthetic application of these complexes as reducing agents for unsaturated

56 —AcCl→ *57* (120-1)

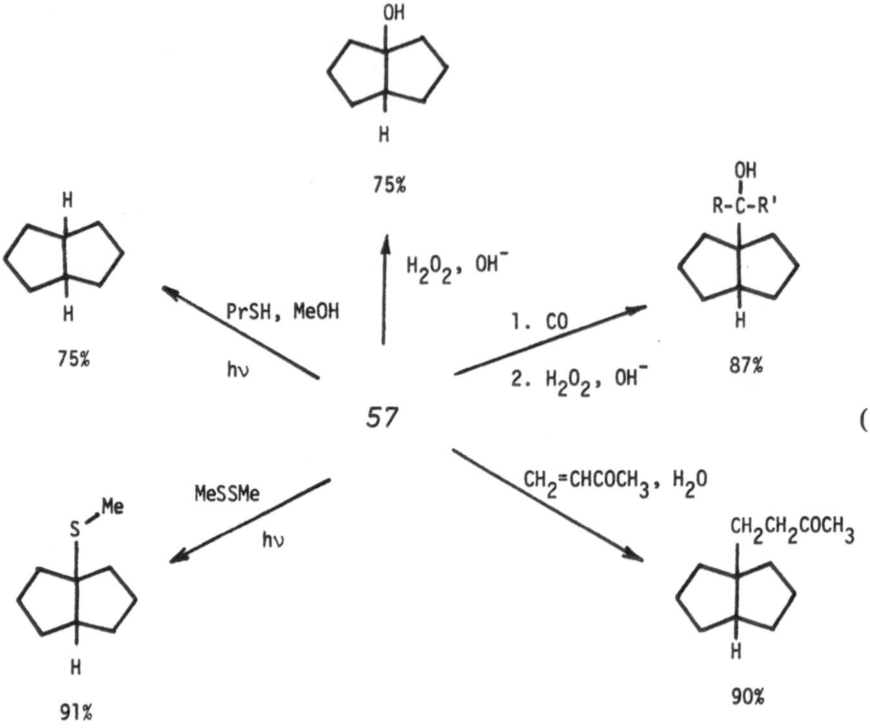

OH

H

75%

H

H

75%

PrSH, MeOH

hν

H_2O_2, OH^-

OH
R-C-R'

H

87%

1. CO

2. H_2O_2, OH^-

57

(120-2)

Me
S

H

91%

MeSSMe

hν

CH_2=$CHCOCH_3$, H_2O

$CH_2CH_2COCH_3$

H

90%

organic compounds has recently been reported. A lanthanide tetrahydroborate has been implicated in the reduction of α, β-unsaturated ketones to allylic alcohols [157] and bis(cyclopentadienyl)chlorotetrahydroborate zirconium(IV) has been shown to reduce aldehydes and ketones [158].

Bis(triphenylphosphine)copper(I) tetrahydroborate (*58*), readily prepared from cuprous chloride, triphenylphosphine and sodium borohydride in chloroform-ethanol solution, reduces aliphatic and aromatic acid chlorides to the corresponding aldehydes in good yields (Eq. 121) [159]. The ideal solvent is acetone, and in most cases,

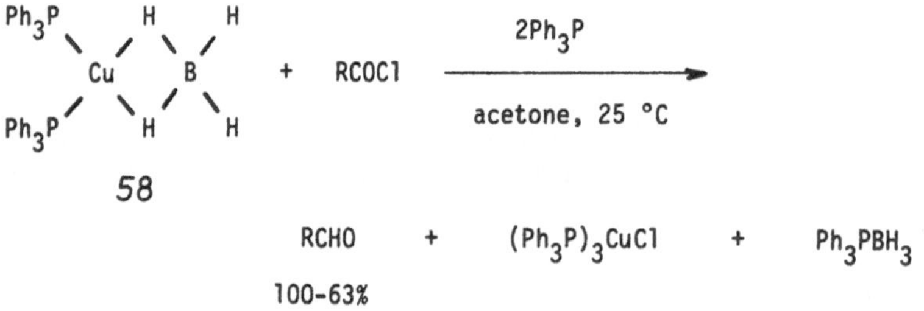

$$(Ph_3P)_2Cu(\mu\text{-}H)_2BH_2 \; (58) + RCOCl \xrightarrow[\text{acetone, 25 °C}]{2Ph_3P} \underset{100\text{-}63\%}{RCHO} + (Ph_3P)_3CuCl + Ph_3PBH_3 \qquad (121)$$

2 additional equivalents of triphenylphosphine are added, one to bind to the copper by-product and the other to trap liberated BH_3.

Saturated ketones and aldehydes are converted via their p-toluenesulfonylhydrazones or trisylhydrazones to the corresponding alkanes by treatment with bis(triphenylphosphine)copper(I) tetrahydroborate (*58*) (Eq. 122) [160]. Neither aromatic nor α,β-unsaturated carbonyl compounds could be converted to the corresponding alkanes in satisfactory yields by this procedure.

$$RR'C{=}O \longrightarrow RR'C{=}NNHTs \xrightarrow[\text{reflux}]{\mathbf{58},\ CHCl_3} RR'CH_2 \quad (84\text{-}56\%) \tag{122}$$

VIII Summary

In this article, a description has been given of the recent development of synthetic applications employing organoborates. The syntheses of the various classes of organic compounds by such organoborate intermediates appear to provide useful, and in some cases, superior alternatives to those using previously established methods. As Professor H. C. Brown [7)] recently commented, a new continent has been discovered, and it requires settlers to develop its riches and contribute them to mankind. With the same theme in mind the author wants to emphasize the immense importance of organoborates, organoboranes in a broad sense, in organic syntheses. The major advantages of these reactions are as follows:

1) Organoboranes with a wide variety of structures are readily available by the hydroboration of alkenes and alkynes.
2) It is possible to synthesize organoboranes with functional groups.
3) Organoborates are readily prepared from various types of organoboranes.
4) The boron-carbon bonds of organoboranes and organoborates are replaced by substituents or by carbon-carbon bonds.
5) The synthetic reactions using organoboranes and organoborates are usually carried out in "one-pot" preparation.
6) The reactions occur under mild conditions.
7) The reactions are generally regio-, stereo-, and chemoselective.

IX Acknowledgements

The author wishes to express his sincere appreciation to Professor Herbert C. Brown of Purdue University, who kindly introduced him to this interesting and attractive field of chemistry. The author is indebted to his active and sedulous co-workers whose names appear in the references, and to the generous support given by the Ministry of Education of Japan, the Japanese Society for the Promotion of Science, Asahi Glass Foundation for Industrial Technology, Mitsubishi Chemical Industries, Mitsui Petrochemical Industries, and Phillips Petroleum Co.

Received May 13, 1982.

X References

1. Negishi, E.: J. Organomet. Chem. *108*, 281 (1976)
2. Weill-Raynal, J.: Synthesis 633 (1976)
3. Cragg, G. M. L., Koch, K. R.: Chem. Soc. Rev. *6*, 393 (1977)
4. Avasthi, K., Devaprabhakara, D., Suzuki, A. in: Organometallic Chemistry Reviews, p. 1, Elsevier, Amsterdam, 1979
5. Brown, H. C., Campbell, J. B.: Aldrichimica Acta *14*, 3 (1981)
6. Suzuki, A.: Acc. Chem. Res. *15*, 178 (1982)
7. Brown, H. C.: Organic Synthesis via Boranes, Wiley, New York, 1975
8. Cragg, G. M. L.: Organoboranes in Organic Synthesis, Dekker, New York, 1973
9. Pelter, A., Smith, K. in: Comprehensive Organic Chemistry, (ed. Barton, D. H. R., Ollis, W. D.) Vol. 3, p. 683, Pergamon, Oxford, 1979
10. Negishi, E.: Organometallics in Organic Synthesis, Vol. 1, p. 286, Wiley, New York, 1980
11. Brown, H. C.: Hydroboration, Benjamine/Cummings, Reading, Massachusetts, 1980[2]
12. Corey, E. J., Albonico, S. M., Koelliker, V., Schaaf, T. K., Varma, R. K.: J. Am. Chem. Soc. *93*, 1491 (1971); Brown, H. C., Kramer, G. W., Hubbard, J. L., Krishnamurthy, S.: J. Organomet. Chem. *188*, 1 (1980)
13. Campbell, J. B. Jr., Molander, G. A.: J. Organomet. Chem. *156*, 71 (1978)
14. Negishi, E., Idacavage, M. J., Chiu, K-W., Yoshida, T., Abramovitch, A., Goettel, M. E., Silveira, A., Bretherick, H. D.: J. Chem. Soc. Perkin II 1225 (1978)
15. Honeycutt, J. B. Jr., Riddle, J. M.: J. Am. Chem. Soc. *83*, 369 (1961)
16. Binger, P., Köster, R.: Tetrahedron Lett. 156 (1961)
17. Hooz, J., Akiyama, S., Cedar, F. J., Bennett, M. J., Tuggle, R. M.: J. Am. Chem. Soc. *96*, 274 (1974)
18. Fletcher, A. S., Paget, W. E., Smith, K., Swaminathan, K.: J. Chem. Soc. Chem. Commun. 573 (1979)
19. Groves, D., Rhine, W., Stucky, G. D.: J. Am. Chem. Soc. *93*, 1553 (1971)
20. Damico, R.: J. Org. Chem. *29*, 1971 (1964); Hart, D. J., Ford, W. T.: ibid. *39*, 363 (1974); Yamamoto, Y., Toi, H., Sonoda, A., Murahashi, S.: Chem. Lett. 1199 (1975); Hall, L. W., Odom, T. D., Ellis, P. D.: J. Am. Chem. Soc. *97*, 4527 (1975); Idem: J. Organomet. Chem. *97*, 145 (1975); Brown, C. A.: ibid *156*, C17 (1978); Hubbard, J. L., Kramer, G. W., ibid. *156*, 81 (1978); Brown, H. C., Hubbard, J. L.: J. Org. Chem. *44*, 467 (1979); Yamamoto, Y., Yatagai, H., Naruta, Y., Maruyama, K.: Tetrahedron Lett. *21*, 3599 (1980); Yanagisawa, M., Yamamoto, O.: Org. Magn. Reson. *14*, 76 (1980)
21. Brown, H. C.: Acc. Chem. Res. *2*, 65 (1969)
22. For reviews, see ref. 8, p. 127; ref. 10, p. 310; Onak, T.: Organoborane Chemistry, p. 100, Academic, New York, 1975
23. De Lue, N. R., Brown, H. C.: Synthesis 114 (1976)
24. Brown, H. C., De Lue, N. R., Kabalka, G. W., Hedgecock, H. C. Jr.: J. Am. Chem. Soc. *98*, 1290 (1976)
25. Kabalka, G. W., Gooch, E. E.: J. Org. Chem. *45*, 3578 (1980)
26. Kabalka, G. W., Gooch, E. E.: J. Org. Chem. *46*, 2582 (1981)
27. Kabalka, G. W., Sastry, K. A. R., Hsu, H. C., Hylarides, M. D.: J. Org. Chem. *46*, 3113 (1981)
28. Kabalka, G. W., Sastry, K. A. R., McCollum, G. W., Yoshioka, H.: ibid. *46*, 4296 (1981)
29. Suzuki, A., Sono, S., Itoh, M., Brown, H. C., Midland, M. M.: J. Am. Chem. Soc. *93*, 4329 (1971)
30. Brown, H. C., Midland, M. M.: ibid. *94*, 2114 (1972)
31. Brown, H. C., Midland, M. M., Levy, A. B.: ibid. *95*, 2394 (1973)
32. Suzuki, A., Tabata, M., Ueda, M.: Tetrahedron Lett. 2195 (1975)
33. Jigajinni, V. B., Pelter, A., Smith, K.: ibid. 181 (1978)
34. Akimoto, I., Suzuki, A.: Synth. Commun. *11*, 475 (1981)
35. Suzuki, A., Ishidoya, M., Tabata, M.: Synthesis 687 (1976)
36. Brown, H. C., Rogić, M. M., Rathke, M. W.: J. Am. Chem. Soc. *90*, 6218 (1968)
37. Hooz, J., Linke, S.: ibid. *90*, 5965 (1968)
38. Tufariello, J. J., Lee, L. T. C., Wojtkowski, P.: ibid. *89*, 6804 (1967)

39. Hughes, R. J., Pelter, A., Smith, K., Negishi, E., Yoshida, T.: Tetrahedron Lett. 87 (1976)
40. Negishi, E., Yoshida, T., Silveira, A. Jr., Chiou, B. L.: J. Org. Chem. *40*, 814 (1975)
41. Midland, M. M., Zolopa, A. R., Halterman, R. L.: J. Am. Chem. Soc. *101*, 248 (1979)
42. Pelter, A., Smith, K., Hutchings, M. G., Rowe, K.: J. Chem. Soc. Perkin I 129 (1975)
43. Murphy, R., Prager, R. H.: Austral. J. Chem. *34*, 143 (1981)
44. Pelter, A., Hutchings, M. G., Rowe, K., Smith, K.: J. Chem. Soc. Perkin II 138 (1975)
45. Pelter, A., Hutchings, M. G., Smith, K.: ibid. 142 (1975)
46. Pelter, A., Hutchings, M. G., Smith, K., Williams, D. J.: ibid. 145 (1975)
47. Zweifel, G., Pearson, N. R.: J. Am. Chem. Soc. *102*, 5919 (1980)
47a. Garst, M. E., Bonfiglio, J. N.: Tetrahedron Lett. *22*, 2075 (1981)
48. Brown, H. C., Sikorski, J. A., Kulkarni, S. U., Lee, H. D.: J. Org. Chem. *45*, 4540 (1980)
49. Kulkarni, S. U., Lee, H. D., Brown, H. C.: bid. *45*, 4542 (1980)
50. Brown, H. C., Hubbard, J. L.: ibid. *44*, 467 (1979)
51. Brown, H. C., Ford, T. M., Hubbard, J. L.: ibid. *45*, 4067 (1980)
52. Bryson, T. A., Pye, W. E.: ibid. *42*, 3214 (1977)
53. Pelter, A., Maddocks, P. J., Smith, K.: J. Chem. Soc. Chem. Commun. 805 (1978)
54. Kawaguchi, T., Ishidoya, M., Suzuki, A.: Heterocycles *18*, 113 (1982)
55. Hara, S., Kishimura, K., Suzuki, A.: Tetrahedron Lett. 2891 (1978)
56. Suzuki, A., Miyaura, N., Abiko, S., Itoh, M., Brown, H. C., Sinclair, J. A., Midland, M. M.: J. Am. Chem. Soc. *95*, 3080 (1973)
57. Midland, M. M., Brown, H. C.: J. Org. Chem. *39*, 731 (1974)
58. Slayden, S. W.: ibid. *46*, 2311 (1981)
59. Pelter, A., Smith, K., Tabata, M.: J. Chem. Soc. Chem. Commun. 857 (1975)
60. Sinclair, J. A., Brown, H. C.: J. Org. Chem. *41*, 1078 (1976)
61. Negishi, E., Lew, G., Yoshida, T.: J. Chem. Soc. Chem. Commun. 874 (1973)
62. Utimoto, K., Kitai, M., Naruse, M., Nozaki, H.: Tetrahedron Lett. 4233 (1975)
63. Negishi, E., Abramovitch, A.: ibid. 411 (1977)
64. Pelter, A., Bentley, T. W., Harrison, C. R., Subrahmanyam, C., Laub, R. J.: J. Chem. Soc. Perkin I 2419 (1976); Idem, ibid. 2428 (1976); Tetrahedron Lett. 1633 (1975); ibid. 3327 (1975)
65. Hooz, J., Mortimer, R.: Tetrahedron Lett. 805 (1976)
66. Köster, R., Hagelee, L. A.: Synthesis 118 (1976)
67. Pelter, A., Rupani, P., Stewart, P.: J. Chem. Soc. Chem. Commun. 164 (1981)
68. Miyaura, N., Abiko, S., Itoh, M., Suzuki, A.: Synthesis 669 (1975)
69. Yamada, K., Miyaura, N., Itoh, M., Suzuki, A.: Tetrahedron Lett. 1961 (1975)
70. Miyaura, N., Yamada, K., Itoh, M. Suzuki, A.: Synthesis 679 (1977)
71. Utimoto, K., Furubayashi, T., Nozaki, H.: Tetrahedron Lett. 397 (1975)
72. Hara, S., Kishimura, K., Suzuki, A.: Chem. Lett. 221 (1980)
73. Pelter, A., Hughes, L.: J. Chem. Soc. Chem. Commun. 913 (1977)
74. Pelter, A., Rao, J. M.: Tetrahedron Lett. *22*, 797 (1981); Pelter, A., Hughes, L., Rao, J. M.: J. Chem. Soc. Perkin I 719 (1982)
75. Midland, M. M.: J. Org. Chem. *42*, 2650 (1977)
76. Zweifel, G., Backlund, S. J., Leung, T.: J. Am. Chem. Soc. *100*, 5561 (1978)
77. Zweifel, G., Pearson, N. R.: J. Org. Chem. *46*, 829 (1981)
78. Merrill, R. E., Allen, J. L., Abramovitch, A., Negishi, E.: Tetrahedron Lett. 1019 (1977)
79. Binger, P., Köster, R.: ibid. 1901 (1965); Binger, P., Benedikt, G., Rotermund, G. W., Köster, R.: Ann. Chem. *717*, 21 (1968)
80. Yogo, T., Koshino, J., Suzuki, A.: Synth. Commun. *9*, 809 (1979)
81. Yogo, T., Koshino, J., Suzuki, A.: Tetrahedron Lett. 1781 (1979)
82. Zweifel, G., Arzoumanian, H., Whitney, C. C.: J. Am. Chem. Soc. *89*, 3652 (1967)
83. Zweifel, G., Arzoumanian, H.: ibid. *89*, 5086 (1967); ibid. *93*, 6309 (1971)
83a. Zweifel, G., Fisher, R. P., Snow, J. T., Whitney, C. C.: ibid. *94*, 6560 (1972)
84. Brown, H. C., Kulkarni, S. U.: J. Organomet. Chem. *218*, 299 (1981); ibid. *225*, 63 (1982)
85. Brown, H. C., Ravindran, N., Kulkarni, S. U.: J. Org. Chem. *44*, 2417 (1979)
86. Kulkarni, S. U., Basavaiah, D., Brown, H. C.: J. Organomet. Chem. *225*, C1 (1982)
87. Brown, H. C., Basavaiah, D., Kulkarni, S. U.: J. Org. Chem. *47*, 171 (1982)

88. LaLima, N. J., Levy, A. B.: J. Org. Chem. *43*, 1279 (1978); Levy, A. B., Angelastro, R., Marinelli, E. R.: Synthesis, 945 (1980)
89. Tagami, H., Miyaura, N., Itoh, M., Suzuki, A.: Chem. Lett. 1411 (1974)
90. Avasthi, K., Baba, T., Suzuki, A.: Tetrahedron Lett. *21*, 945 (1980)
91. Utimoto, K., Uchida, K., Nozaki, H.: Tetrahedron *33*, 1949 (1977)
92. Utimoto, K., Uchida, K., Yamaya, M., Nozaki, H.: Tetrahedron *33*, 1945 (1977)
93. Suzuki, A., Miyaura, N., Itoh, M.: ibid. *27*, 2775 (1971)
94. Akimoto, I., Suzuki, A.: Synthesis 146 (1979)
95. Dickinson, R. P., Iddon, B.: J. Chem. Soc. C 3447 (1971)
96. Akimoto, I., Sano, M., Suzuki, A.: Bull. Chem. Soc. Jpn. *54*, 1587 (1981)
97. Levy, A. B.: J. Org. Chem. *43*, 4684 (1978); Marinelli, E. R., Levy, A. B.: Tetrahedron Lett. 2313 (1979); Levy, A. B.: ibid. 4021 (1979); Sotoyama, T., Hara, S., Suzuki, A.: Bull. Chem. Soc. Jpn. *52*, 1865 (1979)
98. Utimoto, K., Sakai, N., Obayashi, M., Nozaki, H.: Tetrahedron *32*, 769 (1976)
99. Levy, A. B., Schwartz, S. J., Wilson, N., Christie, B.: J. Organomet. Chem. *156*, 123 (1978)
100. Yoshinari, T., Miyaura, N., Itoh, M., Suzuki, A.: Tetrahedron Lett. *21*, 537 (1980)
101. Yogo, T., Suzuki, A.: Chem. Lett. 591 (1980)
102. Koshino, J., Yogo, T., Suzuki, A.: ibid. 1059 (1981)
103. Yamamoto, Y., Yatagai, H., Maruyama, K.: J. Am. Chem. Soc. *103*, 1969 (1981); ibid. *100*, 6282 (1978); Tetrahedron Lett. 3599 (1980)
104. Yamamoto, Y., Yatagai, H., Maruyama, K.: J. Chem. Soc. Chem. Commun. 157 (1979)
105. Evans, D. A., Vogel, E., Nelson, J. V.: J. Am. Chem. Soc. *101*, 6120 (1979); Evans, D. A., Taber, T. R.: Tetrahedron Lett. 4675 (1980); Evans, D. A., Bartroli, J., Shih, T. L.: J. Am. Chem. Soc. *103*, 2127 (1981); Evans, D. A., Nelson, J. V., Vogel, E., Taber, T. R.: ibid. *103*, 3099 (1981); Kuwajima, I., Kato, M., Mori, A.: Tetrahedron Lett. 4291 (1980); Masamune, S., Choy, W., Kerdesky, F. A. J., Imperiali, B.: J. Am. Chem. Soc. *103*, 1566 (1981); Meyers, A. I., Yamamoto, Y.: ibid. *103*, 4278 (1981); Mukaiyama, T., Inoue, T.: Bull. Chem. Soc. Jpn. *53*, 174 (1980); Mukaiyama, T., Murakami, M., Oriyama, T., Yamaguchi, M.: Chem. Lett. 1193 (1981); Thomas, H., Ursula, S., Reinhard, H.: Chem. Ber. *114*, 359 (1981); ibid *114*, 375 (1981); Wada, M.: Chem. Lett. 153 (1981); Yamamoto, Y., Yatagai, H., Naruta, Y., Maruyama, K.: J. Am. Chem. Soc. *102*, 7107 (1980); ibid. *102*, 4548 (1980); J. Chem. Soc. Chem. Commun. 1072 (1980)
106. Midland, M. M., Prestom, S. B.: J. Org. Chem. *45*, 748 (1980)
107. Negishi, E., Chiu, K-W.: ibid. *41*, 3484 (1976)
108. Negishi, E., Idacavage, M. J.: Tetrahedron Lett. 845 (1979)
109. Suzuki, A., Arase, A., Matsumoto, H., Itoh, M., Brown, H. C., Rogić, M. M., Rathke, M. W.: J. Am. Chem. Soc. *89*, 5708 (1967); Kabalka, G. W., Brown, H. C., Suzuki, A., Honma, S., Arase, A., Itoh, M.: ibid. *92*, 710 (1972)
110. Brown, H. C., Rogić, M. M., Rathke, M. W., Kabalka, G. W.: J. Am. Chem. Soc. *89*, 5709 (1967)
111. Suzuki, A., Nozawa, S., Itoh, M., Brown, H. C., Kabalka, G. W.: ibid. *92*, 3503 (1970)
112. Suzuki, A., Miyaura, N., Itoh, M., Brown, H. C., Holland, G. W.: ibid. *93*, 2792 (1971)
113. Suzuki, A., Miyaura, N., Itoh, M., Brown, H. C., Jacob, P.: Synthesis 305 (1973)
114. Sasaki, N., Miyaura, N., Itoh, M., Suzuki, A.: Synthesis 317 (1975)
115. Miyaura, N., Itoh, M., Suzuki, A.: Tetrahedron Lett. 255 (1976)
116. Miyaura, N., Itoh, M., Suzuki, A.: Synthesis 618 (1976)
117. Sasaki, N., Miyaura, N., Itoh, M., Suzuki, A.: Tetrahedron Lett. 173 (1977)
118. Miyaura, N., Itoh, M., Suzuki, A.: Bull. Chem. Soc. Jpn. *50*, 2199 (1977)
119. Yamada, K., Miyaura, N., Itoh, M., Suzuki, A.: ibid. *50*, 3431 (1977)
120. Miyaura, N., Yamada, K., Yano, T., Suzuki, A.: ibid. *52*, 275 (1979)
121. Miyaura, N., Yano, T., Suzuki, A.: ibid. *53*, 1471 (1980)
122. Sasaki, N., Miyaura, N., Itoh, M., Suzuki, A.: Tetrahedron Lett. 3369 (1977)
123. Yamamoto, Y., Yatagai, H., Moritani, I.: J. Am. Chem. Soc. *97*, 5606 (1975); Yamamoto, Y., Yatagai, H., Maruyama, K., Sonoda, A., Murahashi, S.: ibid. *99*, 5652 (1977)
124. Yamamoto, Y., Yatagai, H., Sonoda, A., Murahashi, S.: J. Chem. Soc. Chem. Commun. 452 (1976); Yatagai, H.: J. Org. Chem. *45*, 1640 (1980)

125. Yamamoto, Y., Yamamoto, S., Yatagai, H., Maruyama, K.: J. Am. Chem. Soc. *102*, 2318 (1980); Maruyama, K., Yamamoto, Y.: ibid. *99*, 8068 (1977); Yamamoto, Y., Maruyama, K.: J. Organomet. Chem. *156*, C9 (1978)
126. Yamamoto, Y., Yamamoto, S., Yatagai, H., Ishihara, Y., Maruyama, K.: J. Org. Chem. *47*, 119 (1982); Idem: J. Am. Chem. Soc. *100*, 3240 (1978)
127. Yamamoto, Y., Yatagai, H., Maruyama, K.: J. Org. Chem. 44, 1744 (1979)
128. Campbell, J. B., Brown, H. C.: ibid. *45*, 549 (1980)
129. Brown, H. C., Campbell, J. B.: ibid. *45*, 550 (1980)
130. Brown, H. C., Molander, G. A.: ibid. *46*, 645 (1981)
131. Normant, J. F., Commercon, A., Villieras, J.: Tetrahedron Lett. 1465 (1975)
132. Baba, S., Negishi, E.: J. Am. Chem. Soc. *98*, 6729 (1976)
133. Negishi, E. in: New Applications of Organometallic Reagents in Organic Synthesis (ed. Seyferth, D.), p. 93, Elsevier, Amsterdam, 1976
134. Posner, G. H.: Org. Reac. *22*, 253 (1975); Normant, J. F.: ref. 133, p. 219
135. Dang, H. P., Linstrumelle, G.: Tetrahedron Lett. 191 (1978)
136. Larock, R. C.: Angew. Chem. Int. Ed. Engl. *17*, 27 (1978)
137. Whitesides, G. M., Casey, C. P., Krieger, J. K.: J. Am. Chem. Soc. *93*, 1379 (1971)
138. Okukado, N., VanHorn, D. E., Klima, W. L., Negishi, E.: Tetrahedron Lett. 1027 (1978)
138a. Baba, S., Negishi, E.: J. Am. Chem. Soc. *98*, 6729 (1976); Yatagai, H., Yamamoto, Y., Murahashi, S.: J. Chem. Soc. Chem. Commun. 852 (1977); Larock, R. C., Mitchell, M. A.: J. Am. Chem. Soc. *100*, 180 (1978)
139. Kochi, J. K.: Organometallic Mechanisms and Catalysis, Academic, New York, 1978; Collman, J. P., Hegedus, L. S.: Principles and Applications of Organotransition Metal Chemistry, University Science Books, Mill Valley, Calif. 1980
140. Miyaura, N., Yamada, K., Suzuki, A.: Tetrahedron Lett. 3437 (1979)
141. Miyaura, N., Suginome, H., Suzuki, A.: ibid. *22*, 127 (1981)
142. Miyaura, N., Suzuki, A.: J. Chem. Soc. Chem. Commun. 866 (1979)
143. Yano, T., Miyaura, N., Suzuki, A.: Tetrahedron Lett. *21*, 2865 (1980)
144. Davidson, J. M., Triggs, C.: J. Chem. Soc. A 1324 (1968)
145. Miyaura, N., Yanagi, T., Suzuki, A.: Synth. Commun. *11*, 513 (1981)
146. Miyaura, N., Suzuki, A.: unpublished
147. Miyaura, N., Suzuki, A.: J. Organomet. Chem. *213*, C53 (1981)
148. Miyaura, N., Maeda, K., Suginome, H., Suzuki, A.: J. Org. Chem. *47*, 2117 (1982)
149. Miyaura, N., Suzuki, A.: Chem. Lett. 879 (1981)
150. Rossi, R., Carpita, A., Quirici, M. G.: Tetrahedron *37*, 2617 (1981)
150a. Miyaura, N., Suginome, H., Suzuki, A.: Bull. Chem. Soc. Jpn. *55*, 2221 (1982)
150b. Miyaura, N., Suginome, H., Suzuki, A.: unpublished
151. Dieck, H. A., Heck, R. F.: J. Org. Chem. *40*, 1083 (1975)
152. Yatagai, H.: Bull. Chem. Soc. Jpn. *53*, 1670 (1980); Yatagai, H., Yamamoto, Y., Maruyama, K.: J. Chem. Soc. Chem. Commun. 702 (1978); ibid. 852 (1977)
153. Miyaura, N., Tanabe, Y., Suginome, H., Suzuki, A.: J. Organomet. Chem. *233*, C13 (1982)
154. Yamamoto, Y., Toi, H., Murahashi, S., Moritani, I.: J. Am. Chem. Soc. *97*, 2558 (1975)
155. Yamamoto, Y., Toi, H., Sonoda, A., Murahashi, S.: ibid. *98*, 1965 (1976)
156. Kramer, G. W., Brown, H. C.: J. Org. Chem. *42*, 2832 (1977); J. Am. Chem. Soc. *98*, 1964 (1976); J. Organomet. Chem. *90*, C1 (1975)
157. Luche, J. L.: J. Am. Chem. Soc. *100*, 2226 (1978)
158. Sorrell, T. N.: Tetrahedron Lett. 4985 (1978)
159. Sorrell, T. N., Pearlman, P. S.: J. Org. Chem. *45*, 3449 (1980); Sorrell, T. N., Spillane, R. J.: Tetrahedron Lett. 2473 (1978); Fleet, G. W. J., Harding, P. J. C.: ibid. 975 (1979)
160. Fleet, G. W. J., Harding, P. J. C., Whitcombe, M. J.: Tetrahedron Lett. *21*, 4031 (1980)

Stable Pyridinyl Radicals

Edward M. Kosower

Chemistry Departments, Tel-Aviv University, Ramat-Aviv, Tel-Aviv, Israel 69978
and State University of New York, Stony Brook, New York 11794, USA

Table of Contents

Pyridinyls are pi-radicals (formally, pyridinium ion + e^-) which are highly reactive, some being stable enough to be isolated. Spectroscopy and magnetic resonance reveal a variety of radical species: monomeric radicals, singlet complexes (pi-mers), triplet pairs, and dimers which are thermally and photo-dissociable (see Fig. 27). A new thin film spectroscopic apparatus for the study of air-sensitive species is described. EPR spectra provide details of the electron distribution within pyridinyl radicals. Reactive pyridinyls, non-volatile pyridinyls and pyridinyl cation radicals can be prepared. Pyridinyl radical dimers photodissociate via a $\pi\sigma \rightarrow \pi\sigma^*$ electronic transition. Pyridinyl radicals and halocarbons react via a) solvent insensitive atom transfer or b) solvent sensitive electron transfer processes.

1 Introduction

Stable pyridinyl radicals were discovered while establishing the linear relationship between the lowest charge-transfer transition energy for 1-alkylpyridinium iodides (Eq. 1) and the one-electron reduction potentials of 1-alkylpyridinium ions [1].

$$4^{+}I^{-} \text{ (Py}^{+}I^{-}) \xrightarrow{h\nu} 4^{\cdot}I^{\cdot} \text{ (Py}^{\cdot}I^{\cdot}) \quad (1)$$

COOCH3 / N+ I− / R → COOCH3 / N / R

The pure 1-ethyl-4-carbomethoxypyridinyl radical ($4^{\cdot}$) was isolated by distillation from the mixture produced by sodium amalgam reduction of the corresponding pyridinium iodide (4^{+}) [2]. Flash photolysis of $4^{+}I^{-}$ also yielded the pyridinyl radical $4^{\cdot}$ [3].

The kinetics and mechanism of the reaction of $4^{\cdot}$ with halocarbons (dichloro-, trochloro- and tetrachloromethane, various benzyl halides) have been reported [4,5,6]. Earlier work on pyridinyl radicals has been reviewed [7,8], as well as the organic chemistry of stable free radicals [9]. This chapter will concentrate on developments of the past decade, since new techniques, described specifically, have increased our understanding of pyridinyl radical behavior considerably.

1.1 Nomenclature

The most frequently used radicals will be denoted by the symbols, $4^{\cdot}$, 1-ethyl-4-carbomethoxypyridinyl, or $2^{\cdot}$, 1-methyl-2-carbomethoxypyridinyl. Other alkyl and carbonyl substituents will be specifically shown, e.g. (CH_3, $CONH_2$) $4^{\cdot}$ is 1-methyl-4-carbamidopyridinyl radical.

Two points of nomenclature deserve comment. We introduced the name, pyridinyl, without fanfare in 1963 when the radicals were discovered [10]. Pyridin*yl* radicals are formed by the addition of an electron to pyridin*ium* ions; all numbering problems are resolved at the level of the pyridinium ion [11]. Hanson [12] has adopted the term "1-alkyl-1-hydropyridinyl" (i.e., addition of a hydrogen atom to a pyridinium ring, then its replacement by a 1-alkyl), noting that Chemical Abstracts has used "1-alkyl-1,4-dihydropyridyl" (i.e., 1,4-dihydropyridine less a hydrogen atom, plus a 1-alkyl group). It seems self-evident that our names are created without difficulty, are simple, useful, unambiguous, and do not incorrectly assign a specific double-bond structure to a delocalized π-radical.

"Stable" radicals can be isolated in pure form, e.g., 1-ethyl-4-carbomethoxy-pyridinyl ($4^{\cdot}$) [13]. Griller and Ingold [14] have proposed the term "pensistent" to refer to long-surviving radicals. (Neither "stable" nor "persistent" can be defined precisely since exposure of the radicals to different reactants will lead to different degrees of "stability" or "persistency"). We shall continue to use "stable" with reference to appropriate pyridinyl radicals.

2 Preparation of Pyridinyl Radicals

Pyridinyl radicals are prepared primarily by reduction of the corresponding pyridinium ions, but may also be generated for special types of experiments by dissociation of dimers and disproportionation of pyridinium ions and dihydropyridines (a special case of the oxidation of dihydropyridines).

2.1 Reduction of Pyridinium Ions

Quantities of pyridinyl radicals are best prepared by chemical reduction, defined concentrations of radicals by electrochemical reduction, by photolytic procedures, or by pulse radiolysis in aqueous solution.

2.1.1 Metal Reduction

Pyridinium halides are easily reduced to pyridinyl radicals (e^- in place of hν, eq. 1) with sodium amalgam in oxygen-free acetonitrile [2,15], using a vacuum line and the apparatus shown in Fig. 1 [16]. After filtration, the pure pyridinyl radicals are isolated by several distillations. Magnesium or zinc have been used on a preparative scale; various other metals (e.g., calcium, strontium and barium) are suitable for preparing solutions of pyridinyl diradicals [17].

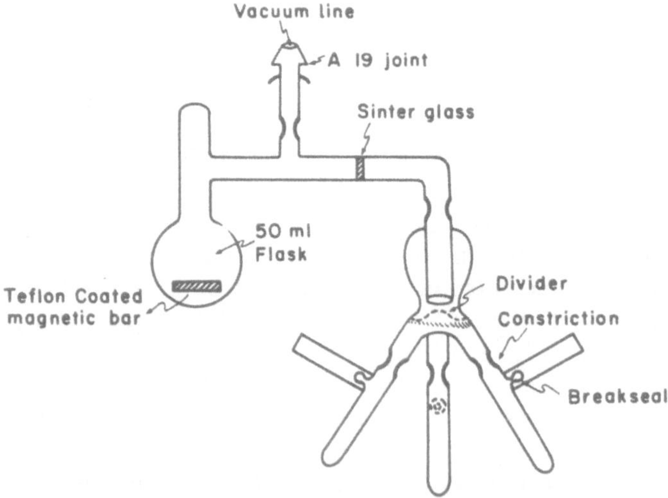

Fig. 1. Apparatus for the preparation of pyridinyl radicals through sodium amalgam reduction of 1-alkylpyridinium iodides in acetonitrile [16]

2.1.2 Electrochemical Reduction

Electrochemical reduction in CH_3CN (containing $LiClO_4$) can be used to generate known quantities of radicals, especially non-volatile 1-alkyl-4-carbamidopyridinyl radicals [(CH_3, $CONH_2$)*4*˙ and (t-Bu, $CONH_2$)*4*˙] from the corresponding pyridinium ions in acetonitrile solution [11]. The cell is illustrated in Fig. 2 [16].

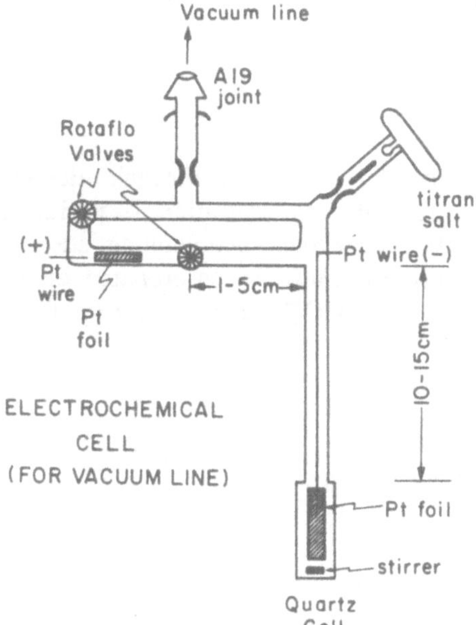

Fig. 2. Spectroelectrochemical cell for the preparation of solutions of pyridinyl radicals [16]. The optical path is parallel to the Pt foil located near one side of the cell

Exhaustive reduction of an exactly known amount of pyridinium salt (weight, absorption spectrum) leads to accurate absorption coefficients which agree with those obtained by alternative means. Electrochemical techniques have been used to probe some of the chemical properties of both mono- and bis-pyridinyl radicals [18,19] (see also Sect. 4.5).

(CH_3, $CONH_2$)*4*˙   (t-Bu, $CONH_2$)*4*˙   (2)

2.1.3 Photochemical Reduction

The flash photolysis of a pyridinium iodide in benzene (Eq. 1), was the first demonstration that a charge-transfer absorption actually led to a charge-transfer process [3].

Photoreduction (253.7 nm) of pyridinium ions in isopropyl alcohol is used to generate pyridinyl radicals for EPR measurements [20,21]. The mechanism of the reaction probably involves formation of the S_1 state of the pyridinium ion followed by electron transfer from the solvent. Dimer formation from the radical is observed in a number of cases (Eq. 3) [22].

$$Py^{+} \xrightarrow{h\nu} Py^{+*}(S_1) \xrightarrow{(CH_3)_2CHOH} Py^{\cdot} \quad (3)$$

$$(2\,Py^{\cdot} \rightarrow Py\text{-}Py)$$

2.1.4 Pulse Radiolysis Reduction

A pulse of 2–10 Mev electrons from a linear accelerator, passed through aqueous solutions of pyridinium ions leads rapidly to substantial concentrations of pyridinyl radicals [23], including those derived from the coenzyme, nicotinamide adenine dinucleotide [24,25]. The pyridinyl radicals disproportionate, are protonated or dimerize. The reducing agent is a solvated electron or the $CO_2^{\overline{\cdot}}$ radical anion (Eq. 4) [26].

$$HOH \xrightarrow{10\ Mev\ e^-} HOH^+ + e^-$$

$$HOH^+ \rightarrow H^+ + {}^{\cdot}OH \quad (4)$$

$${}^{\cdot}OH + HCOO^- \rightarrow HOH + CO_2^{\overline{\cdot}}$$

2.2 Dissociation of Dimers

The reduction of 1-methyl-2-carbomethoxypyridinium iodide (2^+, I^-) with sodium amalgam leads to the pyridinyl radical dimer (*2-2*) (see sect. 4.1). The dimer is in equilibrium with the monomer (*2*˙) (K_{dissoc} ca. $1.3\text{–}2.0 \times 10^6\ M^{-1}$ by titration or EPR measurements) [16,27]. Dissociation of the dimer yields *2*˙, isolated via low temperature distillation (ca. 40–50 °C) with condensation at 77 K (Eq. 5). Dimerization occurs on warming the condensate above ca. 120 K. The spectroscopic properties of the radical *2*˙ are determined with a new thin film spectroscopy apparatus, a device which dramatically simplifies experiments on stable radicals (Sect. 3.1.1).

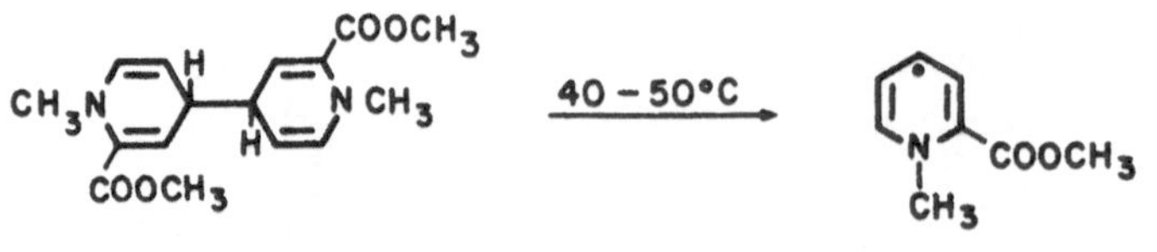

COOCH3 / CH3N / H / H / N CH3 / COOCH3 —40–50 °C→ N / COOCH3 / CH3 (5)

2-2 *2*˙

2.3 Disproportionation

Since the appropriate dihydropyridine anions are not readily available, disproportionation is not a usual reaction for the preparation of pyridinyl radicals. However, the reaction of the anion, 4^-, with the corresponding cation, 4^+, to form the radical $4^{\cdot}$ has been detected electrochemically for acetonitrile solutions (Eq. 6) [18,28]. A parallel reaction is also observed for 2^- and 2^+. Detailed analysis [18] suggests that the disproportionation proceeds via dimerization, i.e., that the anion adds to the cation to form a covalent bond. The dimer subsequently dissociates to form two free radicals, rapidly in the case of the $4^{\cdot}$ radical and less rapidly (1 sec^{-1}) in the case of the $2^{\cdot}$ pyridinyl (for dimer structure, cf. Sect. 4.1). Pyridinyl radicals disproportionate in water to pyridinium ions and dihydropyridines (Sect. 4.4).

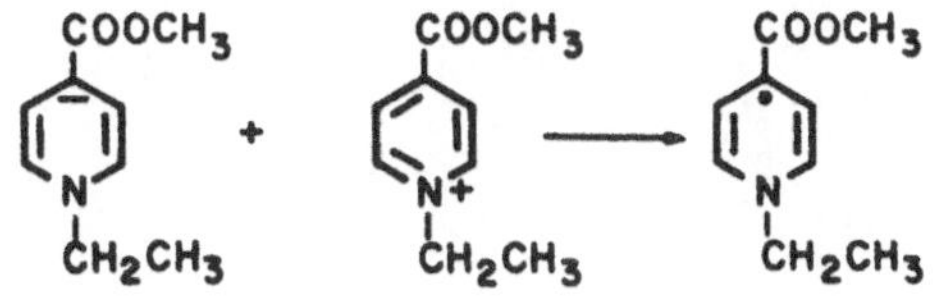

$$4^- + 4^+ \longrightarrow 4^{\cdot} \quad (6)$$

(1-ethyl-4-carbomethoxypyridinyl species: $COOCH_3$ at C-4, CH_2CH_3 on N)

2.4 Oxidation of Dihydropyridines

Although some oxidations of dihydropyridines proceed via pyridinyl radicals, the latter are normally not observable due to the ease of further oxidation to pyridinium ions. By a judicious choice of agents, Land and Swallow succeeded in converting NADH to the radical NAD˙, and so demonstrated that NAD˙ formed by oxidation was identical to that formed by reduction (Eq. 7) [29].

$$\begin{array}{lll} e^- + N_2O & \longrightarrow & {}^{\cdot}OH + N_2 + {}^-OH \\ {}^{\cdot}OH + 2\,Br^- & \longrightarrow & {}^-OH + Br_2^{\dot{-}} \\ NADH + Br_2^{\dot{-}} & \longrightarrow & NAD^{\cdot} + HBr + Br^- \end{array} \quad (7)$$

3 Physical Properties

3.1 Spectroscopic Techniques

Ultraviolet, visible, infrared and EPR spectra are useful in studying pyridinyl radicals. The major recent developments have come from experiments using new apparatus for known techniques. The developments, especially that for thin film spectroscopy, will be useful for other chemical problems. Therefore, the techniques are described in detail.

3.1.1 Thin Film Spectroscopy

The blue film, formed by distillation of 1-alkyl-2-carbomethoxypyridinyl radical, $2^{\cdot}$, onto a 77 K surface, disappears on warming. The thin film spectroscopy apparatus depicted in Fig. 3 allowed a) direct cooling of the condensing surface so that temperature could be altered without delay and b) interferometric measurement of deposition rates on the surface [30].

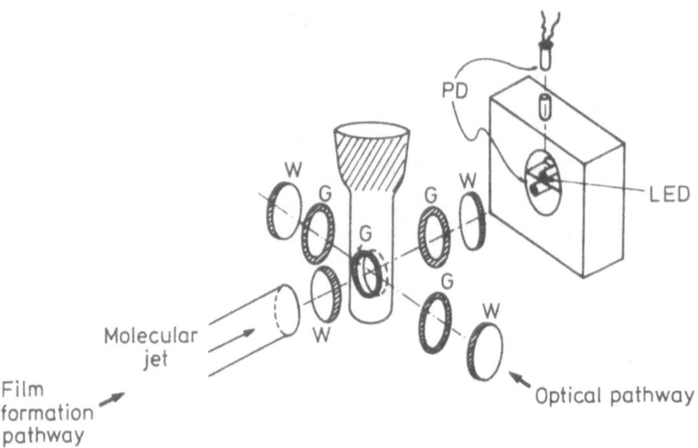

Fig. 3. Exploded view of the thin film spectroscopy apparatus, showing the relative positions of the optical cooling surface, the interferometer, the optical pathway for measurement and the material source for the film. W, window (polished quartz or sapphire), G, ground surface onto which windows are glued, LED, light emitting diode, PD, photodiode [30]

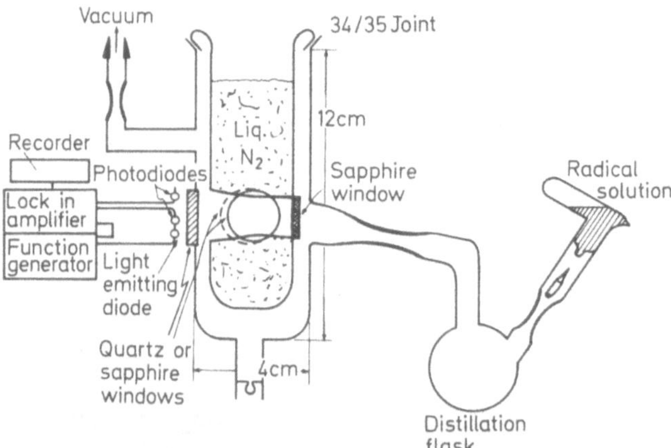

Fig. 4. The thin film spectroscopic apparatus with condenser dimensions. Pyridinyl radicals or dimers are introduced via a breakseal, solvent removed and the radical or dimer transferred to the upper surface of the distilling flask with hot air to facilitate removal of residual volatiles. Liquid nitrogen is added and the distilling flask gently warmed, the interferometer being used to monitor the amount of material deposited on the window. After the desired thickness has been obtained, the distillation is halted, the interferometer detached, the sample section sealed off from the distilling flask, and then from the line. The thin film sample may now be examined by a variety of spectroscopic techniques. The material can also be recovered from the window without exposure to oxygen by attaching the apparatus back to the line, liquid nitrogen; distilling in a suitable solvent, warming and collecting the resulting solution for analysis by chemical or spectroscopic methods [30]

Thermally conductive sapphire served as an optically transparent (200 nm to 6200 nm) condensing surface. The sapphire window is glued to the ground surface of a hollow torus which carries the coolant, assuring good temperature control of the sample. Glued optical windows do not suffer from differences in thermal conductivities and coefficients of expansion between the glass and the window. Film thickness is measured with a interferometer made from light-emitting diode and two photodiodes (number of interference fringes times $^1/_4$ the wavelength of light: 890 nm).

The apparatus has proven to be extremely useful for studies of thin films of viscous liquids and glassy and organized solids over a wide range of temperatures, from 77 K to 577 K. With simple adaptors, the apparatus may be placed in a UV-IVIS spectrophotometer, a spectrofluorimeter and an IR spectrometer. It is small enough (200–300 g) to be carried in one hand. The apparatus for pyridinyl radicals is shown in Fig. 4; an electrical schematic for the interferometer is given in Fig. 5.

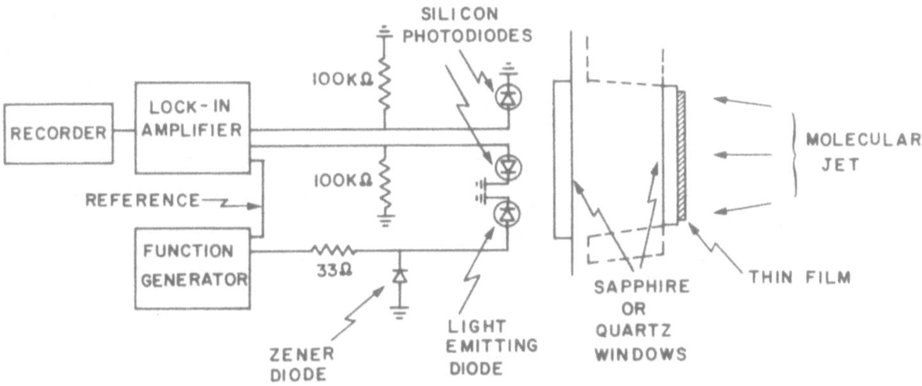

Fig. 5. Electrical schematic for the interferometer used with the thin film spectroscopic apparatus. The windows are included in the diagram to indicate the general location of the LED and photodiodes with respect to the thin films [30]

The condensing surface is an optically polished sapphire window (1.6–2.0 cm diameter) cemented with RTV-108 (General Electric), a silicone glue flexible enough to maintain the bond to the window at 77 K and thermally stable enough to be used at 600 K. The external windows of the apparatus are glued ("Araldite" epoxy, Ciba-Geigy) to ground surfaces on the outside shell. This technique for attaching windows is extremely simple, very rapid, allows different windows to be used (quartz, sapphire, etc.) and makes optical alignment easy. A further advantage of the glue is ease of removal: the RTV-108 is cut with a razor, the sapphire window removed and cleaned thoroughly with lens tissue, detergent, water and organic solvents (CH_3CN, CH_3OH, etc.) then fired in an oven to remove residual organic material. The epoxy glue is burned away by heating the apparatus in an oven at 560 °C, after which the windows are remounted.

Film deposition is followed by a simple interferometer, consisting of a light emitting diode (LED) (Fairchild FPE 104), λ_{max} 899 nm, and two silicon pin photodiodes (Monsanto, MD-1), one for a reference signal directly from the LED, and one for the light reflected from the surfaces of the optical system. As the film thickness reaches 0.25 λ/n (λ = 890 nm), destructive interference of the reflected light decreases the photodiode signal to a minimum, followed by an increase in light signal as the film thickens (Fig. 6). This applies for film material with a smaller refractive index than that of the sapphire (i.e. < 1.758 at 890 nm). The number of maxima and minima are counted during the distillation to obtain the film thickness (Eq. 7). Phase sensitive detection using a lock-in amplifier and a function generator operating at 70 Hz is used to extract the weak signal (200 μv) from the noise. The distillation is stopped by removing the heater from the substrate source.

$$d = 10^{-7} \cdot m/n \cdot \lambda/4 \tag{7}$$

in which d = film thickness in cm; m = the number of maxima and minima detected by the interferometer; λ = the wavelength of the light emitted by the LED (in nm); n = the refractive index of the film.

Interferograms preceding IR spectra of 1-methyl-2-carbomethoxypyridinyl (*2*˙) and 1-methyl-2-acetylpyridinyl radicals ((CH_3CO) *2*˙), are given in Fig. 6. The decreasing signal amplitude with increasing film thickness is due to absorption of the 890 nm light by the film.

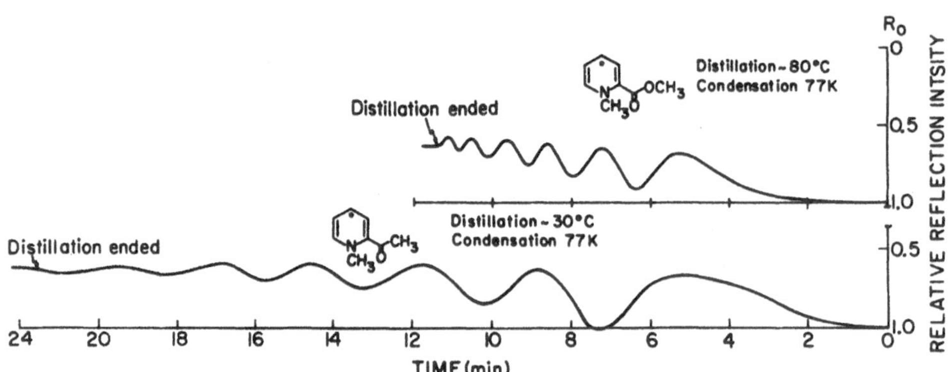

Fig. 6. Interferograms obtained for 1-methyl-2-carbomethoxy-pyridinyl and 1-methyl-2-acetyl-pyridinyl radical films for IR spectra [30]

The absorption coefficient of the species in the film can be obtained from Eq. 8.

$$\varepsilon_\lambda = OD \cdot MW/(1000d \cdot D) \tag{8}$$

in which ε_λ = absorption coefficient at a particular wavelength λ; OD = optical density measured at λ; d = film thickness (in cm); D = estimated density of film [11] (g/cm^3).

The refractive index of the thin film may be estimated from the amplitude of the interferometric signal with Eq. 9, starting from the first maximum, and provided that neither absorption nor dispersion affect the signal intensity too greatly. A refractive index of 1.50 for the pyridinyl radicals at 77 K has been verified.

$$A/R_0 = 4V \cdot W \cdot [1 - X^2]/[2X^2 - X^4] \tag{9}$$

in which $V = (n_s - n)/(n_s + n)$; $W = (n - 1)/(n + 1)$; $X = (n_s - 1)/(n_s + 1)$; n = refractive index of film; n_s = refractive index of sapphire (1.758 at 890 nm); A = amplitude of the signal measured by the interferometer, a quantity proportional to the initial intensity of light reaching the sapphire window, and measured by a photodiode looking directly at the LED; R_0 = a signal proportional to the intensity of light reflected from the clean sapphire window before any sample has been deposited.

3.1.2 Vacuum Variable Path Spectroscopy

In order to measure absorptions of very different intensities with a single solution, or the concentration dependence of the absorption of an oxygen-sensitive compound, a variable pathlength cell with some unusual features was designed and constructed. Mounted on an appropriate base, the cell fits in the Cary model 17 spectrophotometer cell compartment. A detailed drawing is given in Fig. 7.

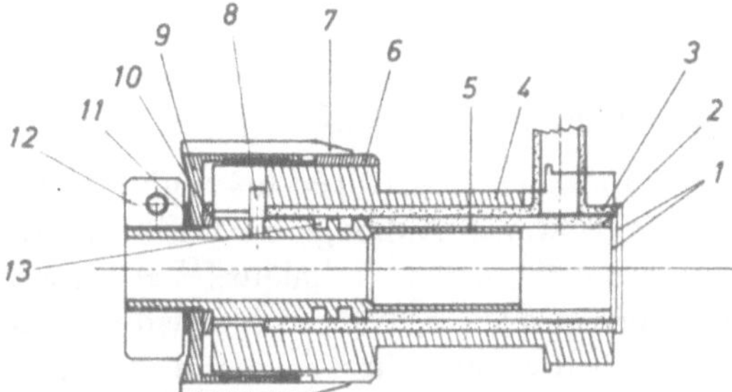

Fig. 7. A detailed drawing of the VV-cell (Vacuum Variable path length cell). The cell mount may be attached to adaptors for the Cary model 17 or an IR spectrophotometer. The optical path goes through the center of the cell, passing through two sapphire windows and an annular opening in the calibrated handle used to turn the micrometer in order to vary the path length.

1) sapphire windows: outside diameters — inner, 2.0 cm; outer, 2.5 cm; 2, 3) precision ground pyrex syringe tubes: diameters — inner, 2.0 cm; outer, 2.5 cm; 4) aluminium structural support; 5) stainless steel support for inner tube and connection to micrometer drive; 6) stainless steel threated sleeve, fixed to aluminum support; 7) calibrated micrometer drum; 8) stainless steel pin stop; 9) bronze nut; 10) stainless steel washer; 11) Delrin washer; 12) aluminum nut; 13) Viton O-rings (Parker 2-017)

The overall length of the cell is 10 cm; the height of cell, not including the glass tube which is connected to the vacuum line or to a manifold carrying solvent and/or solution, is 5 cm. The glass tube protrudes at least 4 cm above the cell, and should extend far enough to permit the necessary glassblowing. The weight of the cell is about 300 g, requiring that the cell be supported for gaassblowing operations.

The cell is connected to a manifold carrying freshly prepared pyridinyl radical solution, additional degassed solvent, a 10 cm quartz cell (for low pyridinyl radical concentrations), a calibrated, graduated cylinder (pipette) for measurement of solvent volumes, and a titrant such as 1,1′-dimethyl-4,4′-bipyridylium dichloride. The radical solution is introduced into the VV-cell, the spectrum measured and the solvent carefully distilled into the graduated cylinder, using liquid N_2-cooled cotton or ice-water to promote condensation. The path length of the VV-cell is decreased after each solvent transfer out of the cell in order to obtain an optimum optical density (ca. 0.9) which is matched from spectrum to spectrum. The concentration of the radical in the cell is calculated from the known total volume and the volume of solvent in the graduated cylinder. The band shapes and absorption coefficients as a function of concentration are thus easily obtained [11]

The basic elements of the VV-cell (Vacuum Variable pathlength cell) are: A) Solution Carrier: 1) Inner and outer cylinders made from a precision pyrex syringe, 2) sapphire windows affixed to the outer end of the outer cyclinder, and to the end of the inner cylinder. B) Sealing Elements: 1) Two Viton O-rings between the outer and inner cylinders. 2) Epoxy glue between the sapphire windows and the cut and ground cylinder ends, the glue being fixed under vacuum from the side away from the solution carrier, so that the solutions used (especially those containing oxidant-sensitive pyridinyl radicals) have minimal contact with the glue. C) Carriage Mechanism: 1) Threaded sleeve 2) Micrometer drum fixed to the inner cylinder by a stainless steel insert. D) Structural Support: 1) Aluminum barrel around the outer cylinder 2) Stainless steel pin to halt motion of inner barrel. E) External Connection: A side arm is sealed to the outer cylinder, and connects the cell to vacuum or to another tube carrying an oxygen-free solution.

The pathlength in the VV-cell is variable from 0.001 cm to 0.8 cm. (Paths as long as 5–10 cm might be considered) Cell volume, including the external tube, may be 1–30 ml. A calibrated pipette is used as a second arm. A dilute solution is concentrated by distilling solvent from the cell, and measuring the volume increase in the pipette. Temperature equilibration of the apparatus is necessary for accurate volume measurement. The temperature coefficient of expansion for CH_3CN is particularly high. Mixing after the removal or addition of solvent is carried out by pouring the solution back and forth at right angles to the calibrated pipette.

3.1.3 Photodissociation Spectra by Reverse Pulse Polarography

Photodissociation of the dimer [*2-2*] to the pyridinyl radical (*2*˙) occurs readily in thin films at low temperatures or in acetonitrile solutions. Although excitation spectra could not be obtained from these experiments [30], a new technique [31] was used: 1) the wavelength dependence of radical formation from dimer generated electrochemically in small amounts, 2) the measurement of the radical produced by a rapid jump to a potential at which reoxidation of the radical takes place. Since the photodissociation spectra of dimers may well determine their practical use, a convenient procedure is useful.

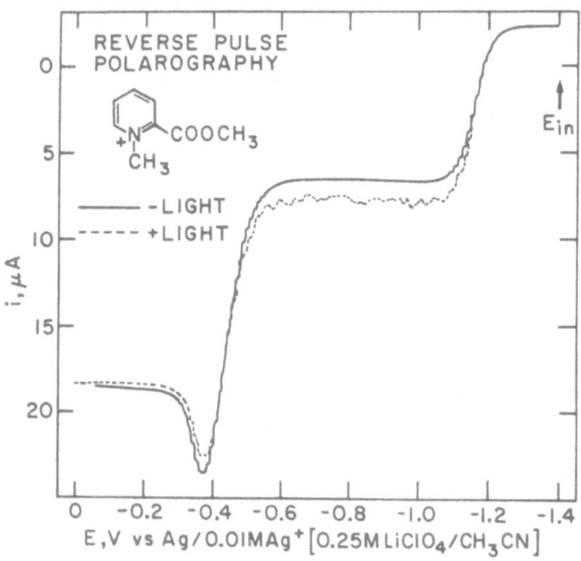

Fig. 8. Reverse pulse polarograms (constant R_{in}-mode) of 1-methyl-2-carbomethoxypyridinium (2^+) perchlorate (0.95 mM) in 0.25 M $LiClO_4$—CH_3CN. Without light, ———; with light, --------- (irradiation with pulsed N_2 laser). E_{in}, −1.4 v vs. Ag/0.01 M Ag^+ (CH_3CN) (t = 0.5 s); E_p, plotted on the X-axis (t_p = 1.7 ms) [31]

The pyridinyl radical dimer (*2-2*) is formed by brief reduction (0.5–8.0 s) of 1-methyl-2-carbomethoxypyridinium ion (2^+) around an Hg drop. A rapid jump (t_p = 0.8–90 ms) to a potential suitable for oxidation of any new species present near the electrode led to an increase during irradiation (Fig. 8) in the amount of monomeric pyridinyl radical (*2*˙) detected. The detection technique is termed reverse pulse polarography, RPP [32] (see below), Variation of the wavelength of irradiation yielded a variation in the "excess" current, corrresponding to a photodissociation spectrum similar to the absorption spectrum of the dimers (Fig. 9).

The photosensitivity of compounds on or near the electrode surface can thus be measured.

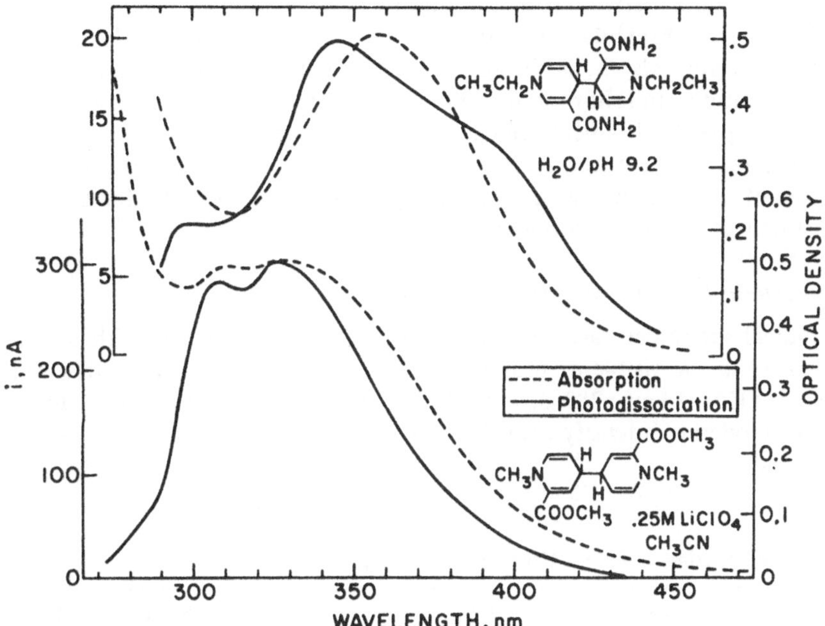

Fig. 9. Photodissociation (———) and absorption (--------) spectra of (above) 1-ethyl-3-carbamido-pyridinyl radical dimer (*3-3*) and (below) 1-methyl-2-carbomethoxypyridinyl radical dimer (*2-2*). The photodissociation curves represent the reverse pulse current of the radical oxidation under irradiation minus the dark current. The dimers were generated near the electrode by RPP as follows: (*2-2*), 1-methyl-2-carbomethoxy-pyridinium (2^+) perchlorate (0.6 mM) in 0.25 M $LiClO_4$—CH_3CN. E_{in}, −1.3 v vs. Ag/.01 M Ag^+ (CH_3CN) (t = 2 s); E_p, −1.0 v (t_p = 1.7 ms). (*3-3*), 1-ethyl-3-carbamidopyridinium (3^+) perchlorate in 0.25 M $LiClO_4$-aq. $NaHCO_3$—K_2CO_3, pH 9.2 buffer. E_{in}, −1.3 v vs. SCE (t = 2 s); E_p, −1.0 v (t_p = 0.8 ms) [31]

Reverse pulse polarography (RPP) has qualitative diagnostic power and combines in one technique the best features of cyclic voltammetry and double potential step chronoamperometry. In this case, RPP has two advantages: a) unstable dimers can be prepared near the electrode and their electrochemical properties characterized, b) irradiation of solutions containing the light-absorbing material (i.e., the dimer) only near the electrode averts a crucial loss of light intensity that would occur if the dimer were distributed throughout the solution. Two modes of RPP were used:

a) Constant E_{in}-mode (see Fig. 8). During the first portion of the drop lifetime ($0.5 \leq t \leq 10$ s), the initial potential is set at a point at which the desired reaction takes place. A single potential pulse ($0.8 \leq t_p \leq 100$ ms) is then applied. The current flowing during the pulse as a result of the electrochemical activity of the products and intermediates accumulated at E_{in} is measured. The usual pulse train of normal pulse polarography is then used to characterized the initial electrode reaction.

b) Constant E_p-mode. The potential of the pulse is held constant, and the current passing through the drop measured as a function of the initial potential, changing from drop to drop. The second mode has been named inverse normal pulse polarography (INPP) [33]. RPP has been used to study 1-methyl-2-, 3- and 4-carbomethoxypyridinium ions [18].

3.2 Optical Spectra

3.2.1 Ultraviolet and Visible Spectra

Pyridinyl radicals and derived species exhibit characteristic absorption spectra, especially in the near ultraviolet and visible regions. These spectra have been useful in identifying the radicals, in decisions about the nature of the species present under glven conditions, in following the kinetics of reactions, and in revealing many subtle features of pyridinyl radical behavior.

A interpretation of the spectra of pyridinyl radical monomers and complexes has been presented [34]. A somewhat different qualitative level scheme for the electronic transitions of isomeric pyridinyls is due to Kosower [7].

3.2.1.1 Monomeric Pyridinyls and Pimers

Pyridinyl radical monomers are weakly colored in dilute solutions, with a weak absorption band in the visible or near IR (640–1250 nm) [35,36] and two strong absorption bands (365–410 nm; 280–309 nm) in the in the near UV (Table 1). A fourth absorption of moderate intensity near 225 nm has been observed for a number of pyridinyl radicals.

Table 1. Absorption Maxima in nm for Pyridinyl Radicals in Solution

Substitution	$\lambda_1(\varepsilon_{max})$	$\lambda_2(\varepsilon_{max})$	$\lambda_3(\varepsilon_{max})$	Solvent	Ref.
1-CH_3-4-$COOCH_3$	309 (12300)	400 (6900)	640 (90)	Isopentane	33)
1-$(CH_3)_3C$-4-$COOCH_3$	308 (14000)	398 (7800)	—	CH_3CN [a]	11)
1-CH_3-4-$CONH_2$	307 (11000)	404 (7100)	650 (—) [b]	CH_3CN [a]	11,35)
1-H-4-COO^-	295 (8000)	391 (5000)	—	HOH [c]	36)
1-CH_3-2-$CONH_2$	307 (6500)	365 (3000)	900 (—)	HOH [c]	35)
1-CH_3-3-$CONH_2$	280 (10200)	410 (4300)	1250 (—)	HOH [c]	35)
1-CH_3			750 (—)	HOH [c]	35)

a Electrochemical preparation in CH_3CN-0.2M $LiClO_4$
b For 1-ethyl derivative
c Pulse radiolysis was used to generate the radical

The strong and striking blue color of distilled pyridinyl radicals is due to a charge-transfer absorption in a pyridinyl-pyridinyl complex (a pimer) (Eq. 10), first identified by Itoh and Nagakura [34]. The electronic transition may be described by Eq. 11.

$$\underset{\text{monomers}}{Py^{\cdot} + Py^{\cdot}} \rightleftharpoons \underset{\text{pimer}}{(Py^{\cdot})_2} \tag{10}$$

$$Py^{\cdot}, Py^{\cdot} \xrightarrow{h\nu} Py^{\pm}Py^{\mp} \tag{11}$$

The visible charge-transfer (pimer) band decreases in intensity and shifts to longer wavelengths with increasing 1-alkyl group size. The maxima for four different

1-alkyl-4-carbomethoxypyridinyls in thin films, (*4*˙) (1-alkyl = CH_3, CH_3CH_2, $(CH_3)_2CH$, $(CH_3)_3C$), are listed in Table 2. The spectra are illustrated in Fig. 10.

Table 2. Absorption Maxima for 1-Alkyl-4-carbomethoxypyridinyl Radicals in Thin Films on a Sapphire Window at 77 K

R =	$\lambda_{max}(\varepsilon_{max})$				
CH_3	630–640 (2800)	400 sh (3500)	378 (7200)	288 (13000)[a]	244 (6850)
CH_3CH_2	640–650 (2550)	400 sh (5100)	380 (6750)	293 (13000)[a]	234 (6050)
$(CH_3)_3CH$	655–665 (1750)	396 sn (5450)	381 (6050)	295 (13000)[a]	232 (4550)
$(CH_3)_3C$	660–670 (1400)	400 (5900)	382 (6250)	297 (13000)[a]	231 (5200)

a absorption coefficients were measured using two methods: (1) film thickness (2) dissolution of film and measurement of radical concentration by titration with paraquat. The coefficients found varied between 9000 and 13000. Since other experiments indicate that the band has an intensity near 13000 relatively insensitive to changes in substitution, the value of 13000 was chosen and all other maxima normalized to this result.

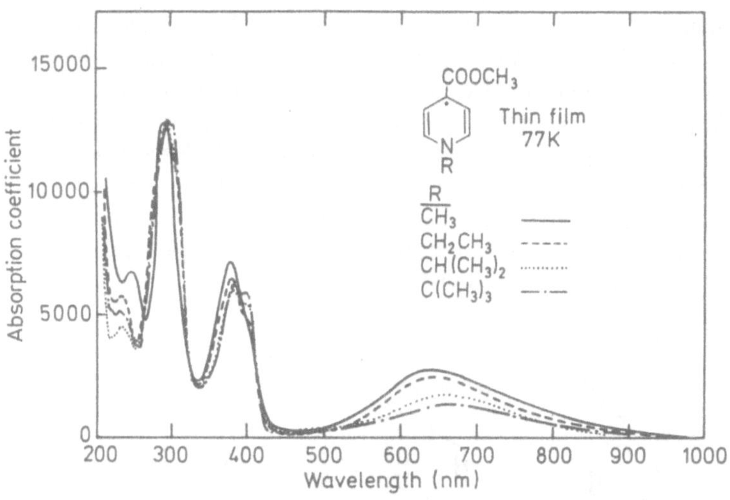

Fig. 10. Absorption spectra of four 1-alkyl-4-carbomethoxy-pyridinyl radicals, (R)*4*˙, in thin films at 77 K. The three main bands are at 650 nm, 400 nm and 300 nm. The 650 nm band (charge-transfer absorption of pimer) shifts from 623 nm for 1-CH_3 to 675 nm for 1-$(CH_3)_3C$, along with a two-fold intensity decrease. The 400 nm band is composed of the pyridinyl radical absorption of the pimer, and absorptions of the two ester conformational isomers of the pyridinyl radical monomer. The 300 nm band varies with the different relative contributions of the pimer and the monomer [11]

1,1′-Dialkyl-4,4′-bipyridylium cation radicals form pimers in polar solvents [37,38]. The charge-transfer band of the pimer for the 1,1′-dimethyl derivative in water (K_{assoc} 385 M) [39] is found at 870 nm and is accompanied by shorter wavelength visible and UV absorption bands than those found for the monomer (monomer → dimer, 600 → 560 nm, 390 → 370 nm). The wide use of the cation radicals in solar energy conversion systems requires a detailed understanding of the spectroscopic properties.

3.2.1.2 Complex 395 nm Absorption Band

Itoh and Nagakura [40] reported two overlapping 395 nm bands for 1-methyl-, 1-ethyl- and 1-isopropyl-4-carbomethoxypyridinyl radicals, in isopentane solution and 77 K matrices, an observation confirmed for MTHF solutions [41] and thin films [11].

Itoh [42] suggested that the 395 nm band splitting is due to excit on splitting in the pimer. However, thin film spectra of 1-methyl-4-carbo-t-butoxypyridinyl radicals (warmed → dimer, irradiated → monomer + pimer, etc.) indicated that a 380 nm band decreased in parallel with the pimer without loss in radical or the 395 splitting.

If the split 395 nm band were due to the monomer, the band shape should be independent of concentration. Spectra can be measured over a concentration range of about 1000 using a "VV-cell". The maxima are matched in height to simplify the comparisons. Absolutely no concentration dependence of the 395 nm band shape was observed for solutions of the 1-methyl-4-carbo-t-butoxypyridinyl radical in both CH_3CN and 2-MTHF over the range from 10^{-4} M to 10^{-1} M.

The splitting has an intramolecular origin, possibly two ground state ester conformational isomers. One conformer is "inner" (alkyl group directed towards the ring) and the second is "outer" (alkyl group directed away from the ring). The conformers would have different ground state energies, somewhat different absorption spectra, and would give rise to overlapping, or "split", absorption bands. Such isomers are now quite well established as the origin of the two different emitting states of methyl salicylate [43–45]. A pyridinyl radical pimer might consist of three isomers, o,o, i,i and o,i (outer-inner), and would exhibit broadened absorption bands; further effects from excition splitting are not excluded.

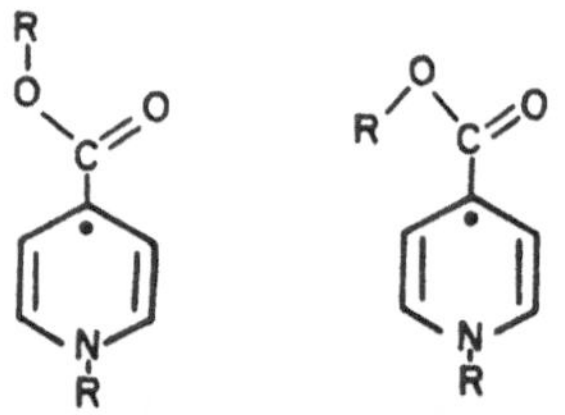

"Outer" conformation "Inner" conformation

Ester conformers

3.2.1.3 Pyridinyl Radical Dimers

The 332 nm absorption for the dimer, *2-2*, in solution resembled that expected for a 1,4-dihydropyridine [46], but the intensity, position, band shape, and variability in position were unusual. The intensity was extremely low (ε_{max} 2000 versus 12000 for two dihydropyridine rings), and the position unusual for a cross-conjugated 2-substituent. A 6,6'-structure was rejected [16], since the spectrum did not exhibit three bands, including one between 250 and 300 nm [46a]. The absorption band at 332 nm in solution was unusually broad, with appreciable intensity well past 400 nm. In the thin film, the breadth of the absorption bands due to the dimers was even greater, extending well past 450 nm for *2-2* and past 600 nm for $(CH_3CO)2$—$(CH_3CO)2$. The dimer absorption was apparently quite sensitive in intensity and position to the geometry around the single bond connecting the two pyridinyl rings, the absorption intensity being higher in the *s-cis* form.

These points suggest an assignment of the absorption to a $\pi\sigma \rightarrow \pi\sigma^*$ transition. Orbital level schemes (not illustrated) for the radical dimer and a radical pair show a) that both photo- and thermal dissociations are allowed and b) that the highest bonding level in the dimer, the σ-bond between the radicals, can conjugate with both π-systems of the dimer. The structure of pyridinyl radical dimers is particularly favorable for πσ interaction as indicated in the formula. "Through-bond" interaction (i.e., conjugation between π-orbitals mediated by σ-bonds [47,48]) probably occurs in many molecules and affects both chemical, e.g., fragmentation [49]) and spectroscopic properties [50–52]. The theory of Hoffmann et al. [53] for the interaction has been extended by Gleiter [54].

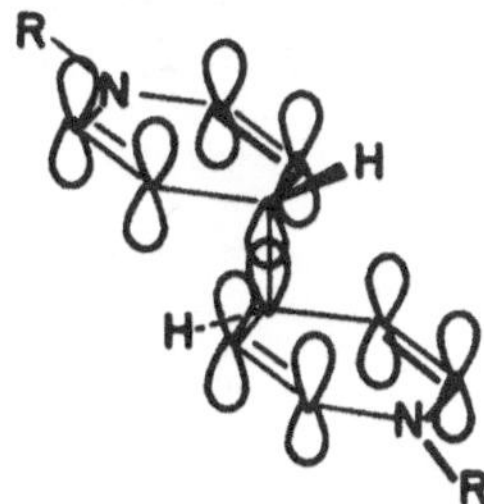

πσ conjugation in pyridinyl radical dimers

The photodissociation of the dimers to radicals (see 4.1) is an important photochemical consequence of the "through-bond" interaction. A photodissociation spectrum for *3-3* is similar to the dimer absorption band [31], as determined by the

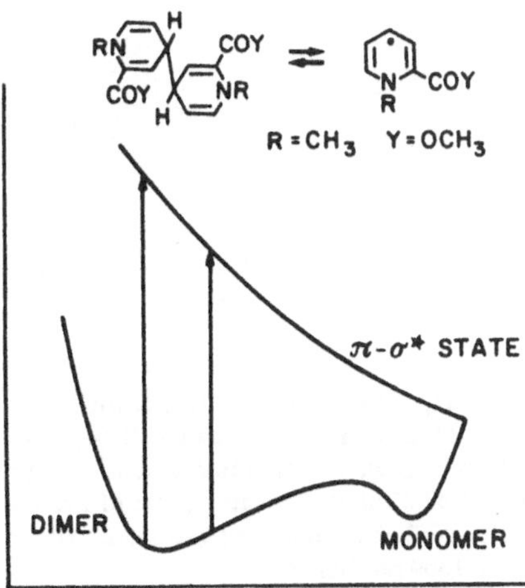

Fig. 11. An energy versus reaction coordinate diagram illustrating the relationship between the 2˙ monomer, the *2-2* dimer and the excited state resulting from the $\pi\sigma \rightarrow \pi\sigma^*$ electronic transition. The observed spectra vary with the particular radical and its local environment. For 1-alkyl-2-carbomethoxypyridinyl radical, the dimer is the predominant form in solution and in the annealed thin film. For the 2-acetyl-pyridinyl radical, the monomer is predominant in solution, but the dimer is the chief component of annealed thin films [30]

technique described in Section 3.1.3. An energy versus reaction coordinate diagram for dimer and monomer (Fig. 11) implies that $\pi\sigma \rightarrow \pi\sigma^*$ transitions should be broad and temperature sensitive.

3.2.1.4 Pyridinyl Diradicals and Cyclomers

Pyridinyl diradicals ($Py^{\cdot}(CH_2)_nPy^{\cdot}$) bearing 4-carbomethoxy groups and separated by two, three, four or five methylene groups have been reported [55,56]. The spectrum of salt-free, distilled diradical *3*$^{\cdot\cdot}$ (thin film) [57] is similar to that of the diradical generated by Na(Hg) or electrochemical ($Bu_4N^+ClO_4^-$) reduction of the bis-pyridinium perchlorate in CH_3CN solution (Fig. 12), with two broad bands observed at 400 nm (3000) and 730 nm (3000) [57]. Neither annealing nor irradiating the film leads to any significant changes in spectrum. Titration confirmed that 2 electrons had been added to the bis-pyridinium cations. The ultraviolet and visible absorption spectra are similar to those expected for pyridinyl diradicals. The weak EPR spectrum (an intensity 1–2% of that expected for *3*$^{\cdot\cdot}$) was attributed to the singlet diradical state ("closed form").

Stable pyridinyl radicals produce dimers with unusual spectroscopic properties (sec 3.2.1.3). The dimer (*2-2*) is the primary form in which *2*$^{\cdot}$ is found in both

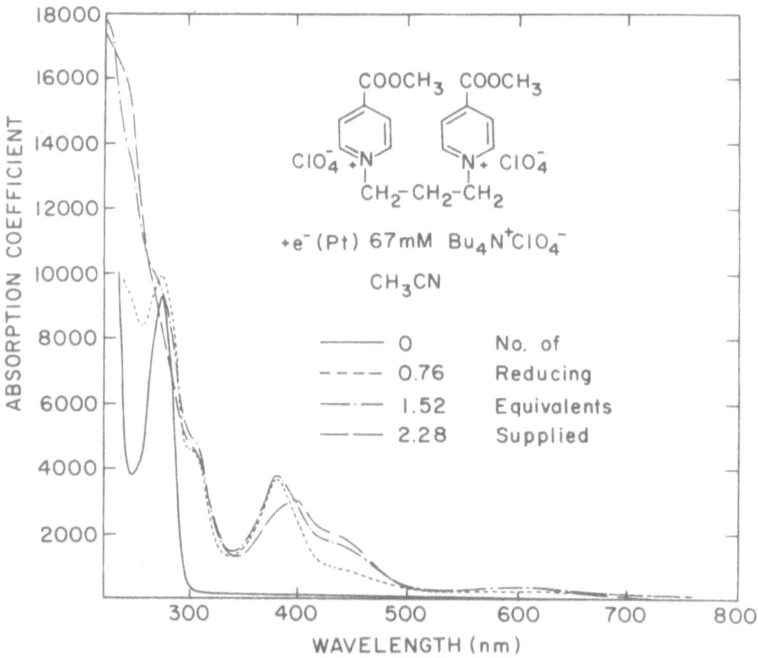

Fig. 12. Spectra for electrochemical reduction of *3*$^{++}$ 2 ClO_4^- (0.99 mM) in CH_3CN-Bu_4NClO_4 (67 mM). Curve *1* (———): Before reduction, maximum at 275 nm (9300). Curve 2 (--------): After passage of 0.76 e$^-$/mole bis-cation, maxima at 381 nm (7100), 275 nm (10000) and 1360 nm (900), near ir not shown. Curve 3 (—·—): After passage of 1.52 e$^-$/mole bis-cation, loss of 275 nm, gain of 440 nm (1900). Curve 4 (— — —): After passage of 2.28 e$^-$/mole bis-cation, maximum at 400 nm (3000), shoulder at 440 nm (2100), no 381 nm or 1360 nm bands [57]

solution and thin films [16]. The dimer of 4˙, 4-4, occurs in thin films from −100 °C to −20 °C (see 4.1.1). Both dimers exhibit the long wavelength and broad absorption assigned to a $\pi\sigma \rightarrow \pi\sigma^*$ transition. Both 2-2 and 4-4 dimers dissociate thermally or on irradiation, and behave chemically as radicals. Reexamination of the spectra of the "closed" diradicals gave rise to the idea that a mixture of several covalently bonded intramolecular "dimers", or *cyclomers*, was present.

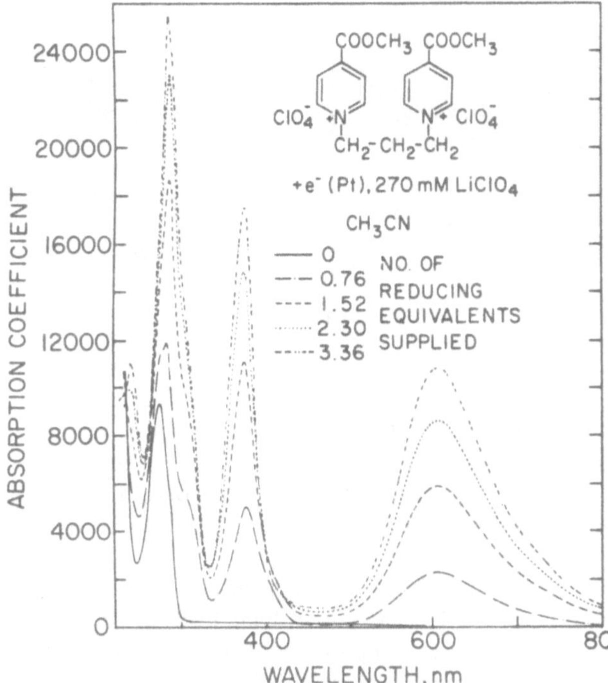

Fig. 13. Spectra for electrochemical reduction of 3^{++} 2 ClO_4^- (0.094 mM) in CH_3CN—$LiClO_4$ (270 mM). The contrast with the spectra shown in Fig. 2 is striking. Passage of 0 (———), 0.76 (—·—), 1.52 (--------), 2.30 (·········) and 3.36 (--·--·--·) e^-/mole bis-cation lead to successive increases in visible absorption at 605 nm and in the pyridinyl radical bands at 377 and 305 nm. No absorption at 1360 nm is seen at any stage of the reduction [57]

The diradical metal complexes˙ $3^{\cdot}M^{+n}$ (Fig. 13), have spectra which are very similar to those of simple pyridinyl radicals, except for an extremely intense pimer band. The $3^{\cdot}MgI_2$ complex has no readily detectable EPR spectrum [17]. Electrochemical studies show that metal ions stabilize pyridinyl radicals bearing carbonyl groups [11]. The weaker and broader absorption bands found in the absence of the metal ion are due to covalent forms ("cyclomers"). At least two cyclomers (maxima at 400 nm and 700 nm) are indicated by the variable ratio of the 400 nm to 700 nm absorptions, ranging from I (400 nm < 700 nm) to I (400 nm ≫ 700 nm). A third cyclomer, with a half-life of about 1 s, has been detected by electrochemical techniques [19]. The 3-cyclomers, like to Mg^{++} complex cited above, show a very weak EPR spectrum [56]. The species contributing to the behavior of $3^{\cdot}$ diradicals are collectively called the ⟨3⟩ system; these together with the metal complexes are shown in Fig. 14.

The structures of the cyclomers are assigned on the basis of probable stability and expected spectrum. The *cis*-A-cyclomer (Fig. 14) should be less stable than the

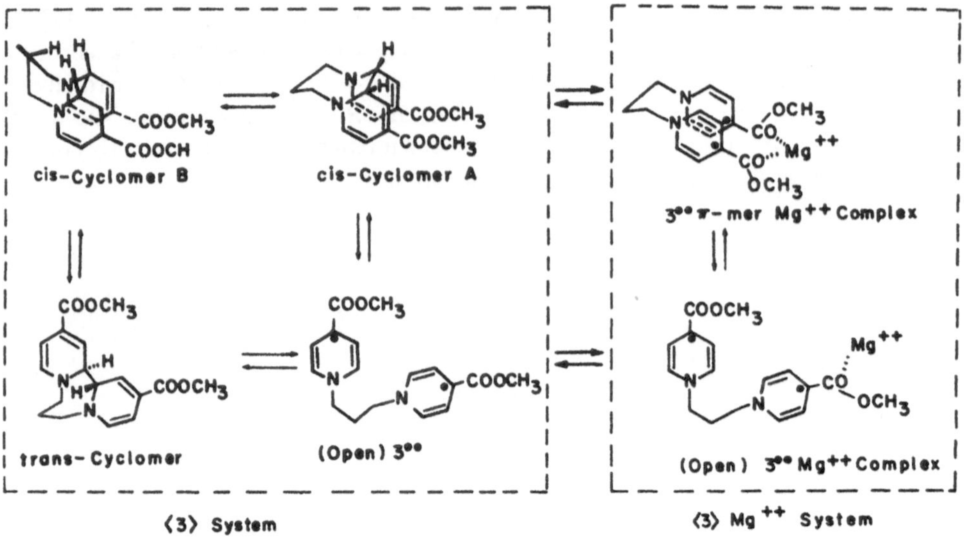

Fig. 14. A scheme illustrating the components of the ⟨*3*⟩ and ⟨*3*⟩ M^{+n} systems [57]

trans-cyclomer, with pimer-like light absorption at 700 nm. The *trans*-cyclomer is the species which absorbs at 400 nm. The third isomer, *cis*-cyclomer B, is presumably the (weakly absorbing) unstable species detected electrochemically. No pyridinyl radical absorption (expected for an "open" diradical) appears near the elctrode during the pulse experiments.

Reduction of the bis-pyridinium ion with various agents (Na(Hg), the *4*˙ radical, or electrochemical reduction with $Bu_4N^+ClO_4^-$ as electrolyte) leads to the *trans*-cyclomer, *trans-3*. Bridged metal ion diradical complexes with Li^+ or Mg^{++} (1:1) are diradical in spectra and chemistry. Extraction or distillation from the metal complexes yields a mixture of *cis*-cyclomer A (*cis*-A-*3*) and *trans*-cyclomer (*trans-3*).

3.2.1.5 Pyridinyl-Pyridinium Intervalence Band

Partial reduction of the trimethylene bis-pyridinium dication, 3^{++} with the pyridinyl radical *4*˙ yields the cation radical, $3^{+\cdot}$ [58]. In addition to modified pyridinyl radical absorption bands at 381 nm (6300) and 305 nm (9000), a near infrared absorption at 1360 nm (1700), characteristic of an intervalence transition, is observed [57]. The transition may be described by the transformation, $Py^+ : Py^\cdot \rightarrow Py^\cdot : Py^+$.

3.2.2 Infrared Spectra

The infrared spectra of pyridinyls are measured conveniently with the thin film spectroscopy apparatus and the changes which ensue on dimerization have been examined. A complete infrared spectrum for *4*˙ at 77 K has been reported [2].

External sapphire windows on the thin film apparatus allow spectra to be measured to ca. 6200 nm. IR absorption coefficients are much lower than those in the visible or ultraviolet regions, so that thicker films are needed (ca. 1.5–2.0μ instead of 0.1 to 0.2μ).

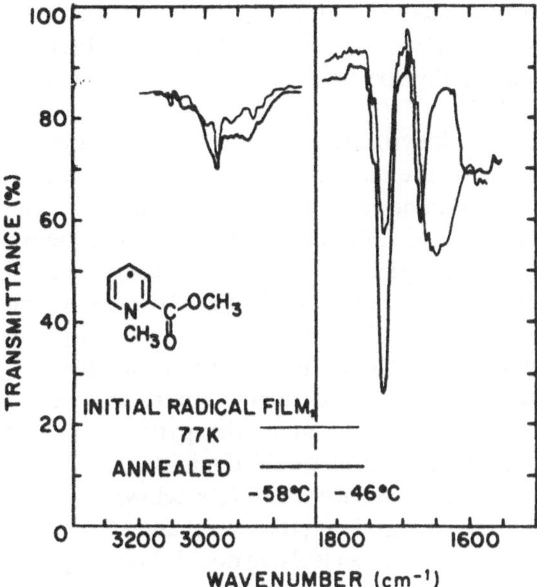

Fig. 15. Infrared spectra of 1-methyl-2-carbomethoxypyridinyl radical, illustrating the change on annealing a 1.5 to 2.0u film at 77 K to between −58 and −46 °C [30]

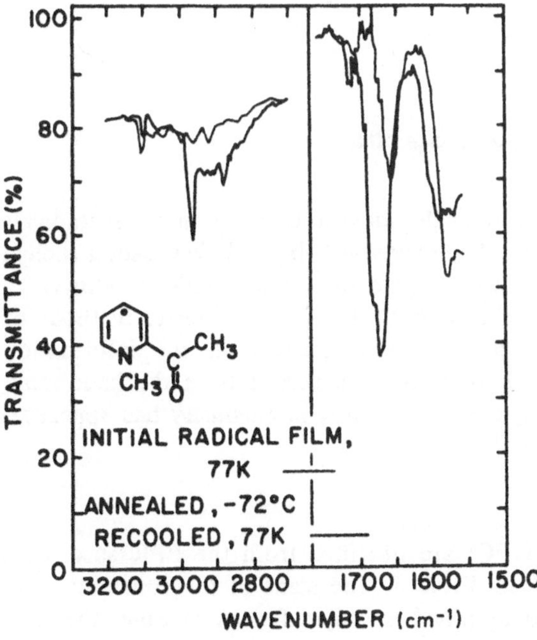

Fig. 16. Infrared spectra of 1-methyl-2-acetylpyridinyl radical ((CH_3CO) *2*˙) showing the changes on annealing a 1.5–2.0 μm film at 77 K at −72 °C (recooled to 77 K for spectrum) [30]

The IR spectrum (between 3300 cm^{-1} and 1570 cm^{-1}) of *2˙* at 77 K shows C—H stretching, C=O stretching (see 3.3.1.1) and C=C stretching vibrations. Warming (—60 °C to —40 °C) forms the dimer and changes the IR spectrum. The C—H and C=O absorptions increase in intensity while the C=C absorption decreases. (Fig. 15). The ir spectrum of (CH_3CO)*2˙* at 77 K changes markedly on dimerization in the region of the C—H, C=O and C=C stretching vibrations (Fig. 16). Photolysis of the dimers causes the spectra to revert to those observed for the pyridinyl radicals.

The ester methyl group is unaffected by dimerization; a pair of bands represents both monomer and dimer. (Methyl groups are characterized by a pair of bands, corresponding to the asymmetrical and symmetrical stretching motions [59]). The N-methyl group stretch at shorter wavelengths (2990 cm^{-1}) than the ester methyl band becomes a strong shoulder in the dimer spectrum. The methyl group bands of the 2-acetylpyridinyl radical are a little more complex, possibly due to two equivalent conformations of the 2-acetyl group. Only one of the three pairs of methyl group bands, that due to the N-methyl group, increases strongly in intensity on dimerization.

The N-methyl stretch absorptions are often weaker in methyl groups attached to electronegative atoms [59]. Spin density on the nitrogen in the pyridinyl radical reflects a positive charge; dimerization of the radical would diminish the transfer of charger into another group, as whon with resonance forms (see below).

IR absorptions due to double bond stretching motions (1663 cm^{-1} and 1655 cm^{-1}) change on dimerization. In the dimer, absorption for an isolated double bond appears, and that for a conjugated double bond moves to longer wavelengths.

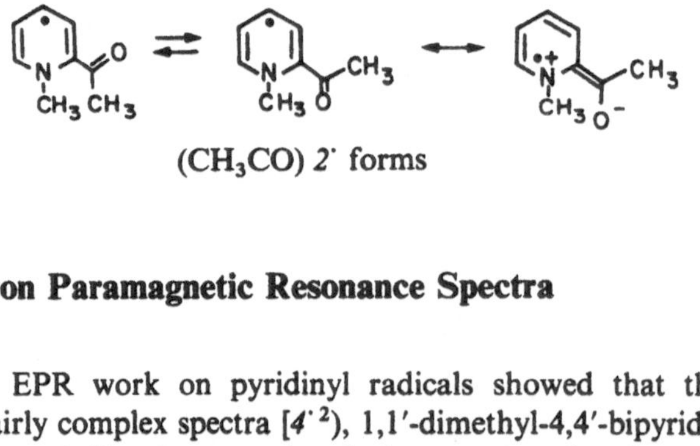

(CH_3CO) *2˙* forms

3.3 Electron Paramagnetic Resonance Spectra

The initial EPR work on pyridinyl radicals showed that the simplest radicals exhibited fairly complex spectra [*4˙* [2]), 1,1′-dimethyl-4,4′-bipyridylium cation radical [37,60])]. Itoh and Nagakura analyzed the spectra of a few simple pyridinyls [40]. Improved methods for generating pyridinyl radicals as well as faster methods of spectroscopic analysis have led to a substantial increase in the information available, including a section on hyperfine coupling constants in the handbook, Landolt-Börnstein [61]. A review by Symons [62] of EPR in chemistry has appeared. Triplet pyridinyl pairs have also been observed.

3.3.1 Hyperfine Coupling Constants

The hyperfine coupling constants (HFC) are obtained from the EPR spectra; a selection of constants are presented in Table 3. The signs of the constants have not been indicated, and have been determined in only a few cases. Since the total

Table 3. Hyperfine Coupling Constants for 1-R-X-Pyridinyl Radicals in Solution at ca. 20 °C

1-R	X	Solv[a]	Coupling constants (G)[b]							Ref.
			N	1-CH (NH)	2-H	6-H	4-H	3,5-H	CH_3	
$C_6H_5CH_2$	4-$COOCH_3$	A	6.30	3.56	3.73	3.83	—	0.56	0.88[c]	d)
4-$CH_3C_6H_4CH_2$	4-$COOCH_3$	A	6.24	3.59	3.70	3.80	—	0.58	0.91	d)
4-$NO_2C_6H_4CH_2$	4-$COOCH_3$	A	6.28	3.51	3.95	4.07	—	0.54	0.92	d)
CH_3	4-$COOCH_3$	A	6.31	5.60	3.57	3.66	—	0.74	0.91	e)
CH_3	4-$COCH_3$	A	5.73	5.40	2.70	3.40	—	0.78	2.52[f]	e)
CH_3	2-$COOCH_3$	A	6.58	5.64	—	2.54	6.28	1.40	0.94	g,h)
								0.94		
CH_3	4-COO^-	W	6.31	5.54	3.68	3.68	—	0.57	—	i)
CH_3	3-COO^-	W	5.20	4.45	2.11	8.36	11.3	0.46	—	i)
H	2-COO^-	P	5.50	4.96	—	2.17	7.97	1.80	—	j)
								1.66		
H	3-COO^-	W	4.46	3.82	2.45	8.47	11.8	1.03	—	i)
H	4-COO^-	W	5.28	4.72	4.14	4.14	—	0.28	—	k)
H	2,6-$(COO^{\dot{-}})_2$	P	6.34	5.38	—	—	8.45	0.95	—	j,l)
H	3,5-$(COO^-)_2$	P	4.48	3.78	5.30	5.30	12.4	—	—	j,m)
H	3-$CONH_2$	P	4.26	3.75	1.77	8.55	11.8	2.22	0.09[o]	p)
H	—	W	5.89	3.45	5.89	5.89	11.6	0.91	—	q)
H	2-CH_3	P	5.31	2.69	—	6.59	11.6	0.54	4.96[r]	s)
								1.36		
H	3-CH_3	P	5.87	3.45	6.16	5.43	11.3	1.16	1.12[r]	s)
H	4-CH_3	P	5.58	2.68	5.98	5.98	—	1.16	12.3[r]	s)
$Si(CH_3)_3$	—	D	4.20	—	6.26	6.26	11.6	1.38	—	t)

a Solvents: A = acetonitrile, W = water, P = isopropyl alcohol (+ water and/or acetone)
b Coupling constants in gauss (G) = 10 × the value in milliTesla (mT)
c CH_3 of $COOCH_3$ group
d Ikegami, Y., Kubota, S. and Watanabe, H.: Bull. Chem. Soc. Japan 52, 1563 (1979)
e Kubota, S. and Ikegami, Y.: J. Phys. Chem. 82, 2739 (1978)
f CH_3 of CH_3CO group
g Hermolin, J., Levin, M., Ikegami, Y., Sawayanagi, M. and Kosower, E. M.: J. Am. Chem. Soc. 103, 4795 (1981)
h Ikegami, Y., Sawayanagi, M. and Kubota, S.: Heterocycles 15, 1027 (1981)
i Neta, P.: Radiat. Res. 52, 471 (1972)
j Zeldes, H., Livingston, R. and Bernstein, J. S.: J. Magn. Res. 21, 9 (1976)
k Zeldes, H. and Livingston, R.: J. Phys. Chem. 77, 2076 (1973)
l Zeldes, H. and Livingston, R.: Radiat. Res. 62, 28 (1975)
m Zeldes, H. and Livingston, R.: Radiat. Res. 58, 338 (1974)
o NH_2 of $CONH_2$ group
p Zeldes, H. and Livingston, R.: J. Magn. Res. 26, 103 (1977)
q Zeldes, H. and Livingston, R.: J. Phys. Chem. 76, 3348 (1972)
r CH_3 substituent on ring
s Rakowsky, T. and Dohrmann, J. K.: Ber. Bunsenges. Phys. Chem. 83, 495 (1979)
t Ponomarev, S. V., Becker, H. P., Neumann, W. P. and Schroeder, B.: Justus Liebig's Ann. Chem. 1895 (1975)

spin density must be 1, and the width of the spectrum is related to the absolute values of the hfc, negative hfc broaden the spectrum [63].

3.3.1.1 Spin Distributions in Pyridinyls

Roughly speaking, the hfc are proportional to the spin densities (ϱ) on the adjacent positions, since the spin polarization parameters (Q) are between 25 and 29 gauss [64] (HFC = $Q\varrho$). In all pyridinyl radicals, a high spin density is present on a) the ring nitrogen, b) the 2- and 6-carbons, c) the carbon adjacent to the ring nitrogen and d) the 4-carbon.

The spin densities at the 2- and 6-positions in 1-alkyl-4-carbomethoxypyridinyls are non-equivalent (Table 3), suggesting that the ester group rotation is slow on the EPR time scale. The low frequency of the ester carbonyl group in the IR spectrum, 1642 cm^{-1}, is consistent with delocalization from the ring [2]. (Normal frequencies for carbomethoxy groups on pyridine or pyridinium rings are 1730–1736 cm^{-1}). The bond from the ester group the ring would have some double bond character ($=C(OCH_3)—O^{\cdot}$), and therefore resist rotation.

The anion radical of pyridinium bis(carbomethoxy)methylide ($P^{\cdot}—CX_2^-$) exhibits non-equivalent ring splittings because the ester groups are non-equivalent [65].

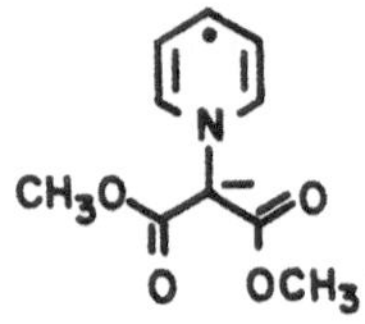

CH_3O O O OCH_3 N

($P^{\cdot}—CX_2^-$)

3.3.1.2 Solvent Effects on Constants

The polarity of the solvent exerts an appreciable effect on some of the hfc constants for pyridinyl radicals [66]. The effects have been correlated with the solvent polarity parameters, Z-value [67] and $E_T(30)$-value [68], and a theory relating the shifts to the permittivity of the solvent has been published [69].

The 2- and 6-coupling constants for (CH_3)*4*˙ vary from 4.27 and 4.35G in n-pentane to 3.57 and 3.66G in CH_3CN (Z 71.3). A good part of the decrease occurs over the range from n-pentane (Z ca. 50) to 2-MTHF (Z 55.4). The data may be consistent with two slopes for correlation with solvent polarity parameters, rather than the one slope suggested [66].

Hydrogen bonding exerts an effect on the hfc coupling constants. For 1-hydro-4-acetylpyridinyl, the hfc for N and NH change only a little from isopropyl alcohol-acetone (PA) to PA:water (4.77 to 4.63; 4.99 to 5.03), but the hfc for the CH_3 of the acetyl group changes from 3.14 to 4.40 [70]. A larger change (1.95G to 5.41G) has been reported for the 1-methyl-4-acetylpyridinyl over the range from THF (Z ca. 56) to water (Z 94.6) [71]. A remarkable change in the hfc for N in *2*˙ in CH_3CN occurs on the addition of $LiClO_4$ [72].

3.3.1.3 Temperature Effects on Constants

The hfc are either unchanged or increase moderately with increasing temperature. The NH coupling constant, however, decreases dramatically. For 1-hydropyridinyl radical, the hfc changes from 4.40 at −80 °C to 3.10 at +40 °C. The out-of-plane bending motion of the hydrogen (estimated frequency, 290 cm^{-1}) is responsible for the change, and also for the "non-transferability" of the spin polarization factor from compound to compound [73−75]. A ring methyl substituent increases the temperature sensitive of the NH hfc, while a 4-cyano group markedly diminishes the temperature sensitivity. The hfc for the nitrogen in the 1-hydro-4-cyanopyridinyl is 5.20G, not very much lower than the 5.56G for 1-hydro-4-methylpyridinyl. However, the hfc for the 2- and 6-hydrogens are 4.10G in the former, and 6.00G in the latter, suggesting that spin density has shifted in the direction of the cyano group. Apparently, the change in electron distribution has also stiffened the NH bond [74].

3.3.2 *Triplet Spectra of Radical Pairs*

Triplet signals were first noted by Kosower and Waits [76] in concentrated solutions of stable pyridinyl radicals (*4*˙) and confirmed for (CH_3)*4*˙ and *4*˙ in MTHF glasses at 77 K [77]. Three pairs of shoulders around the g = 2 signal at 3295G for the monomeric pyridinyl and a ΔM = 2 transition at 1645G suggested the presence of triplet pair. The zero-field parameters were D = 0. 0098 cm^{-1} and E = 0.0011 cm^{-1}, for which a radical-radical separation of 6.5 A was estimated from the relationship for spin-spin dipolar interaction (Eq. 12).

$$D = -(3/2)\, g^2\beta^2 r^{-3} \tag{12}$$

Irradiation of the matrix (420–500 nm, a range *not* corresponding to the pimer charge-transfer transition) produced an additional triplet ($t_{1/2}$ 10 hours) with D = 0.0175 cm^{-1} and E = O, r = 5.4 Å. Careful examination of the spectra of concentrated solutions of *4*˙ revealed absorption in the range of 420–500 nm [78]. Thin film spectroscopy of *4*˙ revealed that covalent dimers (rather than singlet pairs) gave rise to the new triplet via a $\pi\sigma \rightarrow \pi\sigma^*$ transition [16]. Formation of a triplet [D = 0.0146 cm^{-1} and E = O, r = 5.5 Å] from the dimer of *2*˙, *2-2*, has been accomplished by irradiation in an MTHF glass. The equilibria describe the relationship of the triplets and radicals in Eq. 13. The two triplet pairs (A and B) are not readily interconvertible.

$$\begin{array}{ccc} \text{monomeric radical} & \text{pimer} & \text{covalent dimer} \\ Py^{\cdot} + Py^{\cdot} = & Py^{\cdot}, Py^{\cdot} & \rightleftharpoons Py{-}Py \\ & \downarrow\uparrow & \downarrow \\ & Py^{\cdot}, Py^{\cdot} & Py^{\cdot}, Py^{\cdot} \\ & \text{triplet pair A} & \text{triplet pair B} \end{array} \tag{13}$$

4 Chemical Properties

Pyridinyl radicals are highly reactive, reducing species. All pyridinyl radicals, except for the most highly hindered stable radical, 1-t-butyl-4-carbo-t-butoxypyridinyl, dimerize under certain conditions. The dimers are dissociated to the pyridinyl radicals with high efficiencies by light absorption within the $\pi\sigma \rightarrow \pi\sigma^*$ electronic transition.

Pyridinyl radicals react with haloorganic compounds at rates that are very sensitive to the strength of the halogen-carbon bond. Both atom-transfer and electron-transfer mechanisms for the reaction have been detected, with the latter exhibiting a large response to solvent polarity change and to variations in the electron affinity of the organic halide.

Pyridinyl radicals are weakly basic, and disappear in aqueous solution in a reaction which involves either dimerization or disproportionation in the reaction of a radical with a protonated radical. The consumption of protons in disproportionation reactions depends upon the stability of the initial product, varying from one to two.

4.1 Photodissociation and Dimerization

4.1.1 Loss of Pimer and Dimerization

An important observation in identifying the unusual electronic transition of pyridinyl radical dimers was that irradiation of the dimer *2-2* of *2˙* produced dissociation of the dimer into the monomeric free radical. The radical persisted in the thin film at 77 K, but reverted rapidly to dimer in solution (Eq. 14) [2,8,19].

$$2\ \mathrm{Py}^{\cdot} \rightleftharpoons \mathrm{Py\text{-}Py} \tag{14}$$

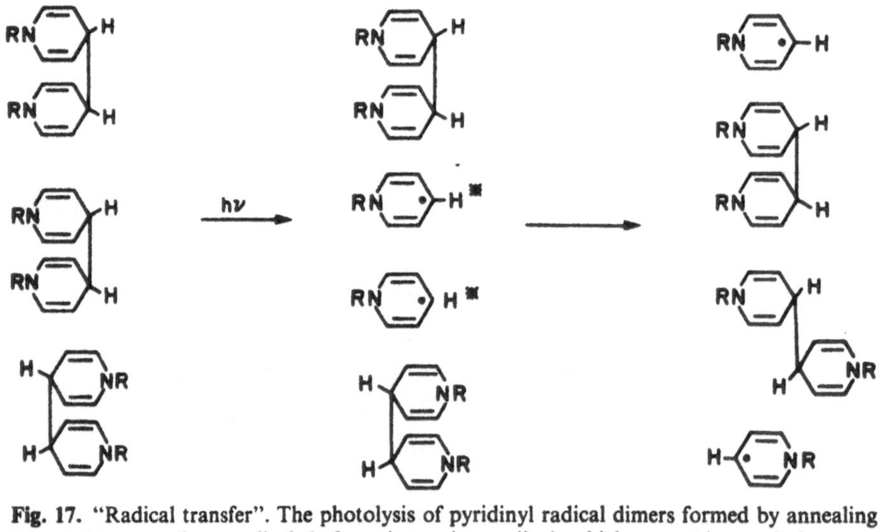

Fig. 17. "Radical transfer". The photolysis of pyridinyl radical dimers formed by annealing at low temperatures, produces radicals before pimers, i.e., radicals which are not in proximity even though pimer formation should be favored in the film at 77 K. A postulated transport process involves the reaction of the "hot" radicals (excess energy, ca. 30 kcal/mole/monomer) with neighboring dimers. New dimers, one moiety from the "hot" radical and the other from the reacted dimer, are produced. The radicals which remain are now separated by two dimer molecules. The mechanism is the radical analogy of the well-known Grotthus mechanism for the mobility of protons through hydrogen-bonded solvents [30]

The visible bands of a thin film of *2*˙ are lost first on warming suggesting that the pimer is prone to dimerization. All radical dimerized on further warming. Initially formed dimer changed in absorption maximum on warming, signifying a change in geometry explained by a "radical transfer" reaction (Fig. 17) analogous to the Grotthus proton transfer mechanism.

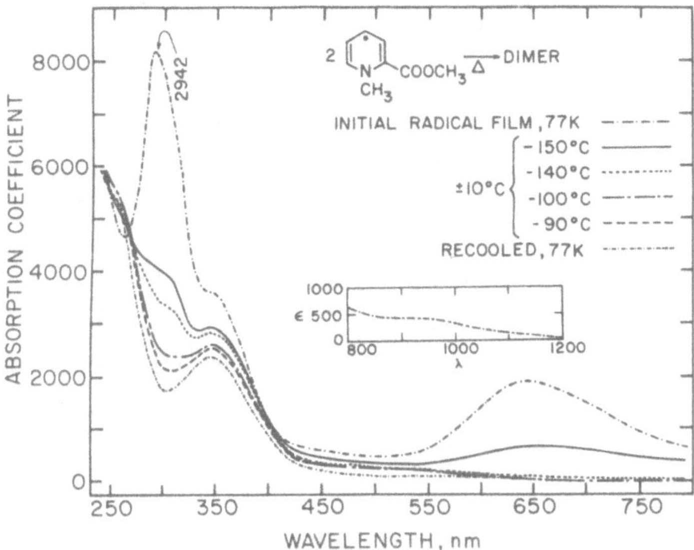

Fig. 18. The spectrum of the 1-methyl-2-carbomethoxypyridinyl radical, 2˙, at 77 K is the "initial radical film". Pimer (*2*˙, *2*˙) (π-complex) absorption at 645 nm overlies the weak monomeric radical absorption in the same wavelength region. The radical bands are lost at high temperatures, with formation of the radical dimer (*2-2*). Successive spectra for —140, —100 and —90 °C are shown [30]

Blue films of *2*˙ have UV absorption at (λ_{max}, ε_{max}) 294 nm (8400) and 345 nm (3600) (Fig. 18). The beautiful blue color at 645 nm (2000) of the 77 K films is lost on warming to ca. —150 °C along with some loss of the UV bands. Warming to —90 °C gives rise to a broad, low intensity band at 345 nm (2400) extending well past 450 nm, with a shoulder at 245 nm. The 345 nm band decreases in intensity and moves towards shorter wavelengths over the range from —113 °C to —8 °C (Fig. 19) accompanied by an increase in the 245 nm shoulder, the final spectrum having a maximum at 340 nm (1900) (ε_{max} 3800 for the dimer) and an e at 245 nm of 7000 (ε 14000 for the dimer). The dimer *2-2* in CH_3CN, 332 nm (2000), sh 245 nm (10000) [16] is not dissociated to *2*˙ by cooling to 77 K.

The green film of the 1-methyl-2-acetylpyridinyl ((CH_3CO)*2*˙) radical [16] has maxima at 310 nm (8100), 390 nm (3200) and 680 nm (2600). Warming the film to —90 °C leads to the dimer, with maxima at 250 nm (5300) (sh), 310 nm (2500) (sh), and 375 nm (2000) (broad, extends well past 600 nm), unchanged by further warming (Fig. 20).

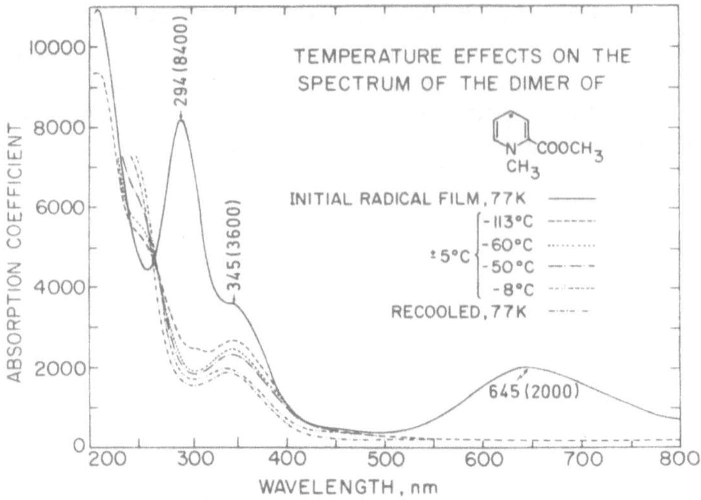

Fig. 19. Temperature effects on the dimer (*2-2*) absorption spectrum: At −113 °C, almost pure dimer is present. At −60 °C, the maximum is at 345 nm. Further temperature increases (up to −8 °C) shift the maximum gradually to 340 nm, and decrease the intensity, the final spectrum being taken after recooling to 77 K [30]

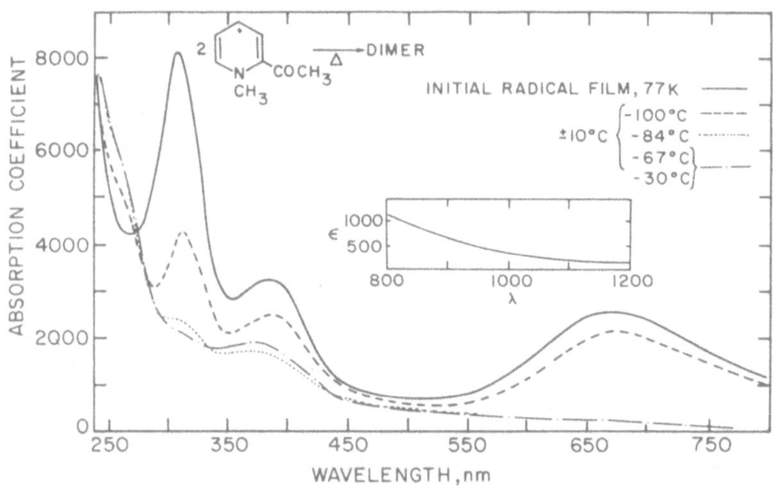

Fig. 20. The spectrum of 1-methyl-2-acetylpyridinyl radical ((CH_3CO) *2*$^{\cdot}$) from 240 nm to 1200 nm in a thin film at 77 K. At ca. −100 °C, there is > 50% loss of ultraviolet absorption along with < 25% loss in the visible. At −67 °C, the radical dimer, (CH_3CO) *2*—(CH_3CO) *2* exhibits a characteristic broad absorption [30]

The changes on warming thin films of 1-alkyl-4-carbomethoxypyridinyl radicals to −60 °C and −25 °C vary with the nature of the 1-alkyl group. Visible absorption disappears at a low temperature, more easily for the 1-t-butylpyridinyl radical than for the 1-methylpyridinyl radical. A striking change in the UV absorption of the 1-methyl derivative is the disappearance of the pyridinyl radical absorption, which is replaced by a broad band near 400 nm with (shoulders at 270 nm and 220 nm). Recooling to 77 K and irradiation produces the pyridinyl radical. However, the 400 nm band has a different ratio of the two overlapping maxima than the initial pyridinyl radical spectrum, and only 50% of the visible band is regained (Fig. 21).

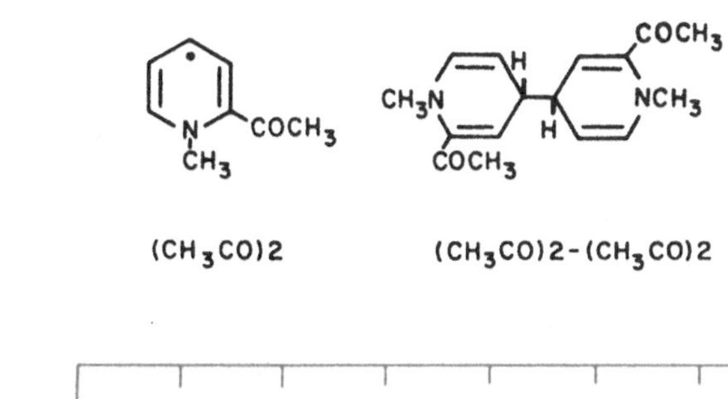

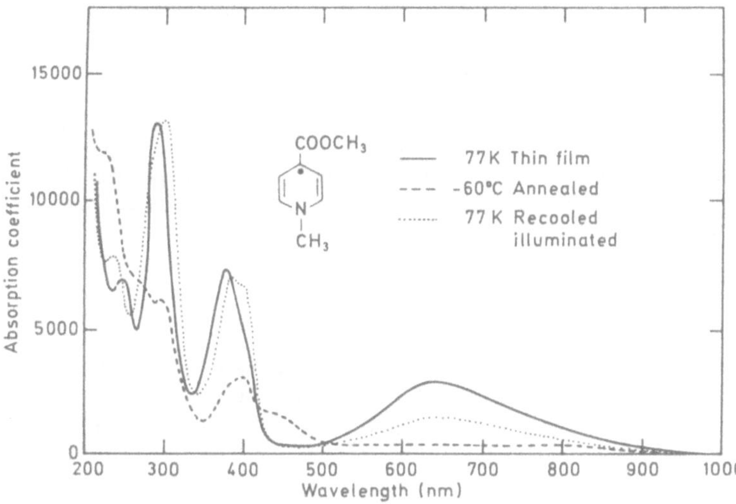

Fig. 21. Spectroscopic changes on warming a thin film of 1-methyl-4-carbomethoxypyridinyl (*4'*). At −60 °C, the radical absorption is replaced by a very broad band in the 400 nm region along with absorptions near 270 nm and 230 nm. Recooling to 77 K and irradiation (350–500 nm) regenerates the radical with about 50% return of the pimer [11]

Thin films of 1-t-butyl-4-carbomethoxypyridinyl radical produce some dimer at −25 °C. The visible absorption is lost, but regenerated partially by irradiation at 77 K. Loss of the pimer band at 380 nm is accompanied a broad absorption in the 400 nm region extending past 500 nm. Steric hindrance to dimerization by a 1-alkyl group implies bonding through the 2-positions.

Thin films of 1-methyl-4-carbo-t-butoxy-pyridinyl radical yield dimers somewhat less readily than the carbomethoxy derivative (Fig. 22). Irradiation at 430 nm rapidly dissociates the dimer to the pyridinyl radical. Thin films of the highly hindered 1-t-butyl-4-carbo-t-butoxypyridinyl radical exhibit a broad weak visible band at

625 nm (100) (no pimer) and maxima at 305 nm (14500) and 394 nm (7500). Covalent dimers are not formed on warming.

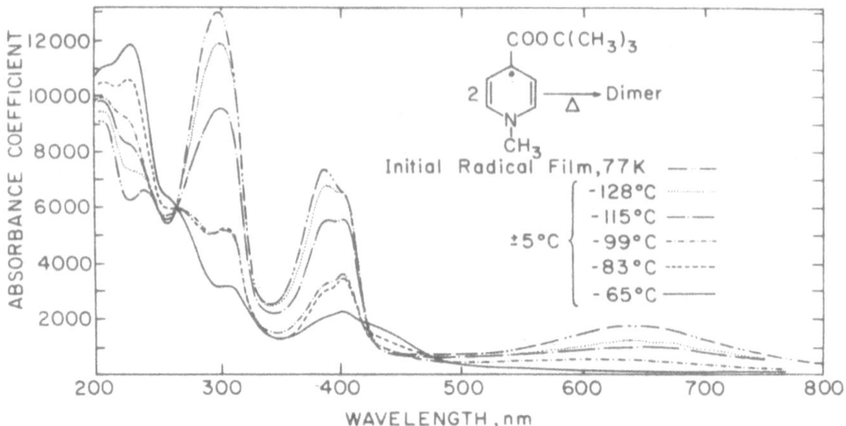

Fig. 22. Spectroscopic changes on warming 1-methyl-4-carbo-t-butoxypyridinyl radical. Pimer absorption is lost completely before loss of all radical absorption in the 400 nm region, and parallels the decrease in the 380 nm absorption. At —65 °C, very little pyridinyl radical is present. The long wavelength absorption ($\pi\sigma \rightarrow \pi\sigma^*$ transition in the dimer) extends from 350 nm past 500 nm [11]

4.1.2 Mechanism of Dimerization

The dimer formed in equilibrium with the 1-methyl-2-carbomethoxypyridinyl radical (*2*˙) is 4,4′. Six dimers (2,2′, 2,4′, 2,6′, 4,6′, 4,4′ and 6,6′) are possible. The 4,4′-dimer (shown below as *2-2*), is considered the most likely on the following grounds.

COOCH$_3$ H CH$_3$N NCH$_3$ H COOCH$_3$

2-2

The 1-alkyl group size has no effect on the association constants, the absorption maxima and the rates of reaction with oxygen and 1,1′-dimethyl-4,4′-bipyridylium dication, implicating dimer bonding at the sterically unhindered 4-position. The radical spin density at the 4-position (hfc 6.54G) is much higher than that at the 6-position (hfc 2.54), favoring coupling at the 4-positions. The UV spectrum of the *2-2* most resembles that of a 1,4-dihydropyridine [46a]. NMR spectra of the dimer in concentrated solution at —20 °C (in which the effects of the radical on the spectrum are minimized) shows four protons of equal intensities (d 4.72 (6-H), 4.26 (3-H), 4.18 (2-H) and 2.91 (4-H)) with signals at 2.73 (NCH_3) and 3.53 (OCH_3) [72].

A 4,4′-structure is also assigned to the dimer of the 1-methyl-2-acetylpyridinyl radical, (CH_3CO)*2*–(CH_3CO)*2* (See previous sect.).

NMR spectra support a 4,4′-structure for the dimer of 1-alkyl-3-carbamidopyridinyls [46b, c]. However, both 4,4′ and 4,6′-dimers have been detected in the dimer derived from 1-methyl-3-cyanopyridinium ion by reduction [46d]. The 4,4′-dimer is formed even when a substituent is present on the 4-position, as with 1-methyl-4-cyanopyridinyl radical dimerization [37]. The rearrangement of a 4,4′-dimer bearing 4 and 4′-methyl groups to a 2,2′-dimer [79] (Eq. 15), and the rearrangement of a 2,4′-dimer to the more stable 4,4′-dimer [80] have been reported.

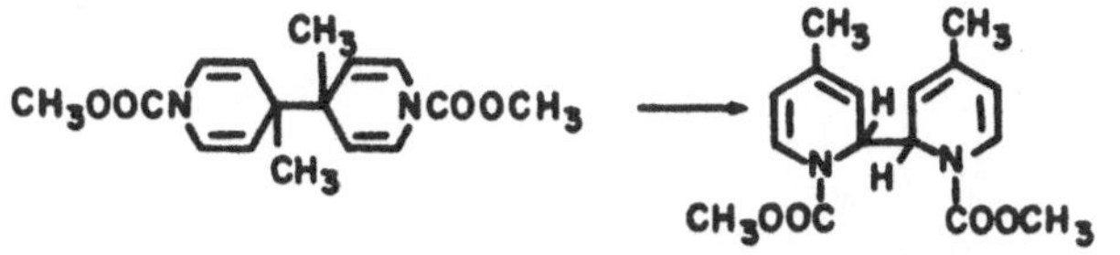

CH₃OOCN⟨4,4′-dimer with CH₃ groups⟩NCOOCH₃ ⟶ 2,2′-dimer (CH₃, CH₃; H, H; N-COOCH₃, N-COOCH₃) (15)

At room temperatures, formation of dimer from pyridinyl radical is a second-order process. However, in many cases (1-methyl-2-carbomethoxypyridinyl, *2*˙, 1-hydro-4-phenylpyridinyl and 1-alkyl-4-phenylpyridinyl [27]), a "slow" first-order process appears after the fast second-order process at low temperatures. Similarily 1-hydro-4-acetylpyridinyl dimerizes to a pinacol [RR(HO)C–C(OH)RR, i.e., a compound in which the dimer bond has been formed through the C=O group of the acetyl] [81]. Pinacol formation has also been noted for 1-trimethylsilyl-4-acetylpyridinyl and 1-trimethylgermylpyridinyl-4-carboxaldehyde [82, 83]. The result has been interpreted in terms of rapid formation of a radical pair followed by "slow" formation of covalent dimer (Eq. 16) [27].

Another possibility is that an unstable dimer is formed rapidly, followed by "slow" rearrangement, presumably by dissociation and recombination within a radical pair complex, to the more stable dimer (Eq. 17). Absorption spectra may help to distinguish these two mechanisms in spite of the similarity of dimer spectra to those of radicals.

$$2\,Py^{\cdot} \rightarrow Py^{\cdot},Py^{\cdot} \rightarrow Py\text{-}Py \qquad (16)$$

$$2\,Py^{\cdot} \rightarrow (Py\text{-}Py)_1 \rightarrow (Py\text{-}Py)_2 \qquad (17)$$

4.1.3 Photodissociation of Dimer

Irradiation (300–450 nm) of a *2-2* dimer film at 77 K produces pyridinyl radical (Fig. 23) (Eq. 18). Only about 50% of the 645 nm pimer band ($\varepsilon_{initial}$ ca. 2000) is regained. The higher the temperature at which the radical is changed into dimer, the lower the amount of pimer produced by irradiation, with none being observed in the limit. The photodissociation quantum yield is about 0.07 (300–400 nm).

$$Py\text{-}Py \rightarrow 2\,Py \qquad (18)$$

Repetition of the cycle for (CH_3CO)*2-2*$(COCH_3)$ dimer (dimerization by warming, photodissociation to (CH_3CO)*2*˙) could be carried out as many as 20 times over a period of 12 hrs, without sign of irreversible thermal or photochemical change.

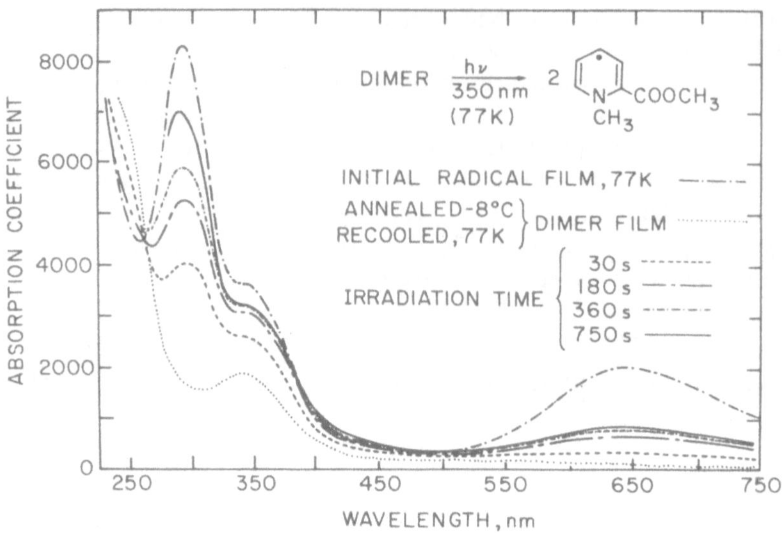

Fig. 23. Photodissociation of 1-methyl-2-carbomethoxypyridinyl dimer (*2-2*) to the monomeric radical, *2*˙. Irradiation at 350 nm for 30, 180, 360 and 750 s rapidly produces radical *2*˙ and some pimer [30)]

Almost all (>98%) of the ultraviolet absorption can be regenerated along with over 90% of the visible absorption band. The quantum yield for the photodissociation process is almost constant at ca. 0.5 from 300 nm to 400 nm. The yields decrease at longer wavelengths, as follows: 425 nm (ca. 0.25), 450 nm (ca. 0.05), 500 nm (ca. 0.004).

Photodissociation of dimer coupled to current measurement of electrochemical oxidation of the pyridinyl radical to the pyridinium ion has been described in Section 3.1.3. Oxidation of the 1-methyl-3-carbamidopyridinyl and NAD˙ (nicotinamide adenine dinucleotide radical) after dissociation of the dimers has been reported [84)], the agents being either oxygen or ˙OH radical. Reasonable mechanisms for the latter are either electron transfer or radical combination, followed by dissociation to Py^+ and ^-OH.

4.1.4 Thermal and Photofission to Pyridine and Radical

The dimer (CH_3CO,H)—(CH_3CO,H), formed by the action of zinc on pyridine and acetic anhydride at low temperatures, decomposes on heating to 1-acetylpyridinyl, the latter yielding products (4-acetylpyridine and 4-(1′-acetoxyethyl)pyridine, derived

CH_3CON(dimer)$NCOCH_3$ ⟶ (pyridinyl radical, N-$COCH_3$) ⟶ (4-$COCH_3$-pyridine) + ... (19)

(CH_3CO,H)—(CH_3CO,H) (CH_3CO,H)Py˙ pyridines

from pyridine and the acetyl radical [85] (Eq. 19). The related dimer, (EtOOC,H)—(EtOOC,H), yields ethyl isonicotinate [79] (Eq. 20).

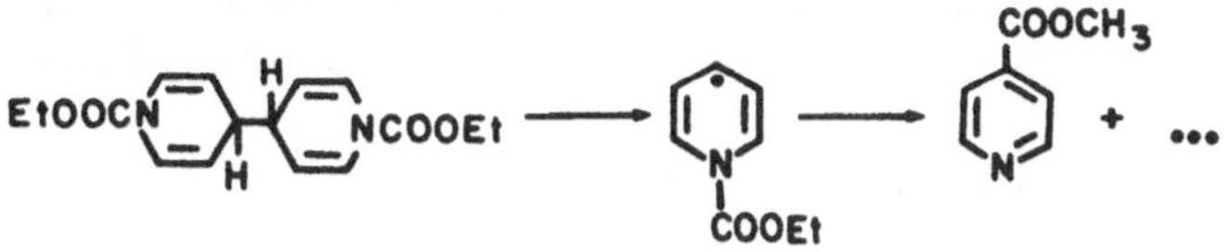

$$\text{(EtOOC,H)—(EtOOC,H)} \longrightarrow \text{(EtOOC,H)Py}^{\cdot} \longrightarrow \text{pyridines} + \ldots \tag{20}$$

Ikegami and Watanabe [86] discovered that 1-benzyl-4-carbomethoxypyridinyl was unstable to light. In a matrix, photolysis of the radical led to methyl isonicotinate and benzyl radical (Eq. 21).

$$C_6H_5CH_2Py^{\cdot} \xrightarrow{h\nu} C_6H_5CH_2^{\cdot} + Py \tag{21}$$

Kosower and Teuerstein [87] showed that o-xylylene bis-pyridinyl and 1,8-biphenylene bis-pyridinyl diradicals could be photolyzed in low quantum yields at 400 nm to yield the corresponding pyridine (methyl isonicotinate) and either a carbomethoxy-azaanthracene or 1,8-dimethylbiphenylene (Eqs. 22).

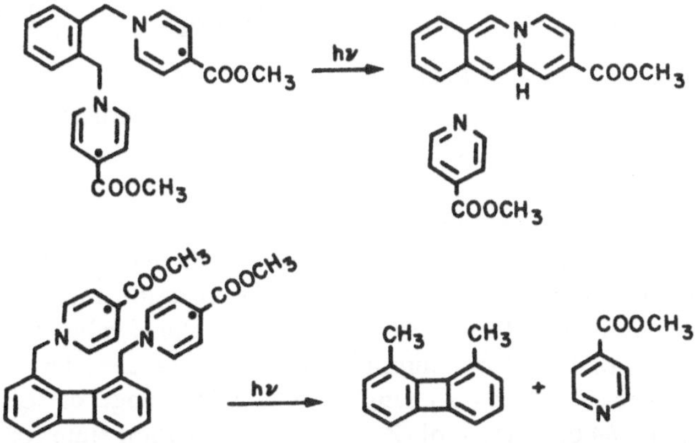

(22)

Although the 1-(4-methylbenzyl)-4-carbomethoxypyridinyl radical was similar to the 1-benzylpyridinyl radical in stability and sensitivity to light, 1-(4-nitrobenzyl)-4-carbomethoxypyridinyl radical was thermally unstable, with $t_{1/2}$ of 15 min in CH_3CN at 20 °C [88].

The thermal instability of the nitrobenzyl radical and the wavelengths required for the photofission reaction suggest a common thermal and photoinduced charge-transfer [89,90] (Eq. 23).

$$ArCH_2Py^{\cdot} \xrightarrow{\Delta \text{ or } h\nu} {}^{-\cdot}ArCH_2Py^{+} \rightarrow ArCH_2^{\cdot} + Py \tag{23}$$

4.2 Reaction with Halocarbons

The discovery [91] that dihydropyridines were oxidized by tetrachloromethane led to the realization that pyridinyl radicals might react rapidly with halocarbons. For this reason, the first kinetic studies of the reactivity of pure 1-ethyl-4-carbomethoxy-pyridinyl, *4*˙, involved halocarbons (Eq. 24). The rate constants varied with the nature of the halocarbon, and, for certain substrates, with the solvent, leading to a clear distinction between atom transfer and electron-transfer mechanisms.

$$2\,Py^{\cdot} + RX \rightarrow Py^{+}X^{-} + PyR \qquad (24)$$

4.2.1 Products of Reaction of Pyridinyls with Halocarbons

After mixing the oxygen-free solutions of halocarbons with solutions of pyridinyl radicals, the rate of the reaction may be followed spectroscopically. Reaction of 1-isopropyl-4-carbomethoxypyridinyl, (i-Pr)*4*˙, with bromochloromethane in acetonitrile gave 1-isopropyl-4-carbomethoxypyridinium bromide, (i-Pr)4^{+}, in 35% yield. The identification of two isomeric, air-sensitive dihydropyridines was facilitated by the characteristic patterns (NMR) for the isopropyl groups [13] (Eq. 25).

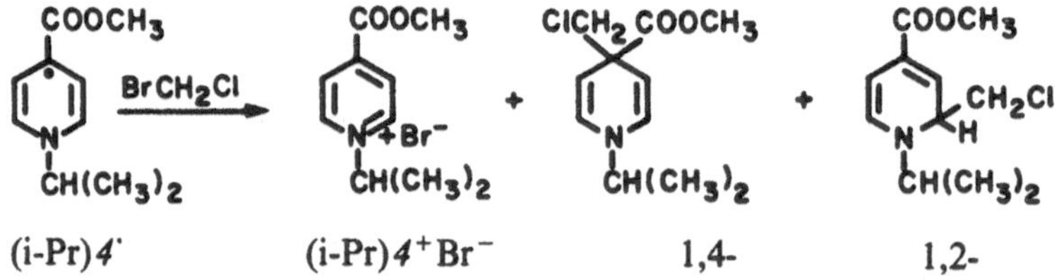

$COOCH_3$ $\xrightarrow{BrCH_2Cl}$ $COOCH_3$ + $ClCH_2$ $COOCH_3$ + $COOCH_3$ CH_2Cl (25)

$CH(CH_3)_2$ — Br^- $CH(CH_3)_2$ — $CH(CH_3)_2$ — H $CH(CH_3)_2$

(i-Pr)*4*˙ (i-Pr)$4^{+}Br^{-}$ 1,4- 1,2-
Dihydropyridines

4.2.2 Mechanism of Pyridinyl-Halocarbon Reactions

The overall reaction of pyridinyl radical and dibromomethane leads to pyridinium bromide and dihydropyridines. The variation in the reaction rate constant with a solvent, using the empirical solvent polarity parameter, the Z-value [4,67] (see below), as a criterion reveals that the rate-limiting step is atom-transfer. The rate constants listed in Table 4 show almost no change over a wide range of solvent polarity (dichloromethane (Z 64.0) to ethanol (Z 79.3). The transition state for the reaction and the initial state have then similar degrees of charge-separation.

Since the pyridinyl radical is soluble in n-hexane, and $BrCH_2Br$ has a dipole moment of 1.0D, the initial state is not very polar and the transition state can thus not be very polar. A free-energy versus reaction coordinate diagram (Fig. 24a) illustrates the point. The rate-limiting step must be the transfer of a bromine atom from the halocarbon to the pyridinyl radical, yielding a bromomethyl radical and one or two bromodihydropyridines. The latter dissociate to the pyridinium bromide, while the former combines with a second pyridinyl radical to form two bromomethyldihydropyridines. The mechanism is shown in Fig. 25 for the reaction with $BrCH_2Cl$.

The rate constants for the reaction of a series of benzyl halides with *4*˙ (Table 4) revealed that the rate for the 4-nitrobenzyl chloride was extraordinarily high, being millions of times faster than expected [5]. A change of mechanism was suspected; solvent polarity changes affected the rate constants very much, with the rate changing by a factor of 10^4 from 2-methyltetrahydrofuran (Z 55.4) to acetonitrile (Z 71.3) [6].

Table 4. Rate Constants for the Reaction of 1-Ethyl-4-carbomethoxypyridinyl Radical with Halocarbons at 25 °C

Halide	Solvent		Rate Constant (M^{-1} sec^{-1})	Ref.
Atom-transfer reactions (element effect)				
$ClCH_2Cl$	CH_3CN		2.6×10^{-8} [a]	64)
$BrCH_2Cl$	CH_3CN		5.0×10^{-5}	64)
ICH_2Cl	CH_3CN		1.3×10^{-1}	64)
ICH_3	CH_3CN		5.0×10^{-6}	64)
Substituent Effect (4-X-benzyl)				
$CH_3OC_6H_4CH_2Cl$	CH_3CN		11.3×10^{-4}	62)
$CH_3C_6H_4CH_2Cl$	CH_3CN		3.68×10^{-4}	62)
$C_6H_5CH_2Cl$	CH_3CN		3.31×10^{-4}	62)
$ClC_6H_4CH_2Cl$	CH_3CN		6.5×10^{-4}	62)
$NO_2C_6H_4CH_2Cl$	CH_3CN		24,000	62)
Atom-transfer (solvent effect)				
$BrCH_2Br$	CH_2Cl_2	(Z 64.0)	0.48×10^{-4}	64)
	CH_3CN	(Z 70.3)	1×10^{-4}	64)
	i-PrOH	(Z 77.4)	0.94×10^{-4}	64)
	EtOH	(Z 79.3)	1.7×10^{-4}	64)
Electron-transfer (solvent effect)				
$NO_2C_6H_4CH_2Cl$	MTHF[b]	(Z 55.4)	1.62	63)
	DME[c]	(Z 59.0)	8.3	63)
	CH_2Cl_2	(Z 64.0)	75	63)
	CH_3COCH_3	(Z 64.0)	450	63)
	DMF[d]	(Z 69)	12000	63)
	CH_3CN	(Z 71.3)	24000	63)

a estimated from data at higher temperatures
b 2-methyltetrahydrofuran
c 1,2-Dimethoxyethane
d Dimethylformamide

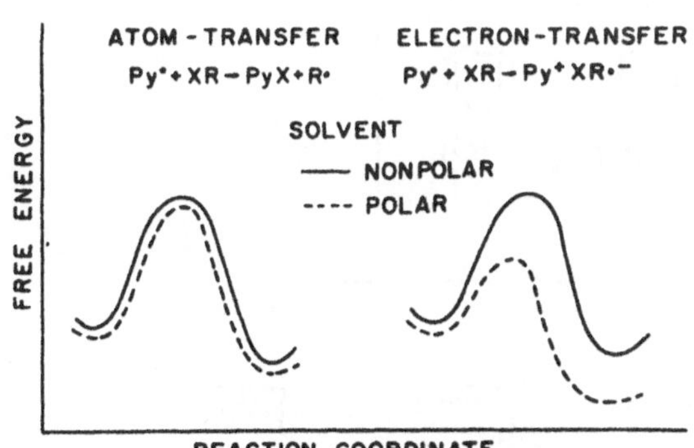

Fig. 24. Free energy versus reaction coordinate diagrams for the reaction of a pyridinyl radical with halocarbons in solvents of low and high polarity.

a) (left side) The initial and transition states for the reaction are lowered to the same (small) extent by solvation, a reflection of the fact that the rate constants are similar in all solvents.

b) (right side) The charge-separated transition and product ion-pair states are lowered much more by solvation than the initial state

The transition state free energy change of the reaction was half of the Z-value change. The Z-value is derived from a charge-transfer process, and a reaction which varies in its energetics in the same way must also be charge-transfer.

The mechanism is shown in Fig. 26. The initial electron-transfer step is very sensitive to the polarity of the solvent, and produces a pyridinium cation:4-nitrobenzyl

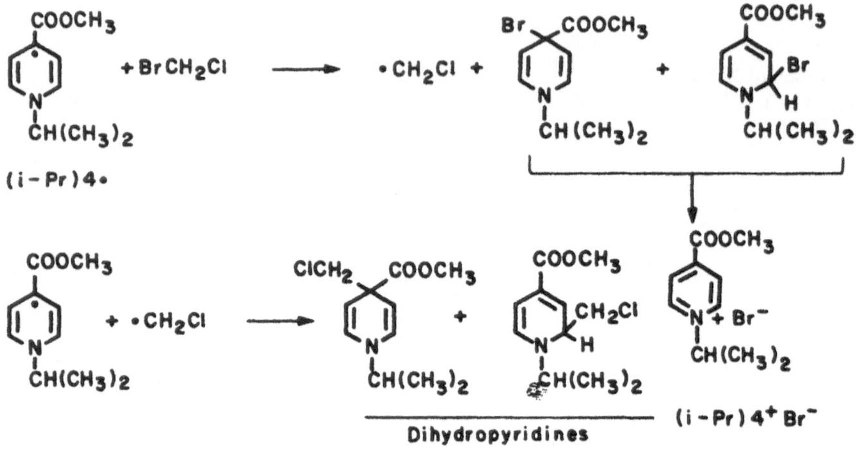

Fig. 25. The mechanism for the reaction of a pyridinyl radical with bromochloromethane. The initial, rate-limiting, solvent-insensitive step involves transfer of the bromine atom to the pyridinyl radical, forming one or more bromodihydropyridines and a chloromethyl radical. The latter reacts rapidly with another pyridinyl radical to yield two isomeric dihydropyridines, while the former dissociates to a pyridinium bromide

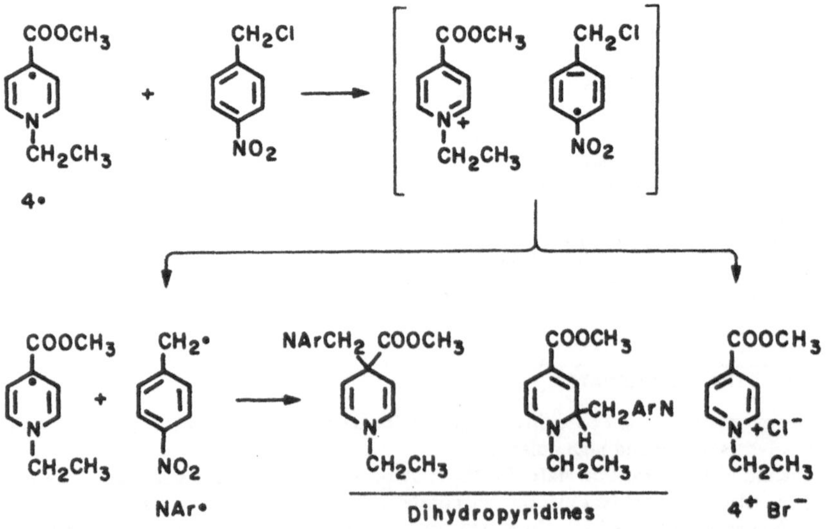

Fig. 26. The mechanism for the reaction of a pyridinyl radical with 4-nitrobenzyl chloride

chloride radical anion pair. The radical anion rapidly dissociates [92] to yield a chloride ion and a 4-nitrobenzyl radical which is consumed in a reaction with another pyridinyl radical. The free energy versus reaction coordinate diagram for the electron-transfer reaction is shown in Fig. 24b.

Evaluation of solvent-sensitive properties requires well-defined reference parameters. A macroscopic parameter, dielectric constant, does not always give interpretable correlations of data. The first microscopic measure of solvent polarity, the Y-value, based on the solvolysis rate of t-butyl chloride [93], is particularly valuable for correlating solvolysis rates. *Y-values* are tedious to measure, somewhat complicated in physical basis, and characterizable for a limited number of solvents. The *Z-value*, based on the charge-transfer electronic transition of 1-ethyl-4-carbomethoxypyridinium iodide [67,94], is easy to measure and had a readily understandable physical origin. However, non-polar solvent Z-values are difficult to obtain because of low salt solubility. The $E_T(30)$-value [95,96], is based on an intramolecular charge-transfer transition in a pyridinium phenol betaine which dissolves in almost all solvents. We have used the $E_T(30)$-value in the studies of ANS derivatives as the measure of solvent polarity. Solvent polarity is what is measured by a particular technique and may refer to different summations of molecular properties in different cases. For this reason, only simple reference processes should be used to derive solvent parameters.

4.3 Basicity of Pyridinyl Radicals

The basicity of pyridinyl radicals is of interest since some pyridinyl radicals react with one another in an acid catalyzed reaction (sect. 4.4). The absorption spectra of radicals generated from the pyridinium ion by pulse radiolysis in aqueous solution at different pH values allow the evaluation of the basicity of the radicals [23,97,98]. The pK_a of the protonated form of 1-methyl-3-carbamidopyridinyl radical, ($(CONH_2)$*3*˙), is 1.43; the protonated radical has absorption maxima at 300 nm and 440 nm, at somewhat longer wavelengths than those for the unprotonated radical at 280 nm and 420 nm. The nicotinamide adenine dinucleotide radical (NAD˙) has a pK_a of 0 or less, with a shift in absorption maximum due to protonation being observed only in 1.5 M $HClO_4$.

The pK_a of the protonated $(CONH_2)$*4*˙ radical is 2.0, with a shift from 425 nm to 400 nm on deprotonation. The change in the absorption maxima for both 3- and 4-carbamidopyridinyl radicals implies protonation on the amide group rather than the ring (Eq. 26). Thus, the pyridinyl radicals are more basic than benzamide derivatives, but very weakly basic compared to dihydropyridines (pK_a ca. 7) [99] or typical tertiary amines (pK_a ca. 10–11).

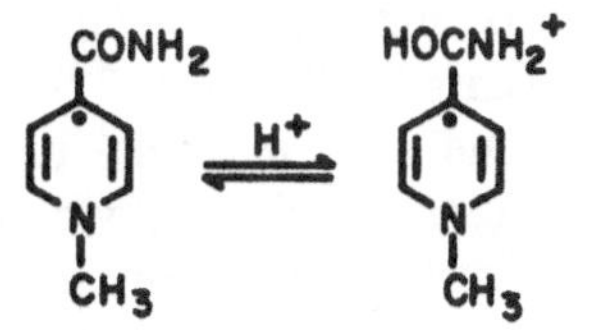

CONH₂ ⇌(H⁺) HOCNH₂⁺ (26)

$(CONH_2)$*4*˙ $(CONH_2)$*4*˙H^+

4.4 Disproportionation of Pyridinyl Radicals in Water

The 1-alkyl-2-, 3- and 4-carbamidopyridinyl radicals may be generated in water from the corresponding pyridinium ions by pulse radiolysis or radiolysis. The rate constant for the disappearance of *3*˙ are pH-independent and close to diffusion controlled (Eq. 27). The rate constants for *4*˙ are pH-dependent. Completely protonated *4*˙ reacts with itself at rates somewhat less than diffusion controlled; the rate constants decrease linearly with increasing pH (slope ca. 1) (Eq. 28). Rates in the range pH 8–11 could be followed using a pyridinyl ester. A minimum rate was found near pH 9.2; at higher pH values, the hydrolysis of the pyridinyl ester to the carboxylate controlled the rate (Eq. 29), since the pyridinyl carboxylate would react with itself in a diffusion controlled process (Eq. 30).

$$3^{\cdot} + 3^{\cdot} \rightarrow 3\text{-}3 \quad (27)$$

$$4^{\cdot} + 4^{\cdot}H^{+} \rightarrow 4^{+} + 4\,H \quad (28)$$

$$4^{\cdot} \rightarrow (COO^{-})4^{\cdot} \quad (29)$$

$$2\,(COO^{-})4^{\cdot} \rightarrow \text{products} \quad (30)$$

The 1-ethyl-4-carbomethoxypyridinyl radical disproportionation products, aside from the pyridinium ion, probably included ethylamine, and a compound thought to be an ester dialdehyde (Eq. 31). The pH changes noted after radiolysis of buffer-free isopropyl alcohol-water solutions of the pyridinium ions revealed that the initial dihydropyridine (*4* H) formed from *4*˙ hydrolyzed and consumed another proton, presumably because of ethylamine.

$$4\,H + 2\,HOH + H^{+} \rightarrow CH_3OOCCH(CH_2CHO)_2 + CH_3CH_2NH_3^{+} \quad (31)$$

The H^+ consumption for the disproportionation of the radicals generated through radiolysis was: a) *3*˙ (none, dimer formed, cf. Eq. 27), b) *2*˙ (one H^+ consumed, cf. Eq. 28), c) *4*˙ (two H^+ consumed, cf. Eqs. 28 and 31).

4.4.1 Disproportionation of Bis-pyridinium Cation Radicals

The two-step reduction of the bis-pyridinium dication, the bis-methyl quaternary derivative of 4,4′-bipyridyl, first to a cation radical, and then to a 2 e^- "quininoid" reduction product, has been known for many years [100,101] (Eq. 32). The system was useful as an oxidation-reduction indicator and as a pyridinyl radical titrant because of the intense blue-violet color of the cation radical ("viologen") (λ_{max} 605 nm, ε_{max} 13000 [16]) (supersedes the lower value previously used [37]). The remarkable effectiveness of the oxidized quaternary salt ("Paraquat") as a herbicide is directly related to the reduction potentials and the stability of the cation radical. Structural variations cause great changes in the reduction potentials and the stabilities of the intermediate cation radicals with respect to disproportionation [102].

$$CH_3\overset{+}{N}C_5H_4\text{—}C_5H_4\overset{+}{N}CH_3 \underset{}{\overset{e^-}{\rightleftharpoons}} CH_3NC_5H_4^{\cdot}\text{—}C_5H_4\overset{+}{N}CH_3 \underset{}{\overset{e^-}{\rightleftharpoons}} CH_3NC_5H_4\text{=}C_5H_4NCH_3 \qquad (32)$$

$4{,}4'^{++}$ $\qquad$ $4{,}4'^{+\cdot}$ $\qquad$ $4{,}4'^{\cdot}$

4.5 Reduction Potentials of Pyridinium Ions

New electrochemical techniques have made possible the measurement of reliable one-electron reduction potentials for pyridinium ions. The shifts in potential introduced by subsequent reactions at the electrode (chiefly, dimerization) have lead to a variety of useful kinetic and thermodynamic properties of the pyridinyl radicals. The potentials are listed in Table 5, along with some values for the one-electron reduction potentials of pyridinyl radicals.

Table 5. Reduction Potentials for 1-Alkylpyridinium Ions and 1-Alkylpyridinyls in Acetonitrile (Formal potentials for the couples $Py^{\cdot}/Py^{+}$ and $Py^{-}/Py^{\cdot}$)

Substituent[a]	$E^{0\prime}$, V[b] $Py^{+} + e^{-} = Py^{\cdot}$				$E^{0\prime}$, V[b] $Py^{\cdot} + e^{-} = Py^{-}$		
	$LiClO_4$[c]	TBAP[c,d]	Method	Ref.	TBAP[c,d]	Method	Ref.
4-$COOCH_3$	−1.081	−1.095	e, f	aa)	−1.910	g	aa)
2-$COOCH_3$	−1.166		g	aa)	ca. −1.83	g	aa)
4-$CONH_2$	−1.154		e, f	bb)			
2-$CONH_2$	−1.358		g	bb)			
4-t-C_4H_9[h]	−1.835		g	cc)			
2-$COCH_3$	−1.05	−1.106	f	dd)	−1.750	f, g	dd)
4-CN	−0.982	−0.965	e, f	ee)	−1.900	g	dd)
2-CN	−1.030		f, g	dd)			
2^{++} [i]	−0.949[j]	−0.960[k]	f	ff)			
		−0.945[l]	e				
3^{++} [i]	−0.978[j]	−1.035[k]	f	ff)			
		−1.025[l]	e				
4^{++} [i]	−1.052[j]	−1.025[k]	e, f	ff)			
5^{++} [i]	−1.066[j]	−1.078[k]	e, f	ff)			
6^{++} [i]	−1.071[j,m]	−1.089[k]	e, f	ff)			
7^{++} [i]	−1.070[j,m]	−1.095	e, f	ff)			
8^{++} [i]	−1.074[j,m]	−1.095	e, f	ff)			
9^{++} [i]	−1.074[j,m]	−1.095	e, f	ff)			
10^{++} [i]	−1.075[j,m]	−1.092	e, f	ff)			

a 1-Alkyl substituent is CH_3, unless otherwise indicated
b versus the reference electrode: Ag/Ag^{+} (0.01 M), 0.25 M $LiClO_4$ in CH_3CN
c 0.25 M electrolyte
d tetra-n-Butylammonium perchlorate
e Direct current (DC) polarography
f Normal pulse (NP) polarography

Table 5. (continued)

g Reverse pulse (RP) polarography
h 1-CH_3, 1-C_2H_5 and 1-$(CH_3)_2CH$ derivatives gave similar results
i n^{++} = 1,1'-n-methylene-bis(4-carbomethoxypyridinium) ions, $Py^+(CH_2)_nPy^+$, for which n = number of methylene groups connecting the nitrogens
j The electrode reaction is: $n^{++} + 2e^- = n^{\cdot\cdot}$; $n^{\cdot\cdot} + Li^+ = n^{\cdot\cdot}Li^+$
k The electrode reaction is: $n^{++} + 2e^- = n^{\cdot\cdot}$ (for n = 2 to 10); $n^{\cdot\cdot} = n_{ccB}$ (ccB = *cis*-cyclomer B) (only for n = 3 or 4)
l The electrode reaction is: $n^{++} + 2e^- = n^{\cdot\cdot}$; $n^{\cdot\cdot} = n_{tc}$ = *trans*-cyclomer) (for n = 2 or 3)
m Spectroscopic evidence for pimer (i.e., pyridinyl-pyridinyl interaction) is the enhanced absorption coefficient in the visible region of the optical spectrum (cf. Table 1, Kosower, E. M., Hajdu, J. and Nagy, J.: J. Am. Chem. Soc. 100, 1186 (1978). The potential measurements (10 mV more positive than the parent 1-methyl-4-carbomethoxypyridinium ion) suggest a cyclic Li^+ complex without pimer absorption being evident
aa Kashti-Kaplan, S., Hermolin, J. and Kirowa-Eisner, E.: J. Electrochem. Soc. 128, 802 (1981)
bb Hermolin, J., Talmor, D., Kashti-Kaplan, S. and Kirowa-Eisner, E., to be submitted
cc Talmor, D., Hermolin, J. and Kirowa-Eisner, E.: J. Electroanal. Chem. 139, 81 (1982)
dd Talmor, D., Hermolin, J. and Kirowa-Eisner, E.: unpublished results
ee Webber, A., Osteryoung, J., Hermolin, J. and Kirowa-Eisner, E., J. Electrochem. Soc. 129, 2725 (1982)
ff Hermolin, J., Kashti-Kaplan, S. and Kirowa-Eisner, E.: J. Electroanal. Chem. 123, 307 (1981)

5 Biological Role

5.1 NAD· and Related Radicals

The 3-carbamidopyridinium ring is the chemically active portion of the enzymatic cofactors, NAD and NADP (nicotinamide adenine dinucleotide and its phosphate). A typical reaction involving NAD is the stereospecific (with respect to both cofactor and substrate) oxidation of ethanol to acetaldehyde catalyzed by the enzyme, alcohol dehydrogenase (Eq. 33).

$$\text{NAD (3-CONH}_2\text{-pyridinium, N-RPPRA)} + CH_3CH_2OH \xrightarrow{-H^+} \text{NADH (4,4-H,H-1,4-dihydro, 3-CONH}_2\text{, N-RPPRA)} + CH_3CHO \quad (33)$$

NAD NADH

The overall reaction is currently regarded as an apparent hydride-transfer (for review, see [103]) because a) hydrogen is transferred directly between the reactants in most cases, without exchange with the solvent, 2) the base-catalyzed disproportionation of benzaldehyde to benzyl alcohol and benzoic acid (Cannizzaro reaction) seems to be a hydride transfer (3) stereospecificity can be observed in model reactions. In addition, Verhoeven and coworkers [104, 105] have analyzed the thermodynamics of the photoinduced reaction between 1-benzyl-3-carbamido-1,4-dihydropyridine and 1,1-dimethyl-4,4'-bipyridylium dication, and have concluded that a *thermal* 1 e^- reaction between a 1-alkyl-1,4-dihydronicotinamide and a carbonyl compound is unlikely.

Nevertheless, pyridinyl radicals or a protonated pyridinyl radical may still be worth consideration, since the way in which the enzymes catalyze reactions of molecules with NAD (or NADH) is still not understood. A complex conformational change takes place in alcohol dehydrogenase on binding the substrate to a complex of NAD and the enzyme; a hydrogen bond between threonine and the 3-carboxamido group of the pyridine ring is required [106]. The amide group is not coplanar with the pyridine ring [102,108] and rotates unusually rapidly in NADH [109]. If the amide group were rotated out of the plane in the reaction complex, the "deformed" 1,4-dihydropyridine would be a much better donor (less conjugation) and would form a protonated radical through 1 e^- transfer to a substrate. The electron-affinity of the carbonyl substrate might be promoted by hydrogen bonding within the active site. The fate of the intermediate is not so clear, but one possibility is that a hydrogen atom is transferred to the radical anion of the carbonyl substrate (Eq. 34). Thus, a free pyridinyl radical need not be formed, and the protonated pyridinyl so reactive that the reaction is completed with stereospecificity within the active site. Even in model compounds, one could expect considerable stereospecifity and direct cofactor-carbonyl hydrogen transfer.

Numerous examples of radical formation during model reductions with 1,4-dihydropyridines are known [110]. The overall reaction would then be (e^-, $H^{\cdot}$). The reverse reaction, between an alcohol and NAD, would produce the equivalent of an alkoxide, as is also required in the "hydride-transfer" mechanism (Eq. 35).

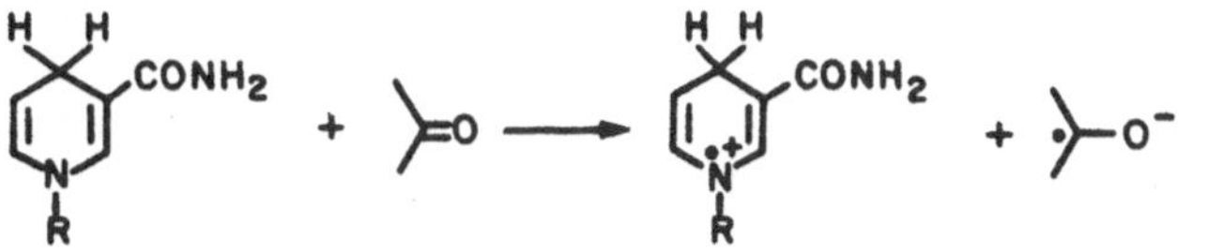

$$\text{(H,H)C}_5\text{H}_3\text{N(R)-CONH}_2 + \;>\!=\!O \longrightarrow \text{(H,H)C}_5\text{H}_3\overset{\cdot +}{\text{N}}\text{(R)-CONH}_2 + \;>\!\cdot\!-O^- \tag{34}$$

$$\text{NADH} + \;>\,=0 \rightleftharpoons \text{NAD}^+ + \;>(\text{H})\text{—O}^- \tag{35}$$

5.2 "Paraquat" Cation Radical

Novel approaches to agriculture have been made practical by the use of new herbicides (like Paraquat). It is necessary to explain not only the useful activities but also the considerable human toxicity encountered in a number of cases.

The herbicidal activity of 1,1-dimethyl-4,4′-dimethylbipyridylium dication ("Paraquat") seems to depend on the formation of the cation radical through reaction with a component of photosystem I in the chloroplast, followed by reaction with oxygen to form superoxide ion (Eq. 36) (for a summary of references, see [111]). The rate constant for the latter reaction has been reported [112].

$$\text{Py-Py}^{+\cdot} + O_2 \rightarrow \text{Py-Py}^{++} + O_2^- \tag{36}$$

Superoxide ion is not highly reactive towards hydrocarbons (i.e., lipids) and is readily removed from biological systems by superoxide dismutase, forming hydrogen peroxide and oxygen. However, the protonated superoxide ion or hydroperoxyl radical, $HO_2^{\cdot}$, reacts with NADH [113] and alkanes. Although such a reactive radical might be responsible for cell damage, even in small concentrations, it does not seem likely that the range of pH (7.1–7.6) attained in respiratory pathology could lead to the conversion of a significant fraction of the superoxide ion (pK_a 4.8) into the more reactive radical.

Another plausible route for the creation of a reactive radical might be the carbonic anhydrase catalyzed addition of superoxide ion to carbon dioxide (Eq. 37). An enzymatic reaction is consistent with the variable response of different individuals to paraquat exposure. There are at least two common genetic variants of carbonic anhydrase, and additional variation can occur through posttranslational modification of the protein [114].

$$O_2^{\cdot-} + O{=}C{=}O \rightarrow {}^{-}O{-}C({=}O){-}O{-}O^{\cdot} \text{ (carboxyperoxyl radical)} \quad (37)$$

The carboxyperoxyl radical anion thus produced should be similar in reactivity to the hydroperoxyl radical, $HO_2^{\cdot}$. The nucleophilic activity of the superoxide ion towards carbonyl groups in acid chlorides, esters and ketones is well documented [115, 116].

The reaction between superoxide ion and the Py-Py$^{+\cdot}$ cation radical, which leads to destruction of the latter, would seem more likely to mitigate the long-term effects of the Py-Py$^{+\cdot}$ rather than promote damage to components of the cell [117]. The occurence of Rh(bipy)$_3^{+2}$-mediated photoreduction of alkenes with NADH models and proceeding via pyridinyls indicates that other, more complex pathways may exist [118].

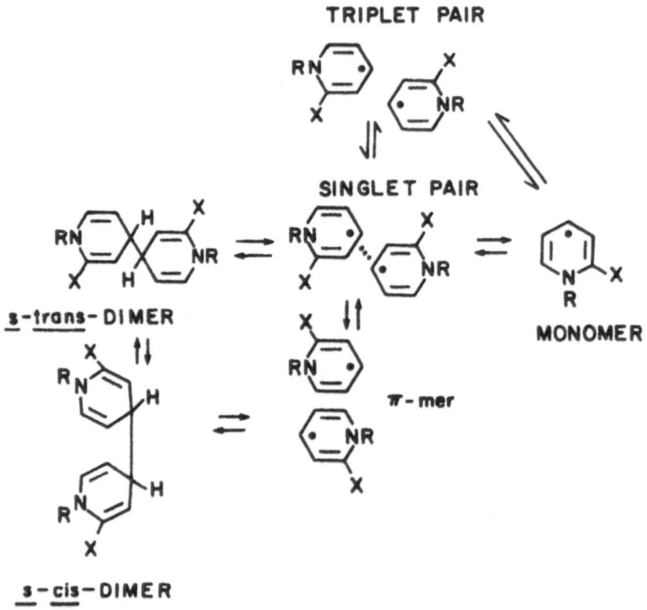

TRIPLET PAIR

SINGLET PAIR

s-trans-DIMER

MONOMER

π-mer

s-cis-DIMER

Fig. 27. A scheme illustrating many of the species identified for the 1-methyl-2-carbomethoxypyridinyl radical 2· system [30]

6 Conclusions

1. Pyridinyl radicals yield detailed information on radical spectra, radical complexes and radical chemistry. (See Fig. 27).
2. Pyridinyl radicals are among the best examples of molecules exhibiting "through-bond conjugation", with unusual UV-VIS spectra ($\pi\sigma \rightarrow \pi\sigma^*$ transition) and chemistry (thermal- and photodissociation).
3. Pyridinyl radical studies have stimultated the development of useful equipment, such as those for thin film spectroscopy, for variable path measurements under oxygen-free conditions (the "VV-cell") and a spectroelectrochemical cell. In addition, a new technique for measuring the photodissociation spectra of unstable species which can be generated electrochemically was introduced.

7 Acknowledgements

Those who have contributed to this subject deserve special and warm acknowledgement for their intense effort, for their creativity, and for their perserverance. Specific mention should be made of Bill Schwarz, who discovered the stability of pyridinyl radicals during electrochemical experiments, of Ed Poziomek, who discovered how to prepare pyridinyl radicals and distill them, of Irv Schwager, Mahboob Mohammad, Harold Waits, Joe Hajdu and Avraham Teuerstein who made the difficult kinetic measurements, of Yusaku Ikegami and Michiya Itoh who began the subject of pyridinyl diradicals, and of Joshua Hermolin whose herculean efforts with thin film spectroscopy very much advanced our knowledge of pyridinyl radical behavior. It has been a pleasure to collaborate on several aspects of the work with John Swallow (Paterson Laboratories, Manchester, England) and Emilia Kirowa-Eisner of Tel-Aviv University. Support for the work has been provided by the National Institutes of Health, Edgewood Arsenal, and the Israel Academy of Sciences,

8 References

1. Schwarz, W. M., Kosower, E. M., and Shain, I.: J. Am. Chem. Soc. *83*, 3164 (1961)
2. Kosower, E. M., Poziomek, E. J.: ibid. *86*, 5515 (1964)
3. Kosower, E. M., Lindqvist, L.: Tetrahedron Letters 4481 (1965)
4. Kosower, E. M., Schwager, I.: J. Am. Chem. Soc. *86*, 5528 (1964)
5. Mohammad, M., Kosower, E. M.: ibid. *93*, 2709 (1971)
6. Mohammad, M., Kosower, E. M.: ibid. *93*, 2713 (1971)
7. Kosower, E.M.: Pyridinyl radicals in biology, in: Free Radicals in Biology, Vol. II, p. 1, (ed. Pryor, W. A.) Academic Press, New York, 1976
8. Kosower, E. M.: Pyridinyl paradigm proves powerful, in: Amer. Chem. Soc. Symp. *69*, (ed. Pryor, W. A.), p. 447 (1978)
9. Forrester, A. R., Hay, J. M., Thomson, R. H.: The Organic Chemistry of Stable Free Radicals, Academic Press, New York 1968, 405 pp.
10. Kosower, E. M., Poziomek, E. J.: J. Am. Chem. Soc. *85*, 2035 (1963)
11. Hermolin, J., Levin, M., Kosower, E. M.: ibid. *103*, 4808 (1981)
12. Hanson, P.: Adv. Heterocycl. Chem. *25*, 205 (1979)
13. Kosower, E. M., Waits, H. P., Teuerstein, A., Butler, L. C.: J. Org. Chem. *43*, 800 (1978)

14. Griller, D., Ingold, K. U.: Accts. Chem. Res. *9*, 13 (1976)
15. Kosower, E. M., Waits, H. P.: Org. Prep. Proced. Int'l. *3*, 261 (1971)
16. Hermolin, J., Levin, M., Ikegami, Y., Sawayanagi, M., Kosower, E. M.: J. Am. Chem. Soc. *103*, 4795 (1981)
17. Kosower, E. M., Hajdu, J., Nagy, J. H.: ibid. *100*, 1186 (1978)
18. Kashti-Kaplan, S., Hermolin, J., Kirowa-Eisner, E.: J. Electrochem. Soc. *128*, 802 (1981)
19. Hermolin, J., Kashti-Kaplan, S., Kirowa-Eisner, E.: J. Electroanal. Chem. *123*, 307 (1981)
20. Livingston, R., Zeldes, H.: J. Chem. Phys. *44*, 1245 (1966)
21. Dohrmann, J. K., Becker, R.: J. Magn. Resonance *27*, 371 (1977)
22. van Bergen, T. J., Kellogg, R. M.: J. Am. Chem. Soc. *94*, 8451 (1972)
23. Kosower, E. M., Teuerstein, A., Burrows, H. D., Swallow, A. J.: ibid. *100*, 5185 (1978)
24. Land, E. J., Swallow, A. J.: Biochim. Biophys. Acta *162*, 327 (1968)
25. Bielski, B. H. J., Chan, P. C.: J. Am. Chem. Soc. *102*, 1713 (1980)
26. Swallow, A. J.: Radiation Chemistry, Longman, London 1973, 275 pp.
27. (a) Tero-Kubota, S., Sano, Y., Ikegamik Y.: ibid. *104*, 3711 (1982)
    (b) Akiyama, K., Tero-Kubota, S., Ikegami, Y.: Abstracts, 22nd Symp. on Organic Radical Reactions, Tsukuba, Japan October 1981
    (c) Akiyama, K., Kubota, S., Ikegami, Y.: Chem. Lett. 469 (1981)
    (d) Sano, Y., Tero-Kubota, S., Ito, O., Ikegami, Y.: Chem. Lett. 657 (1982)
28. (a) Mohammad, M., Iqbal, R., Khan, A. Y., Zahir, K., Riffat: J. Electroanal. Chem. *124*, 139 (1981)
    (b) Mohammad, M., Khan, A. Y., Iqbal, M., Iqbal, R., Razzaq, M.: J. Am. Chem. Soc. *100*, 7658 (1978)
    (c) Mohammad, M., Iqbal, R., Khan, A. Y., Bhattl, M., Zahir, K., Johan, R.: J. Phys. Chem. *85*, 2816 (1981)
29. Land, E. J., Swallow, A. J.: Biochim. Biophys. Acta *234*, 34 (1971)
30. Bermolin, J., Levin, M., Kosower, E. M.: J. Am. Chem. Soc. *103*, 4801 (1981)
31. Hermolin, J., Kirowa-Eisner, E., Kosower, E. M.: ibid. *103*, 1591 (1981)
32. Osteryoung, J., Kirowa-Eisner, E.: Anal. Chem. *52*, 62 (1980)
33. Cummings, T. E., Bresnahan, W. T., Suh, S. Y., Elving, P. J.: J. Electroanalyt. Chem. *106*, 71 (1980)
34. Itoh, M., Nagakura, S.: J. Am. Chem. Soc. *89*, 3959 (1967)
35. Kosower, E. M., Swallow, A. J., Land, E. J.: ibid. *94*, 986 (1972)
36. Ebert, M., Simic, M.: Int. J. Rad. Phys. Chem. *3*, 259 (1971)
37. Kosower, E. M., Cotter, J. L.: J. Am. Chem. Soc. *86*, 5524 (1964)
38. (a) Evans, A. G., Evans, J. C., Baker, M. W.: J. Chem. Soc. Perkin II, 1310 (1975)
    (b) J. Am. Chem. Soc. 99, 5882 (1977)
39. Schwarz, W., jr.: Ph. D. Thesis, Univ. of Wisconsin, 1961
40. Itoh, M., Nagakura, S.: Bull. Chem. Soc. Japan *39*, 369 (1966)
41. Ikegami, Y., Seto, S.: J. Am. Chem. Soc. *96*, 7811 (1974)
42. Itoh, M.: Chem. Phys. Lett. *2*, 371 (1968)
43. Klöpffer, W., and Naundorf, G.: J. Luminesc. *8*, 457 (1974)
44. Kosower, E. M., Dodiuk, H.: ibid. *11*, 249 (1975/6)
45. Klöpffer, W., Kaufmann, G.: ibid. *20*, 283 (1979)
46. (a) Eisner, U., Kuthan, J.: Chem. Rev. *72*, 1 (1972)
    (b) Biellmann, J., Lapinte, C.: Tetrahedron Lett. 683 (1978)
    (c) Carelli, V., Liberatore, F., Casini, A., Mondelli, R., Arnone, A., Carelli, I., Rotilio, G., Mavelli, I.: Bioorg. Chem. *9*, 342 (1980)
    (d) Carelli, I., Cardinali, M. E., Casini, A., Arnone, A.: J. Org. Chem. *41*, 3967 (1978)
47. Cookson, R. C., Henstock, J., Hudec, J.: J. Am. Chem. Soc. *88*, 1060 (1966)
48. Heilbronner, E., Schmelzer, A.: Helv. Chim. Acta *58*, 936 (1975)
49. Grob, C. A.: Angew. Chem. Int. Ed. *8*, 535 (1969)
50. Kuthan, J., Palecek, J.: Z. Chem. *8*, 229 (1968)
51. Dekkers, A. W. J. D., Verhoeven, J. W., Speckamp, W. N.: Tetrahedron *29*, 1691 (1973); Verhoeven, J. W.: Rec. Trav. Chim. Pays-Bas *99*, 369 (1980)
52. Bartetzko, R., Gleiter, R., Muthard, J. L., Paquette, L. A.: J. Am. Chem. Soc. *100*, 5589 (1978)
53. (a) Hoffmann, R., Imamura, A., Hehre, W. J.: ibid. *90*, 1499 (1968)
    (b) Hoffmann, R. A.: Accts. Chem. Res. *4*, 1 (1971)

54. (a) Gleiter, R.: Angew. Chem. Int. Ed. *13*, 696 (1974)
(b) Spanget-Larsen, J., de Korswagen, C., Eckert-Maksić, M., Gleiter, R.: Helv. Chim. Acta *65*, 968 (1982)
55. Kosower, E. M., Ikegami, Y.: J. Am. Chem. Soc. *89*, 461 (1967)
56. Itoh, M., Kosower, E. M.: ibid. *89*, 1843 (1968)
57. Hermolin, J., Kosower, E. M.: ibid. *103*, 4813 (1981)
58. Itoh, M.: ibid. *93*, 4750 (1971)
59. Bellamy, L. J.: The Infrared Spectra of Complex Molecules, 3rd ed., Chapman and Hall, London 1975
60. Johnson, jr., C. E., Gutwosky, H. S.: J. Chem. Phys. *39*, 58 (1963)
61. Landolt-Börnstein, Vol. 9b, Magnetic Properties of Free Radicals, Springer-Verlag, Berlin, 1977
62. Symons, M.: Chemical and Biochemical Aspects of Electron Spin Resonance Spectroscopy, Van Nostrand Reinhold, New York, 1978, 190 pp.
63. Wertz, J. E., Bolton, J. R.: Electron Spin Resonance, McGraw-Hill, New York, 1972, p. 117
64. Rakowsky, T., Dohrmann, J. K.: Ber. Bunsenges. Phys. Chem. *83*, 495 (1979)
65. Fujita, H., Yamauchi, J., Matsumoto, K., Ohya-Nishiguchi, H., Deguchi, Y.: J. Magn. Res. *35*, 171 (1979)
66. Kubota, S., Ikegami, Y.: J. Phys. Chem. *82*, 2739 (1978)
67. Kosower, E. M.: J. Am. Chem. Soc. *80*, 3253 (1958)
68. Reichardt, C., Dimroth, K.: Fortschr. Org. Chem. *11*, 1 (1968)
69. Abe, T., Tero-Kubota, S., Ikegami, Y.: J. Phys. Chem. *86*, 1358 (1982)
70. Dohrmann, J., Kieslich, W.: J. Magn. Res. *31*, 69 (1978)
71. Grossi, L., Minisci, F., Pedulli, G. F.: J. Chem. Soc. Perkin *2*, 943 (1977)
72. Ikegami, Y., Sawayanagi, M., Kubota, S.: Heterocycles *15*, 1027 (1981)
73. Sander, U., Dohrmann, J. K.: Ber. Bunsenges. Phys. Chem. *83*, 1258 (1979)
74. Sander, U., Dohrmann, J. K.: ibid. *84*, 270 (1980)
75. Guerra, M., Bernardi, F., Pedulli, G. F.: Chem. Phys. Lett. *48*, 311 (1977)
76. Cited in Kosower, E. M.: An Introduction to Physical Organic Chemistry, John Wiley, New York, 1968, p. 438
77. Ikegami, Y., Watanabe, H., Seto, S.: J. Am. Chem. Soc. *94*, 3294 (1972)
78. Ikegami, Y., Seto, S.: ibid. *96*, 7811 (1974)
79. Atlani, P., Biellmann, J. F., Briere, R., Rassat, A.: Tetrahedron *28*, 5805 (1972)
80. McNamara, F. T., Nieft, J. W., Ambrose, J. F., Huyser, E. S.: J. Org. Chem. *42*, 988 (1977)
81. Krohn, H., Leuschner, R., Dohrmann, J. K.: Ber. Bunsenges. Phys. Chem. *85*, 139 (1981)
82. Neumann, W. P., Reuter, K.: Chem. Ber. *112*, 936 (1979)
83. Neumann, W. P., Reuter, K.: ibid. *112*, 950 (1979)
84. Czochralska, B., Szweykowska, M., Shugar, D.: Arch. Biochem. Biophys. *199*, 497 (1980)
85. Atlani, P., Biellmann, J. F., Briere, R., Lemaire, H., Rassat, A.: Tetrahedron *28*, 2827 (1972)
86. Ikegami, Y., Watanabe, H.: Chem. Lett. 1007 (1976)
87. Kosower, E. M., Teuerstein, A.: J. Am. Chem. Soc. *100*, 1182 (1978)
88. Ikegami, Y., Kubota, S., Watanabe, H.: Bull. Chem. Soc. Japan *52*, 1563 (1979)
89. Kosower, E. M.: Progr. Phys. Org. Chem. *3*, 81 (1965)
90. Mochida, K., Kochi, J. K., Chen, K. S., Wan, J. K.: J. Am. Chem. Soc. *100*, 2927 (1978)
91. Kurz, J. L., Hutton, R., Westheimer, F. H.: ibid. *83*, 584 (1961)
92. Mohammad, M., Hajdu, J., Kosower, E. M.: ibid. *93*, 1793 (1971)
93. Grunwald, E., Winstein, S.: ibid. *70*, 846 (1948)
94. Kosower, E. M.: An Introduction to Physical Organic Chemistry, J. Wiley, New York, 1968
95. Dimroth, K., Reichardt, C.: Fortschr. Organ. Forsch. *11*, 1 (1968)
96. Reichardt, C.: Solvent Effects in Organic Chemistry, Verlag Chemie, Weinheim, 355 pp., 1979
97. Neta, P., Patterson, L. K.: J. Phys. Chem. *78*, 2211 (1974)
98. Bruhlmann, U., Hayon, E.: J. Am. Chem. Soc. *96*, 6169 (1974)
99. Kosower, E. M., Sorensen, T. S.: J. Org. Chem. *27*, 3764 (1962)
100. Weitz, E., Fischer, K.: Angew. Chem. *38*, 1110 (1925); Ber. Deut. Chem. Ges. *59*, 432 (1926)

101. Weitz, E.: Angew. Chem. *66*, 658 (1954)
102. Hünig, S., Gross, J., Schenk, W.: Liebig's Ann. Chem. 324 (1973)
103. Walsh, C.: Enzymatic Reaction Mechanisms, W. H. Freeman, San Francisco, 978 pp., 1979
104. Martens, F. M., Verhoeven, J. W., Gase, R. A., Pandit, U. K., De Boer, Th. J.: Tetrahedron *34*, 443 (1978)
105. Martens, F. M., Verhoeven, J. W.: Rec. Trav. Chim. Pays-Bas *100*, 228 (1981)
106. Brånden, C.-I., Eklund, H.: Coenzyme-induced conformational changes and substrate binding in liver alcohol dehydrogenase in: Molecular Interactions and Activity in Proteins, Ciba Foundation Symp. 60 (New series), pp. 63–80, Excerpta Medica, Amsterdam, 1978
107. Reddy, B. S., Saenger, W., Mühlegger, K., Weimann, G.: J. Am. Chem. Soc. *103*, 907 (1981)
108. Van Lier, P. M., Donkersloot, M. C. A., Koster, A. S., Van Hooff, H. J. G., Buck, H. M.: Rec. Trav. Chim. Pays-Bas *101*, 119 (1982)
109. Tropp, J., Redfield, A. G.: J. Am. Chem. Soc. *102*, 534 (1980)
110. Yasui, S., Nakamura, K., Ohno, A., Oka, S.: Bull. Chem. Soc. Japan *55*, 196 (1982)
111. (a) Dodge, A. D.: Biochem. Soc. Trans. *10*, 73 (1982)
(b) Fridovich, I.: ibid. *10*, 67 (1982)
112. Farrington, J. A., Ebert, M., Land, E. J., Fletcher, K.: Biochim. Biophys. Acta *314*, 372 (1973)
113. Nadezhdin, A., Dunford, H. B.: J. Phys. Chem. *83*, 1957 (1957)
114. McKusick, V. A.: Mendelian Inheritance in Man, Jones Hopkins Univ. Press, Baltimore, 975 pp., 1978[5]
115. (a) Lee-Ruff, E.: Chem. Soc. Revs. *6*, 195 (1977)
(b) Sawyer, D. T., Valentine, J. S.: Acc. Chem. Res. *14*, 393 (1981)
116. Lissell, M., Dehmlow, E. V.: Tetrahed. Lett. 3689 (1978)
117. Nanni, jr., E. J., Angelis, C. T., Dickson, J., Sawyer, D. T.: J. Am. Chem. Soc. *103*, 4268 (1981)
118. Pac, C., Ihama, M., Yasuda, M., Miyauchi, Y., Sakurai, H.: ibid. *103*, 6495 (1981)

Labeled Proteins, Their Preparation and Applications

Jan Káš and Pavel Rauch

Department of Biochemistry and Microbiology, Institute of Chemical Technology, Suchbátarova 5, 166 28 Prague 6, CSSR

Table of Contents

The review describes the character and properties of labels used in protein labeling (radio- and stable nuclides, chromophores, fluorophores, enzymes, biospecific, affinity and other types of labels). The ways of labeling are described and evaluated in relation to a certain label. A lot of attention is paid to the individual areas in which labeled-proteins are applied (enzymology, investigations of protein structure and mechanism of enzyme action, studies on cell morphology and organisation, metabolic studies, imunochemistry, analysis, medical diagnosis and research and the possible industrial separations of enzymes).

The survey concludes with a proposed definition of a protein label and with an account of present trends in protein labeling.

Preface

In the "Dictionary of Biochemistry" [1] a labeled compound is defined as "a compound containing a label" and this is further specified as:

a) A radioactive or stable isotope that is introduced into a molecule.

b) A group of atoms or a molecule that is linked covalently to another molecule for purposes of identification.

"Double-label experiment" is defined as an experiment using either one compound labeled with two different isotopes, or two compounds, each labeled with different isotopes [1].

All these definitions follow the generally widespread meaning that isotope labeling is the most important one (or at least the most widely used) and that a covalent bond is necessary for the attachment of a label (reporter molecule, detector, probe, marker, tracer, i.e. other terms having the meaning "label" which can be found in literature).

The aim of this review is to show that the term label must be understand in a much broader sense. Although the review is limited to proteins, the high number of labeling techniques and areas of application does not allow the authors to cover all the aspects of this problem in detail. Nevertheless, we believe that it will clearly show the diversity and wideness of application of labeled proteins in research and even in common practice.

1 Introduction

Science provides increasing support for the opinion that the majority of the regulated metabolic processes, which take place in living organisms are joined to some kind of a function, are realized by means of specifically labeled compounds.

The protein labeling taking place in organisms makes possible the interaction with a definite type of molecule (both on the surface of cells or organelles and in the biological fluids), the transport of this molecule over membrane barriers, the starting or stopping of the particular processes, the defence of the organism against specific external factors (e.g. foreign proteins, bacteria, viruses).

The majority of biologically active substances which have an important function seem to be labeled in some way. The labeling is genetically included in the structure of certain macromolecules (i.e. RNA, enzymes, hormones) or the molecules are labeled in relation to the requirement of various metabolic processes according to well established rules.

Proteins play very important role in this respect, performing a lot of fundamental functions (as enzymes, hormones, antibodies, translocators, receptors, etc.) and thus in some ways their labeling can be considered as a common phenomenon.

All enzymes are in fact labeled, at least by their active binding site for substrates, and by coenzymes (catalytic sites) which both together enable them to fulfil their very specific catalytic function. The same can be said about hormones and antibodies which are adapted for a specific interaction with a target cell or antigen by some definite mode of labeling.

In this respect protein labeling by glycosylation can be remembered as a universal feature of eukaryotic organisms. Present knowledge supports the idea that it is essential for the development of organisms of increasing complexity [2]. The most frequent labels seem to be especially the oligosaccharides with asparagine-N-acetylglucosamine type linkages or N-glycans [3].

For instance, the liver Kupffer cells and macrophages probably have on their surfaces lectin-like components which recognize a manose determinant of lysosomal glycosidase (e.g. glycoprotein enzyme) and remove them from the blood circulation [4]. A similar effect which is applied in affinity chromatography [5].

Also immunoglobin has arginin-linked oligosaccharide which enables the interaction with antigen to take place [6]. Carbohydrate speciticity also appears to play a significant role in the recognition of whole cells, which leads to many varied and more complex biological phenomena [2].

Another form of protein labeling in vivo seems to be a label consisting of a part of a protein molecule which is split off under certain circumstances, usually by specific hydrolysis. This "quite simple" way of labeling, not obvious at first sight, enables biological activity to be kept latent, hinders the formation of an oligomeric structure [7] and most probably helps the particular protein to cross the membrane barrier (mitochondrial membrane, yeast cell wall, etc.) [8]. This "temporary" labeling seems to be a very important aspect of many regulatory mechanisms occuring in organisms. From a wider point of view, all conjugated proteins should be considered as "labeled proteins", provided this "labeling" predetermines them for a specific type of a function.

This short introduction should show that protein labeling in organisms is really a common phenomenon whose importance is still beyond our full understanding. On the other hand, it is evident that suitable protein labeling is an important experimental feature which will bring many outstanding discoveries in the future. There is, of course, one exceptionally difficult task: to find suitable labels and ways of "attaching" them.

In spite of these difficulties, this survey produces a lot of arguments which show that protein labeling has been already applied in many diverse fields. It enables us to understand different metabolic events, mechanisms of protein interactions, and allows us to determine with precision many compounds which are present in very low quantities (in nanogram levels or even less). Briefly, the labeling of proteins enables us to follow them "in action" under in-vivo conditions (turnover, distribution in the body, etc.) and in vitro (i.e. assays of enzymatic activities) or "only" to facilitate their detection in tissues (e.g. histochemistry) or directly in a live body (scintigraphy).

Protein labeling has become an important tool in everyday practical life (i.e. medical diagnosis) and in pure and applied research (immunochemistry, membranology, histochemistry, enzymology, etc.). In spite of the wide range applications of this method, up to now its possibilities have by no means been fully utilized.

Further applications can be expected, mainly in the tracing of proteins in order to elucidate their behaviour in relation to different functions (in the broadest sense), and in the field of analysis (immunochemical and enzymatical assays).

2 Radionuclides Applied for Protein Labeling

G. Hevesy was the first to label proteins with radionuclides as early as 1938. The Nobel Prize was awarded to him in 1943 for his discoveries in this field [9]. Hevesy fed the radionuclide of phosphorus to sheep per os and isolated ^{32}P-casein from their milk.

The next label, first used in vitro experiments, was ^{35}S and was applied by the chemists of American army in the course of World war II. They labeled proteins with ^{35}S-mustard gas, nowadays considered a rather curious reagent. These secret results were published, after some delay, in the Biochemical Journal in 1946 [10]. At about the same time, the first productive atomic reactor was put into operation at Oak Ridge.

Table 1. Basic properties of radioactive labels of proteins [120,121]

Nuclide	Radiation	Max. energy (MeV)	Half life[a]	Type of counting	Counting efficiency (%)
^{3}H	beta$^-$	0.018	12.26 y	LSC	60
^{14}C	beta$^-$	0.158	5700 y	LSC	90
^{32}P	beta$^-$	1.71	14.3 d	IM, LSC	50, 95
^{35}S	beta$^-$	0.167	87 d	LSC	90
^{51}Cr	EC	0.76	27.8 d	LSC	15–30
	gamma	0.32		NaI	2– 6
^{57}Co	EC	0.122	270 d	LSC	60–70
	(gamma)	0.122		NaI	60–80
^{55}Fe	EC	0.22	1.7 y	LSC	20
^{59}Fe	beta$^-$	0.46 0.27 1.56	45.1 d	LSC, IM	80
	gamma	1.10 1.29		NaI	40
^{63}Ni	beta$^-$	0.067	120 y	NaI	70
^{75}Se	EC				
	gamma	0.265	120 d	NaI	65–90
^{77}Br	EC		57 h	NaI	65
	beta$^+$	0.34			
	gamma	0.52			
^{99m}Tc	gamma (isomeric transition)	0.140	5.9 h	NaI	60–80
^{111}In	EC	0.247 0.173	2.81 d	NaI	40–60
^{125}I	EC (gamma)	0.035	57.4 d	NaI, LSC	70–80
^{131}I	beta$^-$	0.61	8.05 d	IM, LSC	50–80
	gamma	0.364		NaI	60–70
^{211}At	EC	0.062	7.2 h	NaI	40–60
	(gamma)	0.67			
	alpha	5.86			

a y — year, d — day, h — hour, LSC — liquid scintillation couting, IM — ionization methods, NaI — crystal scintillation

From that point, on the first artificial radionuclides ^{3}H, ^{14}C, ^{32}P, ^{35}S, ^{125}I, ^{131}I and some others became commercially available and provided the material basis for protein labeling. An almost exponential increase in the study of protein labeling with radionuclides followed.

Such a brief historical diversion can provide, besides some peculiarites, interesting information. Jean Loiseleur [11] who studied, first of all, in 1935 the interactions of proteins with radionuclides, found that ammonia, hydrogen sulphide and urea are released as a result of this interaction. We point it out here because the majority of subsequent studies, right up till the present, try to demonstrate successful results in protein labeling (i.e. little or no alternation in the protein structure and the maintenance of biological activity). In fact, the process of labeling is always very risky in view of possible protein denaturation and degradation.

The great number of radionuclides that have been applied in protein labeling is summarized and characterized in Table 1. The majority of radioactive labels are beta and gamma emitters. Radionuclides with gamma emission are more convenient from the radiometric point of view (i.e. from the standpoint of experimental evaluation). The main advantage of gamma emitters is a relatively simple way of measuring their radioactivity. Special sample treatment is usually not necessary. The measurement is very convenient especially when a high number of samples is assigned for counting. A lot of simple counters are now available at prices which are not higher than those of the quite common laboratory devices, such as photometers, fluorometers and so on. On the other hand, the radioactivity measurement of beta emitters requires, in general, time-consuming sample preparation, liquid scintillators as well as expensive liquid scintillation counters. These are the reasons why the most widely used radionuclides are those of iodine in biochemistry and ^{99m}Tc in medicine.

2.1 Radionuclides of Iodine

Radionuclides ^{125}I and ^{131}I have short half-lives (60 and 8 days respectively). Higher specific radioactivity can be therefore achieved more easily with iodinated preparations than with those labeled with "more natural label" such as ^{3}H and ^{14}C. The specific radioactivity of iodinated proteins is up to 75 times higher than that of tritiated proteins and up to 35000 times higher compared with carbon labeling [12]. Radionuclide ^{125}I is the more suitable of the two above-mentioned iodine radionuclides because the isotopic enrichment is about 5 times higher (about 95%). It also has a 7.5-fold longer half-life and it also has the advantage of being a limited health hazard.

^{123}I has been applied recently in medicine [13–16] where it causes lower radiation injury to the organism compared to the other iodine isotopes.

The main advantage of all iodine radionuclides is their relatively easy introduction into the protein molecule that is to be labeled, whose structure is only slightly changed as a result of the labeling. Iodine binds by electrophilic substitution onto tyrosyl and histidyl residues.

Bolton [12] describes five methods of protein iodination: with chloramine T (N-monochlorderivative of p-toluensulfonamide) [17], monoiodochloride [18], electrolytically [19,20], with other oxidizing agents such as Iodogene and with enzymes [21–23].

The most widespread method is chloramine T oxidation. Chloramine T is slowly decomposed in aqueous solutions forming hypochloric acid which acts as a mild oxidizing agent. Sodium iodide in presence of chloramine T is oxidized at pH 7.5 to I^+. The phenyl group of the tyrosyl side chain is ionized only very slightly (pK is higher than 10), but sufficiently for the course of electrophilic substitution of I^+ onto phenyl residue [12]. Monoiodotyrosine has lower pK (about 8.5) and thus it is more ionized at the pH of the reaction. Therefore, it is also less reactive for diiodo-substitution. The reaction course is also affected by the microenvironment of the individual tyrosyl residues in the labeled protein molecule. Tyrosyl residues situated on the molecule surface are more accessible to iodination. One iodine atom per one mole of labeled protein is generally considered as suitable and quite satisfactory labeling. It means, of course, that with the increasing relative molecular weight the specific radioactivity decreases.

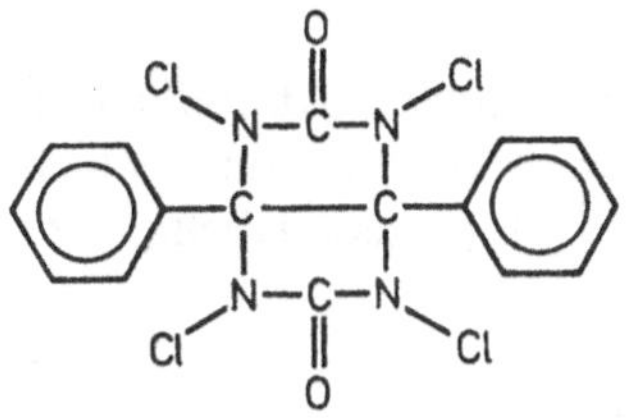

Fig. 1. Formula of IODO-GEN
1,3,4,6-Tetrachloro-3α,6α-diphenylglycouril

Iodogene, a very mild oxidizing agent (1,3,4,6-tetrachloro-3-alfa,6-alfa-diphenyl-glycouril) has been used since 1978 for iodine labeling of isolated proteins and proteins of cell membranes (Fig. 1). It is insoluble in water and enables fast iodination with iodine in the solid phase and the protein in aqueous solution. The side reactions are very limited. The iodination is terminated by simply decanting of the solution of iodinated protein from the solid reagent. A reducing agent is not used. This labeling method yields lower specific radioactivity, but high biological activity. By means of gel chromatography, isoelectric foccusing and immunological reactivity, Paus et al. [236] showed that protein iodinated with Iodogen appeared to be structurally unaltered. Their experience is in agreement with other authors [237–239]. Thus Iodogen application seems to be very advantageous for the labeling of proteins used in RIA.

Electrolytic iodination has also been developed and the procedure has been miniatured. Micro-electrolytic iodination was recently described [24]. Neither oxidation changes of the protein nor changes in biological and immunological activity occur. This procedure also makes it possible to reach high specific radioactivity and high reproducibility.

The iodination with lactoperoxidase is also very gentle and very often used [25]. Because the application of an enzyme for oxidation is more expensive than the use of common oxidation reagents, procedures involving immobilized lactoperoxidase have been worked out [21,22,26].

Commercial kits for protein labeling based on the application of immobilized lactoperoxidase have appeared recently (Fig. 2). For instance, Iodo-beads of Pierce Eurochemie B. V. or Enzymo-beads of Bio-Rad Laboratoires are on the market [27,28].

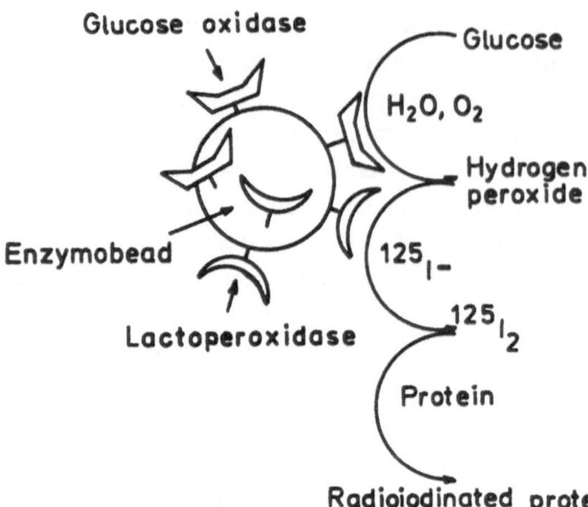

Fig. 2. Schematic diagram of protein iodination with Enzymobeads (by kind permission of Bio-Rad-Laboratories)

These kits bring a lot of advantages for protein labeling with iodine. All iodination procedures can be easily controlled, product contamination is eliminated (which occurs when soluble enzyme is used) and higher yields of protein labeling are achieved (more than 90%).

Table 2. Examples of proteins labeled with iodine radionuclides

Labeled protein	Labeling technique	Ref.
^{125}I (nuclide[a])		
fibrinogen	electrolytic	19,47)
fibrinogen	chloramin T	42,44,46)
fibrinogen	iodine monochloride	30,44)
fibrinogen	Iodogen	30)
fibrinogen	enzymic	26,47)
fibrinogen	comparison of various methods	45,48)
human serum albumin	electrolytic	51)
polypeptide hormones	micro-electrolytic	24)
secretin porcine	comparison of various methods	53)
ferritin	comparison of various methods	29)
transferrin	comparison of various methods	54)
human thyroxine-binding alpha globulin	enzymic	59)
bovine parathormone	enzymic, chloramine T	60)
bovine growth hormone	electrolytic	61)
protein hormones	enzymic	63,64)
insulin	comparison of different methods	65)
globulins	comparison of various methods	67)
antibody	chloramin T	68,72,73)
collagen	Bolton-Hunter reagent	70)
apolipoprotein A—I	chloramin T	71)
transmembrane protein	iodosulfanilic acid	75)
transmembrane protein	enzymic	76)
proteins (generally)	iodo-p-hydroxybenzimidate	74)

Table 2. (continued)

Labeled protein	Labeling technique	Ref.
^{131}I (nuclide[b])		
human serum albumin	chloramine T	77)
fibrinogen	chloramine T	80)
fibrinogen	iodine monochloride	81)
^{123}I (nuclide[c])		
fibrin	iodine monochloride	13)

a fibrinogen [41,43,49,50)], gastrin [55)], chromatin [56,], concanavalin A [57)], cyclosporin A [58)], tissue thromboplastin [62)], angiotensin I [66)], alpha foetoprotein [69)]
b human serum albumin [78,79)], orosomucoid [15)], fibrin canine [82)]
c fibrinogen [14)], orosomucoid [15)]

Note: For other examples see Bolton [12)] and commercial catalogues

Examples of proteins labeled with iodine radionuclides are given in Table 2. Many authors have been concerned with a mutual comparison of the individual methods of protein iodination [29-34)]. Their results show that the enzymatic method (with lactoperoxidase) gives the best results for the majority of proteins tested. A very important step in protein iodination is the separation of labeled protein from unreacted iodine. This problem has been studied in detail by Bolton [12)]. Gel and ion-exchange chromatography are generally recommended [35)]. Recently rapid centrifugation [36)] and high-voltage electrophoresis [37)] were also found very useful. For all labeling procedures, i.e. labeling and separation of labeled protein, standardized equipment has been developed [38,39)].

2.2 Radionuclide ^{99m}Tc

Technetium is a very suitable radionuclide for medical applications. Owing to its short half-life, radiation doses to the human organism are very low. The problems of protein labeling with technetium have been reviewed by several authors [83-91)]. The following methods of technetium reduction are used in protein labeling: by ascorbate alone [84,85)]; by ascorbate and Fe^{3+} [92)]; by Fe^{2+} [84,85)]; by Sn^{2+} [84,85,93-95)]; electrolytic reduction [99-101,84,85)]; by Sn^{2+}-tartarate [102,103)]; by Sn^{2+}-citrate [104)] (Tab. 3).

Table 3. Examples of proteins labeled with ^{99m}Tc

Labeled protein	Way of TcO_4^- reduction	Ref.
bovine trombin	Sn^{2+} chloride	109)
plasmin	Sn^{2+} chloride	109,115)
human gamma globulin	Sn^{2+} chloride	111)
fibrinogen	electrolytic	98,112,116)
human serum albumin	electrolytic	100,101,117,118)
human serum albumin	Sn^{2+} chloride	113,114,119)
transferrin	Sn^{2+} citrate	110)
human serum albumin	Sn^{2+} citrate	104)
human serum albumin	Sn^{2+} tartrate	102,103)

The most common methods are Sn^{2+} and electrolytic reduction. The presence of impurities causes the main difficulties in the preparation of proteins labeled with technetium. Thus the quality control and the assay of product contamination with tin colloid is the subject of several papers [93,94,105]. Cífka [106] studied the contamination of technetium which does not react with protein. He found that pertechnate is not the only possible impurity in the case of colloids, but may be present as a soluble tartrate complex in the case of antimony sulphide colloid or as a soluble chlorocomplex in the case of the sulphur colloid.

Commercial kits for technetium labeling are also available [107,108]. The whole problem of labeling has been very well worked out [83,84,88–90,96,97,107,108].

2.3 ^{3}H and ^{14}C Radionuclides

Besides gamma emitters of iodine and technetium ^{3}H and ^{14}C radionuclides, representative of β-emitters, are often utilized in protein labeling.

2.3.1 ^{3}H Radionuclide (Tritium)

Tritium as a low energy beta-emitter is most conveniently measured by liquid scintillation counting. In general, the tritium measurement is the most complicated of all the radionuclide labels. The difficulties are related to the character of the sample which is tritium labeled. Tritium labeling is not suitable in all cases in which the system yields high chemiluminiscence; the samples are coloured or turbid. Recently, however, new types of liquid scintillation counters appeared which make it possible to count tritium even under such complicated conditions (Beckman, Packard, Kontron, Berthold).

Proteins are relatively easy to label by exposure to tritium gas. This method was first described by Hembree et al. [122]. According to their procedure, the protein is situated in a thin porous membrane filter of polymer material (cellulose ester) 0.5 μm thick. Protein solution in the membrane filter is then exposed to $^{3}H_2$ using microwave discharge. Isotopic exchange (H → T) occurs. The labeled protein is then purified by chromatography.

An alternative procedure has been described by Tscheske et al. [123]. The protein sample (2–3 mg) is irradiated by an electron beam or ultra-violet light. The isotopic exchange is carried out at the temperature of boiling point of nitrogen for about 20 min. The labeled protein is purified by gel chromatography and retains about 60–100% of the original biological activity. Approximately one in a hundred or a thousand molecules is ^{3}H labeled. This method was adopted for labeling many organic compounds of biological interest [124–126]. It was used for introducing ^{3}H into compounds of molecular weight between 10^2 and 10^7 daltons, resulting in a specific activity between 3.7 and 18.5 KBq/μg. For sensitive biological molecules a more gentle tritiation procedure has been described [127]. This method is performed by depositing the substance to be tritiated on a supporting substrate in an evacuated vacuum chamber near, but not in the path of, an electron beam which traverses the chamber, by admitting tritium gas into the chamber, and subjecting it to the electron beam. Vibrationally excited tritium gas collides and reacts with the substance, thus incorporating tritium atoms into the substance.

Another tritium labeling method is the reductive methylation [128,129]. ^{3}H-formaldehyde is allowed to combine with amino groups of a protein and is then reduced using sodium borohydride. The details will be given later in a paragraph concerned with ^{14}C labeling.

Some important methods of protein labeling utilize various conjugation reagents [130-136] (Tab. 7). The common advantage of these methods is that they are relatively easy to perform. Special devices such as a microwave discharge system are not necessary for the course of the chemical reaction, as they are in the above-described methods. On the other hand a new functional group (i.e. conjugation reagent) is introduced into the protein molecule and thus the molecular structure is altered to some extent.

Table 4. Examples of ^{3}H-labeled proteins

Labeled protein	Way of labeling	Ref.
beta$_2$-micro globulin	microwave discharge	124)
protein hormones	microwave discharge	125)
galactose oxidase	reductive methylation	128)
galactose oxidase	acetic anhydride	128)
luteinizing hormone	reductive methylation	129,139,140)
collagen	acetic anhydride	130)
lipoproteins	acetic anhydride	131)
bovine fibrinogen	acetic anhydride	132)
histones, antibodies	N-succinimydyl-(2,3-3H)-propionate	133,134,136)
intracellular proteins	N-hydroxysuccinimide ester	134)
proteins (generally)	methyl acetimidate	135)
glycoproteins	fucose, glucoseamine	137)
bradykinin	microwave discharge	138)
peptides	^{3}H-phenylalanine	141)

Other possibilities of protein labeling have been described. For instance, glycoproteins can be labeled with ^{3}H-fucose or ^{3}H glucosamine [137]. Examples of ^{3}H-labeling methods are included in Table 4.

It must be noted that ^{3}H is a relatively labile label and can be reversibly exchanged with hydrogen atoms. The results obtained with the ^{3}H-label must therefore be carfully verified.

2.3.2 ^{14}C Radionuclide

^{14}C-Carbon is now utilized more often in accordance with the development of new labeling techniques which involve so-called natural labeling, i.e. ^{14}C replaces ^{12}C by isotopic exchange in molecules of the proteins to be labeled. The application of ^{14}C is more convenient than tritium labeling. Radiocarbon may be measured by various gas flow counters and semiconductor detectors, but liquid scintillation counting is still preferred.

Table 5. Examples of proteins labeled with ^{14}C

Labeled protein	Way of labeling	Ref.
gliadin	acetic anhydride	[143]
proteins, viruses	acetic anhydride	[142]
proteins	formaldehyde	[144]
luteinizing hormone	reductive methylation	[145]
initiation factor-3, elongation factor Tu	reductive methylation	[146, 147]
proteins	ionized CO gas	[151, 152, 153]
proteins with methionine residues	methylation	[154]
collagen	acetic anhydride	[159]
glykoproteins	glucoseamine, galactose, mannose, fucose	[155]
algal protein	CO_2	[156]
proteins	2,4-dinitrophenol	[157]
proteins	irradiation + ^{14}C-tryptophan	[158]

Protein labeling with ^{14}C radionuclide is most often performed in vivo by utilizing ^{14}C-amino acids (see 4.1) or in vitro by using various conjugation reagents (Table 5). The method most frequently applied is protein acetylation because consequent alterations of protein structure and biological activity are limited [142, 143].

Reductive methylation is another important method, and one in which changes occurring in protein molecules as a result of labeling are also insignificant. This method is very fast and thus enables protein labeling to be carried out even with the short-lived ^{11}C radionuclide [144, 145]. The protein to be labeled is treated with ^{14}C formaldehyde and sodium borohydride at 22 °C for about 2 hours. This method of labeling is one of the extremely mild procedures [149] and enables a product with high specific radioactivity to be obtained [146–149]. Reductive methylation has been studied in detail by Koch et al. [150].

Natural protein labeling is achieved by a method which utilizes ^{14}CO ionized gas. This method is also very fast and applicable to protein labeling either with ^{14}C [151, 152] or with ^{11}C [153]. The apparatus used consists of a high vacuum system in which a small amount of ^{14}CO is ionized by electron impact [151]. The resulting species drift towards a targer where they interact with the protein molecule to produce ^{14}C-labeled compounds. Since the reaction time is only two minutes, the method is particularly promising for producing labeled proteins with short-lived ^{11}C at high specific radioactivity. Chromatographic purification of the labeled protein is necessary. The yields usually vary in the range of 370–3700 KBq/mol. For details see, for example, Sanche and Van Lier [153].

Another new and fast procedure which has recently been described utilizes the ^{11}C-methylation of methionine residues of protein molecules [154].

For the introduction of ^{14}C into proteins other, less frequent, methods have been applied, e.g.: glycoprotein labeling using saccharide components such as glucosamine, mannose, galactose and fucose [155], in vivo labeling of algae proteins with $^{11}CO_2$ [156]; protein labeling with ^{14}C-2,4-dinitrophenol [157]; or by irradiation in presence of ^{14}C-tryptophane [158].

2.4 Other Radionuclides

Besides iodine, technetium, tritium and carbon, a number of other radionuclides have been utilized in protein labeling (see Table 6).

Table 6. Examples of proteins labeled with less frequent radionuclides

Radionuclide	Labeled protein	Ref.
^{32}P	brain phosphoproteins	160)
	proteins from rat liver nuclei	161)
	proteins from rat ovarian plasma membranes	162)
^{35}S	caerulein	164)
^{51}Cr	chicken erythrocytes	165)
	platelet	166)
^{57}Co	serum albumin	167)
^{59}Fe	rabbit hemoglobin	169)
^{67}Ga	albumin microspheres	172)
^{68}Ga	albumin microspheres	171, 173, 174)
	albumin	171)
^{75}Se	lymphocyte membrane antigens	176)
^{77}Br	albumin, thyroglobulin	177)
^{99}Mo	human erythrocyte membrane	183)
^{111}In	human serum albumin, bovine fibrinogen	186, 187)
	rabbit serum albumin, bleomycin	186)
	human platelets	189)
	antibodies to canine cardiac myosin	190)
	albumin microspheres	188)
^{113}In	albumin macroaggregates	188, 191, 193, 194, 195)
	albumin microspheres	192)
^{152m}Eu	albumin microspheres	116)
^{197}Hg	fibrinogen	197)
^{203}Hg	phosphate-transporting protein	198)
^{211}At	proteins	199, 200)
	globulins	202)
	concanavalin A	203)

^{32}P was the first radioactive label used for labeling proteins in which phosphorus regularly occurs. Although a lot of other methods of labeling are at present available, ^{32}P is still used as a protein label in certain cases, e.g. it has been applied for labeling brain proteins [160)], proteins from rat liver nuclei [161)] and rat ovarian plasma membranes [162)].

^{35}S radionuclide is used as protein label in the form of ^{35}S-methionine for labeling in vivo [163)], and in the form of various conjugation reagents, e.g. ^{35}S-pyridine-N-sulfonate [164)].

^{51}Cr is considered a very firm label. The way of labeling is relatively easy. ^{51}Cr has been used for labeling erythrocytes [165)], platelets [166)], etc.

^{57}Co has been used for the labeling of bovine serum albumin. As a result of labeling, a complex consisting of a peptide bond, cobalt and ethylene diamine or triethylene-tetramine is formed [167, 168)].

[^{59}Fe]-Iron is most frequently used for hemoglobin labeling [169].

Binding of ^{63}Ni to rabbit serum α_1-macroglobulin in vivo and in vitro has been studied [170].

^{67}Ga and ^{68}Ga is applied for labeling of albumin [171] or albumin microspheres [172]. Albumin microspheres have been synthesized by binding EDTA and DTPA chelating groups covalently to their surfaces. The microspheres may be labeled with ^{67}Ga at high yield (97 ± 2%) by transcomplexation from a 0.1 M ^{67}Ga-acetate solution [172]. The use of short-lived ^{68}Ga radionuclide (half-life 68 min) is very advantageneous for medical applications because of reducing the health hazard from irradiation [173,174].

^{75}Se: Because the proteosynthetic system is able to utilize methyl-^{75}S-cysteine and/or methyl-^{75}S-methionine for protein synthesis, both these amino acids have been applied for protein labeling in vivo. [175,176]

^{77}Br: Iodination of proteins may be replaced by bromation. The ^{77}Br-label is considered to be more stable than those of the radionuclides of iodine [177–180]. It splits off by positron decay and has a short half-life (Table 1). Proteins are labeled enzymatically either by bromperoxidase from the microorganisms *Pecillus capitatus* and *Bonnemaisonia hamifera* [181] or by myeloperoxidase [182].

^{99}Mo has been used for labeling of human erythrocyte membrane [183,184].

^{110}Ag: Serum albumin has been labeled with $^{110}AgNO_3$. The radionuclide combines to SH-groups of protein [185].

[^{111}In] and [^{113m}In] Indium, like technetium, was first employed in medicine. Labeling with indium is facilitated by the application of bifunctional chelates [186]. The use of a new class of chelating agents for the binding of metal ions to macromolecules as a novel approach to radiopharmaceutical labeling was described by Goodwin and Diamanti [187]. Human serum albumin and bovine fibrinogen labeled with ^{111}In or ^{133m}In, using this technique, were relatively stable in vitro and in vivo, showed little functional alteration, and have potential as tumor localizing agents in humans [188]. ^{111}In-8-hydroxyquinoline and ^{111}In-diethylentriamine pentaacetic acid may be applied for indium labeling, too [189,190]. The preparation of albumin macroaggregates, as new lung scanning agents, containing ^{113m}In-sulfide has been described [191] as consisting of albumin microspheres labeled with ^{113}In [192–195].

[^{152m}Eu]: Europium is being utilized as a label in a new method which has been developed for producing labeled albumin microspheres. This method is based on the incorporation of stable elements into the microspheres with subsequent neutron activation. In this way microspheres with high specific activity (up to 185 GBq/g for ^{152m}Eu microspheres) are obtained. The achievable specific activity is higher by one order of magnitude in comparison to other methods. This new method not only enlarges the number of isotopes applicable to the labeling of proteins or microspheres, but also simplifies the preparation and reduces the radiation doses to the manufacturing personnel [196].

^{197}Hg and ^{203}Hg: The procedure for fibrinogen labeling with a radionuclide of mercury was described [197]. Human fibrinogen, both commercial and from own production, has been labeled with ^{197}Hg. The effects of various physico-chemical factors on the labeling yield were assessed. An optimum yield of 90 to 95% was obtained by incubating 10 mg fibrinogen with 74 KBq of $^{197}HgCl_2$ at room temperature and pH 7.4 for 30 minutes.

In mitochondrial research the phosphate-transporting protein from rat liver mitochondria has been labeled with ^{203}Hg-mersalyl [198].

^{211}At: For protein labeling with astatine (alpha emitter) the following procedures may be used: reaction of the protein with p-astatobenzoic acid [199]; condensation reaction with peptide bond and protein acetylation [200]. While labeling by the above procedures seems to be sufficiently stable [201], a remarkable instability of the ^{211}At-label was observed when astatinated protein was prepared electrophoretically [202]. The results of these authors indicate that the tyrosine-astatine bond is unstable. The conclusion of Vaughan et al. [203] that astatinated proteins lose as much as 50% of their biological activity and, in addition, are extremely toxic, is very serious.

3 Stable Nuclides in Protein Labeling

Besides radionuclides, stable nuclides are also utilized in protein labeling. The reason is that some elements do not have a suitable radionuclide. In such cases, a stable nuclide must be applied for protein labeling in spite of difficulties with its detection.

Mass spectrometry or nuclear magnetic resonance are expensive methods and thus the use of stable nuclides is a privelege of well-situated, usually large, institutions. The most important studies in this field have been performed in the Argonne National Laboratory [204]. For protein labeling, mainly in vivo, nuclides D, B, ^{13}C, ^{15}N are applied (see 7.4). Proteins with incorporated amino acids labeled with deuterium [205], 4-boronophenylalanine [206] and ^{13}C-amino acids were prepared in this way.

4 Ways of Protein Labeling with Radionuclides

Proteins may in principal be labeled in three different ways (in vivo, "natural" and "foreign" procedures in vitro), which will be briefly described.

4.1 Labeling in Vivo (Naturally)

The most "natural" form of protein labeling is labeling in vivo, i.e. to achieve this amino acids or their precursors labeled with ^{3}H, ^{14}C and ^{35}S radionuclides are most frequently utilized. For instance, of the tritiated amino acids ^{3}H-Leu [207,208], ^{3}H-Lys [209], and ^{3}H-Pro [210] have been used. ^{14}C-Leu [211,212], ^{14}C-Lys [213] and ^{14}C-Met [214,215], as well as acetyl-L-(U-^{14}C) aspartate [216] are the most common representatives of ^{14}C-labeled amino acids. The production of labeled proteins and peptides using radioactive amino acids was studied by Zaidenberg et al. [217]. Of the sulfur-containing amino acids ^{35}S methionine is the most frequently used [163,218]. Radionuclide ^{75}Se may be also used as a component of methyl selenocysteine, which incorporates itself into proteins and peptides where cysteine is normally present. Similary ^{32}P may be incorporated into phosphoproteins (e.g. casein).

In general, all 18 biogenic L-amino acids may be used for labeling proteins. They may be combined for so-called double labeling, and even triple labeling is sometimes used as well. (The differentiation of ^{14}C and ^{35}S is possible, e.g. with a semiconductor detector)[219,220].

Besides amino acids, labeled saccharides are suitable for the labeling of glycoproteins. In this manner, glycoproteins have been labeled with ^{3}H-mannose[221], ^{3}H-glucosamine[222], N-acetylmanosamine[223]. Saccharides may also be allowed to react with proteins after periodate oxidation[222].

Labeling of proteins in vivo has also been performed by asimilation of $^{14}CO_2$[224,225] or by the using ^{14}C-acetate[226] or ^{35}S-sulfate[227–229]. Labeling of lymphocytes with various radioisotopes (^{3}H, ^{14}C, ^{125}I, ^{131}I, ^{35}S, ^{75}Se, ^{51}Cr and ^{99m}Tc) for in vivo tracer studies was reviewed by van Rooijen[230]. To the above radionuclides the application of stable nuclides such as D, ^{13}C and ^{15}N must be added[231].

Besides specific labeling also nonspecific labeling may be used under in vivo conditions. Proteins have been labeled in this way by ^{99m}Tc[232], ^{125}I[232,233] and ^{63}Ni[170]. Nonspecific labeling is carried out either by labels in inorganic form or by use of nonspecific reagents such as ^{14}C-trichlorethylene[234]. By means of the above-described methods, proteins labeled with radioactive or stable nuclides are formed in the organism.

In vivo labeling procedures yield a large spectrum of labeled compounds, not only proteins. Thus, this method of labeling is not suitable for the preparation of single-labeled proteins, because their isolation would be difficult. In vivo labeling creates some other problems, e.g. protein labeled with ^{14}C is subjected to "destructive" decay if the carbon atom, participating in peptide bond, is split off ($^{14}C \rightarrow {}^{14}N$). On the other hand, the main advantage of protein labeling in vivo is that the original structure and biological activity may be kept. There are no noticeable differences between labeled and unlabeled proteins.

4.2 "Natural" Labeling in Vitro

Labeling in vitro may be also natural. Suitable methods have been developed in last ten years.

A protein can be labeled by isotopic exchange under cooling either with tritium gas or ^{14}CO[153]. Another possible way to label a protein is by irradiating it with a proton beam which has an energy of 185 MeV. This activation enables protein to be labeled with ^{11}C and ^{13}N at about 20% yield[235]. These methods are, of course, relatively risky, nevertheless they have the great advantage that any foreign functional group can introduced into the protein molecule. This is the reason why these methods of labeling in vitro are called "natural".

4.3 "Foreign" Labeling in Vitro

"Foreign" labeling in vitro is certainly very widely used. The most important labeling methods which have been applied for labeling almost all important proteins are halogenation (iodination or bromation)[12] and conjugation (reductive methylation)[148], alkylation and acylation[130–136].

Halogenation has been already described (see 2.1). In conjugation methods a radionuclide is included in the molecule of a conjugation reagent which then reacts with the protein to be labeled. The label introduced into the protein molecules affects their behaviour to different degrees. Protein isolation is not necessary, but, some form of separation of the protein from the reactants used for the labeling is almost always required. Radiolysis is, under conditions of this method of labeling, limited. All these considerations must be taken in account when the results obtained with these radiolabeled proteins are interpreted.

For instance, if the protein has several biological activites, labeling affects them in different ways; while the substrate binding site is damaged the immunological activity may be maintained or reversed. If the breakdown of a given protein is to be followed, a similar rate of decomposition of both labeled and unlabeled proteins is required. On the other hand, if the same protein is assayed by some sensitive immunological technique (e.g. RIA) then the rate of their catabolism is of no interest, but total retention of the immunological response (i.e. no changes in the immunological determinants) is completely necessary.

It can be concluded that the application of labeled proteins for certain kinds of research requires the condition that both the labeled and the unlabeled proteins behave, under the same circumstances, in similar (or better identical) ways. It is not so easy to fulfil this condition as it might seem at first sight.

Conjugation reagent (if it is not commercially available) must first be labeled with a suitable radionuclide. The most frequently applied radionuclides are again ^{125}I, ^{131}I, ^{3}H, ^{14}C and ^{35}S. The other radionuclide labels are used only ocasionally. The reagent, after labeling, is purified by chromatography, and only then may it be used for protein labeling. The yields achieved with conjugation methods are lower compared with those of direct labeling and thus preparations with high specific radioactivity are difficult to obtain. It is possible, however, to prepare a conjugation reagent containing several radionuclide atoms in its molecule. These methods are used for labeling proteins which do not contain tyrosine or when direct iodination of the tyrosyl group would lead to a loss of biological activity [12]. From chemical viewpoint the conjugation methods are very mild and thus any serious damage of the labeled protein cannot occur. The presence of a radiolabel outside of the peptide chain (i.e. the radionuclide is present in side chains of amino acid residues) eliminates the possibility of destructive decay.

Conjugation reagents bind to protein functional groups: NH_2-, SH-, OH-, phenol, etc. Reagents which react with the amino groups are usually preferred. The utilization of SH-groups could be risky because they can be included in the catalytic site of an enzyme or participate in some other way in the biological activity of protein to be labeled protein cannot occur. The presence of a radiolabel outside of the peptide active site.

Some reagents bind to several functional groups of proteins, e.g. dansylchloride, acetic anhydride. The list of the most frequently used conjugation reagents is given in Table 7. The most important ones are briefly characterized below.

Acetic anhydride

Acetic anhydride is a nonspecific acylating reagent that may be used to label serine, lysine, threonine, cysteine and N-terminal residues [240–245].

Bolton-Hunter reagent

First prepared by Rudinger and Ruegg [246], the Bolton-Hunter reagent is one of the most frequently used conjugation reagents. Its iodinated form has been prepared by Bolton and Hunter [247]. Chemically it is N-succinimidyl-3 (4-hydroxy,5-^{125}I iodophenyl) propionate. The N-succinimidyl group of this reagent condenses with the free amino group (therminal or lysyl) of the protein and forms a conjugate containing a radionated fenyl group covalently bound to the protein by an amide bond. The Bolton-Hunter reagent has been recommended for the iodination of sensitive proteins and proteins not containing tyrosine residues. Labeling with this reagent takes

Table 7. Survey of the most frequent conjugation reagents used for protein labeling

Reagent	Radionuclide	Ref.
Reacting with NH_2-groups:		
acetic anhydride	^{3}H, ^{14}C	240)
Bolton-Hunter reagent	^{125}I	247)
dansyl chloride	^{3}H, ^{14}C	253)
diazobenzenesulphonate	^{35}S	245)
dimethylsuberimidate HCl (1,8-^{14}C)	^{14}C	254)
ethyl acetimidate	^{14}C	255)
1-fluoro-2,4-dinitrobenzene	^{3}H, ^{14}C	257)
fluoro-3-nitrophenylazide, 4-(2,6-^{3}H)	^{3}H	259)
formaldehyde	^{14}C	258)
iodosulfanilic acid	^{125}I	260)
isethionyl acetimidate	^{14}C	260)
maleic anhydride	^{14}C	261)
methyl-3,5-diiodohydroxybenzimidate	^{125}I	262)
phenyl isothiocyanate	^{14}C, ^{35}S	263)
potassium borohydride	^{3}H	264)
sodium borohydride	^{3}H	265)
succinic anhydride	^{14}C	266)
N-succinimidyl propionate	^{3}H	267)
succinimidyl-4-azidobenzoate, N-(benzoate-3,5-^{3}H)	^{3}H	268)
succinimidyl-3-N-(4-azido-2-nitrophenyl)-2-aminoethyl-dithio-N-propionate(cysteamine-^{35}S)	^{35}S	269)
Reacting with SH-groups:		
acetic anhydride	^{3}H, ^{14}C	240)
bromoacetic acid	^{14}C	270)
chloroacetic acid	^{14}C	271)
p-chloromercuribenzene sulphonic acid	^{203}Hg	272)
p-chloromercuribenzoic acid	^{203}Hg	273)
dansyl chloride	^{3}H, ^{14}C	253)
N-ethyl maleinimidate	^{3}H	256)
iodoacetamide	^{125}I	256)
iodoacetic acid	^{125}I, ^{14}C	256)
Reacting with phenolic hydroxyl groups:		
acetic anhydride	^{3}H, ^{15}C	240)
dansyl chloride	^{3}H, ^{14}C	253)
iodine	^{125}I, ^{131}I	12)
Reacting with imidazole groups:		
dansyl chloride	^{3}H, ^{14}C	253)
iodine	^{125}I, ^{131}I	12)
Reacting with aliphatic hydroxylgroups:		
acetic anhydride	^{3}H, ^{14}C	240)
(1,3-^{3}H)di-isopropyl phosphorofluoridate	^{3}H	274)

place under mild conditions and avoids exposure of the protein to the reactive reagents commonly used in radioiodinations. The chemical or antigenic behaviour of the labeled protein may not therefore be identical to that of the same protein labeled by traditional methods [248-252].

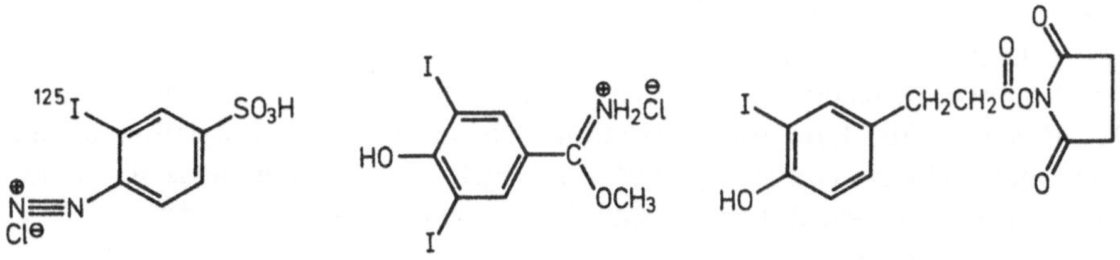

Fig. 3. Formulae of most common conjugation reagents [^{125}I]-Iodosulfanilic acid (left)
Methyl 3,5,-di-[^{125}I]-iodohydroxybenzimidate (middle)
Bolton and Hunter reagent:
N-succiimidyl 3-(4-hydroxy,5-[^{125}I]-iodophenyl)propionate (right)

Dansyl chloride

Dansylchloride reacts with amino, thiol, imidazole, phenolic and hydroxy groups. In general, the reaction with aliphatic hydroxyl groups is very slow. This reagent is used widely to detect very small amounts of protein by means of the intense fluorescence of the sulfonamide formed when dansyl chloride reacts with the terminal amino group of protein [253].

Dimethylsuberimidate. HCl (1,8-^{14}C)

It is a homobifunctional, noncleavable reagent, which has been in use as a cross-linker since 1970 and is commonly used in determining the quarternary structure of proteins [254].

Ethyl acetimidate

Ethyl acetimidate reacts specifically with free amino groups of proteins under relatively mild conditions. It penetrates cells without impairing membrane function, and labels the protein under physiological conditions. This reagent is rapidly hydrolysed by water, but the products do not interfere with the main reaction [255]. On the other hand, a sulfonated derivative of the same compound labels the outer side of the membrane and does not penetrate inside.

Formaldehyde

Formaldehyde is used primarily with a reducing agent such as sodium cyanoborohydride in reductive methylation of free amino group [258].

Fluoro-3-nitrophenylazide,4-(2,6-^{3}H)

This is a protein labeling reagent which readily reacts with amines and other nucleophilic functional groups. This reagent has been used, for example, to study the glucagon receptor in hepatocyte plasma membranes [259].

Iodosulfanilic acid

Iodosulfanilic acid, when diazotized, may be used for labeling cellular membrane proteins under conditions guaranteeing minimal damage to membrane components. This reagent can be used if it is undesirable to label amino groups. It labels histidine and tyrosine residues [256].

Isethionyl acetimidate

This reagent is unable, unlike ethyl acetimidate, to penetrate intact cells and may therefore be used to label proteins on the outer surfaces of the membranes [260].

Methyl 3,5-di(^{125}I)iodohydroxybenzimidate

It is a protein labeling reagent which preserves the charge on the protein and is relatively stable in aqueous media [261].

Succinic anhydride

Succinic anhydride reacts with free amino groups in mildly alkaline solution under conditions similar to those for acetylations. However, while acetylation of the cationic group yields an electrically neutral product, succinylation yields an anionic product, and thus one which is more soluble than the acetylated ones [266].

N-succinimidyl propionate

N-succinimydyl propionate reacts in an analogous manner to the Bolton-Hunter reagent. It has the advantage of being a smaller molecule than the Bolton-Hunter reagent and hence causes less alterations to the protein structure [267].

5 Stability of Proteins Labeled with Radionuclides

The stability of proteins labeled with radionuclides is affected primarily by physical and secondarily by chemical processes [275,276].

Physical processes

The radioactive decay follows first order kinetics with a half-life of the radionuclide. Theoretically, after one half-life, 50% of the labeled protein should remain intact in the stored sample. In practice, this is not the case. Some radioactive molecules of the protein change their identity and in this manner they are converted to impurities during the preparation of the radioactively labeled protein. Radiochemical conversions (^{3}H → ^{3}He, ^{14}C → ^{14}N, ^{35}S → ^{35}Cl, ^{125}I → ^{125}Te, ^{131}I → ^{131}Xe) cause destruction of the respective covalent bonds of these radionuclides in the labeled molecules. The transmutation of elements and the breakdown of the peptide bonds are the reasons for the destructive decay. The stability of the labeled proteins affects not only the radioactive decay, but also other secondary processes. For instance, ^{3}H-labeled protein is more stable than that one labeled with ^{125}I [277]. The shelf life of the labeled protein may be significantly affected by the way it is purified [278,279].

Molecules of a labeled protein may be also degraded during interactions of radiation emitted by adjacent molecules in the preparation. The interaction of a beta particle or a gamma quantum with a protein molecule produces various ionizations and even disruption of chemical bonds. The number of different fragments generated increases with the complexity of the original molecule, but the concentration of each remains negligible. The loss of the labeled compound due to this process is much lower than that due to the radioactive decay process.

Chemical processes

Chemical processes are caused by chemical reactions between the radiolytically produced ions and radicals and the molecules of the labeled proteins. They are very significant and rapidly diminish the useful life of many labeled compounds. The degree of decomposition caused by secondary radiation is affected by the storage conditions of the protein. For instance, proteins stored in aqueous solutions are exposed to a variety of reactive intermediates such as hydrogen peroxide, hydrogen and hydroxyl free radicals, hydrated electrons, and so on, produced by the effect of radiation on molecules of water. To diminish the effect of these reactive intermediates, labeled proteins in aqueous solutions are often stored at low temperatures. The preparations may also contain "scavenger impurities" (e.g. cysteamine, mercaptoethanol, benzylalcohol, ethanol, etc.) which deactivate reactive intermediates and reduce their influence on the labeled compound. A similar protective effect is achieved by the presence of serum albumin or gelatine [280]. When radioactive solutions containing scavenger impurities are used in tracer studies, the possible influence of the impurities on the experimental results should be considered. The differences in the stability of labeled proteins or peptides may, of course, be remarkable. While ^{125}I-insuline in phosphate buffer was useless for RIA after 2–3 days ^{125}I-glucagon in n-propanol: water (1 : 1) can be used for RIA even after six weeks [281]. Ramsey et al. [282] determined the loss of ^{125}I from the labeled protein to about 3% per day. The stability of labeled proteins in aqueous solutions ranges on average from one month for ^{125}I-label to several months for ^{14}C-label. Proteins labeled with technetium are stable for about 5–6 hours [107]. Preparations which are stable for longer period tend to be exceptions. Studies on the effect of the storage temperature proved that the stability of labeled compounds at lower temperatures is generally better. Thus the storage temperature usually recommended is —20 °C. It is advantageous to store labeled proteins in diluted solutions because the reaction rates of the undesired reactions are reduced.

Finally, all proteins labeled or unlabeled, are decomposed to some extent during their storage by mechanisms such as oxidation, hydrolysis, and biological reactions. These mechanisms are sometimes accelerated in labeled compounds, particularly when the compounds are supplied in a very small volume. Kutas et al. [283] recommended that labeled proteins be stored in lyophilized form. Under these conditions the stability of ^{131}I-human serum albumin and ^{125}I-fibrinogen is enhanced.

6 Other Labeling Techniques

Apart from radioactive and stable nuclides, many other labels have been applied for similar reasons and also for other purposes. The advantages and disadvantages of the individual labels will be discussed in the appropriate paragraphs.

6.1 Labeling with Chromophores

Attempts to label proteins with different dyes and to use the conjugates obtained for the colorimetric assay of proteolytic activity are very old. The first experiments concerning the direct determination of proteolytic activity were performed by Grützner as early as 1874. He studied the hydrolysis of fibrin by pepsin and used coagulated fibrin with adsorbed carmin as a substrate [284]. The amount of dye released into solution was evaluated as a relative measure of proteolytic activity. Scharvin [285] modified wool, casein, ovoalbumin, gelatine and other proteins with p-benzochinon and obtained the corresponding aminochinon derivatives of the modified proteins. Several authors have studied the character of dye-protein interactions under various experimental conditions (Rakuzin [286], Gilbert [287]) in order to differentiate between covalent binding and physical sorption. Attempts to use another protein conjugate, congo-fibrin, for proteolytic activity assessment were made by Shackell [289]. Nelson et al. [288] applied indigo-carmine-fibrin and congo-red hide powder.

Besides the effort to apply dyed proteins for the measurement of proteolytic activity, attempts were carried out to use them in immunochemistry as well [290].

Great progress in the development of chromogenic proteins gave rise to the method of preparation of azoproteins (nitroproteins). The method does in fact represent an, optimized form of the well-known xanthoprotein reaction (protein solutions boiled with nitric acid yield a yellow colour). Nitrated protein is relatively stable because aromatic nitro groups can react only with very strong reducing reagents. This is a great advantage of azoproteins which are used as substrates for proteases under moderate conditions. On the other hand, it is necessary to point out that the nitration of proteins also causes certain degradation of the protein molecules. Suitably chosen nitration conditions are thus absolutely necessary for the preparation of azoproteins having the required properties. Such a method has been worked out by Pechmann [291]. Pechman has also made clear the course of the protein nitration. Besides casein, he nitrated edestin. Nitrodestin, however, gives lower colouring owing to the lower content of tyrosine in edestin molecules. Hitherto, many different azoproteins have been prepared; a lot of them are now commercially available and recommended for assay of the proteolytic·activity of various proteases (according their substrate specifity and pH optima).

The common advantage of azoproteins as substrates for assaying proteolytic activity can be seen in the fact that they represent natural protein substrates which are altered only very slightly. The unequal distribution of nitrogroups in the azoprotein molecule and the indirect measurement of proteolytic activity should be remembered as their disadvantages. Only the protein fragments containing tyrosine and tryptophan residues are taken into account as a measure of proteolytic activity. These objections are, however, valid in general for all kinds of labels. In the case of azoproteins they become consequential only when the protein under consideration is deficient in tyrosine and tryptophan.

Many other dyes (Congo Red [292], Indigo Carmine [292], Orcein Red [293, 294], Bromsulfalein [295], Remazolbrilliant Blue [296]) and reagents forming dyed proteins have been applied for protein labeling for different purposes.

Experience has shown that the chromophore label must be chosen with regarding to the properties of the protein to be dyed and the purpose of the labeling. Among the main properties of the proteins to be labeled, the number and character of the interacting sites with the labeling agent must first be considered. A few examples

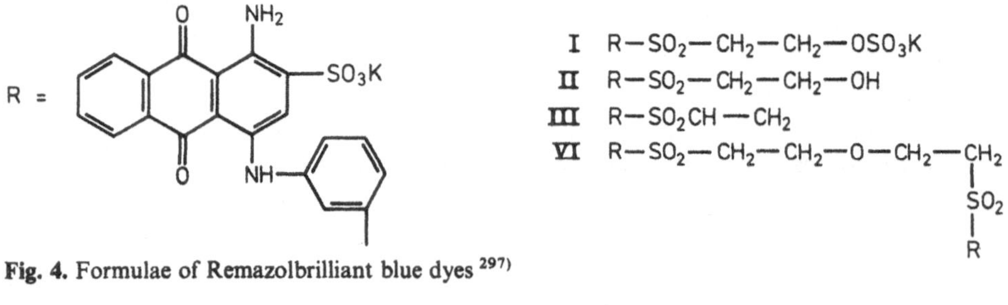

R = (1-amino-4-anilino-anthraquinone-2-sulfonate, with NH_2, SO_3K, NH substituents)

I $R{-}SO_2{-}CH_2{-}CH_2{-}OSO_3K$

II $R{-}SO_2{-}CH_2{-}CH_2{-}OH$

III $R{-}SO_2CH{-}CH_2$

VI $R{-}SO_2{-}CH_2{-}CH_2{-}O{-}CH_2{-}CH_2{-}SO_2{-}R$

Fig. 4. Formulae of Remazolbrilliant blue dyes [297]

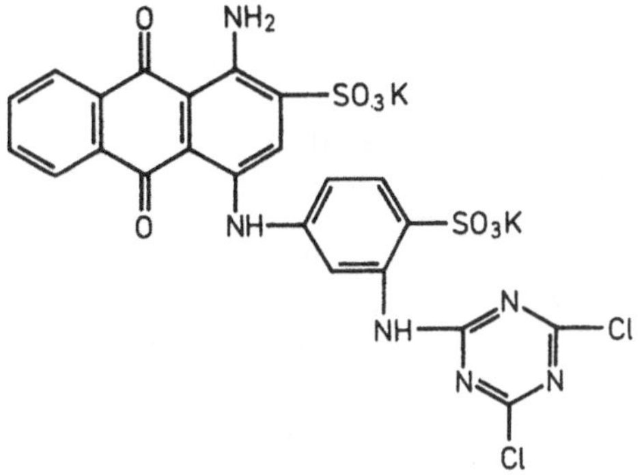

Fig. 5. Formula of Procionbrilliant blue RS [301)]

can be given. Remazolbrilliant Blue (Fig. 4) is a most suitable marker for collagenous proteins (e.g. hide powder) because of its high content of hydroxyproline which offers an abundance of sites for the formation of ether linkages between the dye and hydroxyl groups [297)]. For the labeling of amino groups of proteins the generally-known dabsylchlorid (4-dimethylaminoazobenzene-4-sulphonyl chloride, a derivative of methylorange) may be used, as well as, different dyes containing dichlorotriazynal groups [298, 299)].

Apart from the application of dyed proteins as substrates of proteolytic activity assay so far mentioned (see also 7.2), proteins are labeled with chromophores for studying membrane surfaces, for detecting proteins after several different separation procedures (i.e. chromatographic and electrophoretic techniques), for the determination of proteins in the mixture with many other substances, etc.

A very special case of "temporary" protein labeling is the presently developing branch of affinity chromatography; the proteins are separated from a protein mixture by various dyes covalently bound to insoluble supports. The protein chosen for separation binds reversibly to the immobilized dye and the remaining proteins can be easily removed by washing (usually in the column). The bound protein (in fact labeled for the time of separation) is then released by interrupting the specific interaction between the protein and dye. A few examples of the commonest dyes are given below.

Cibacron Blue is a blue, polyaromatic, sulfonated dye (Fig. 6). It can be attached, as an affinity ligand, to solid matrix supports (e.g. dextran, agarose) by the reaction of the triazine ring with free hydroxyl groups of the supports. The conditions of this triazine coupling method have been described by Bohme [302)]. Such a dye affinity sorbent is also produced commercially, e.g. under trade name Blue Sepharose CL-6B (Cibacron Blue F3G-A covalently bound to the cross-linked agarose gel "Sepharose CL-6B") by Pharmacia, Sweden.

The Procion dyes (Fig. 5 and 6) which contain reactive chlortriazine, consists of various aromatic chromophores linked to the triazine ring via —NH— bridges [301)]. Procion MX dyes, containing dichlortriazinyl groups, are very reactive, whereas the later Procion H range contains monochlortriazines which are less reactive [300)]. The Procion dyes are structurally very similar to the Cibacron dyes.

Congo Red is a dye related in structure to Cibacron Blue. According to the results obtained by Edwards and Woody [303)], Congo Red seems to be more

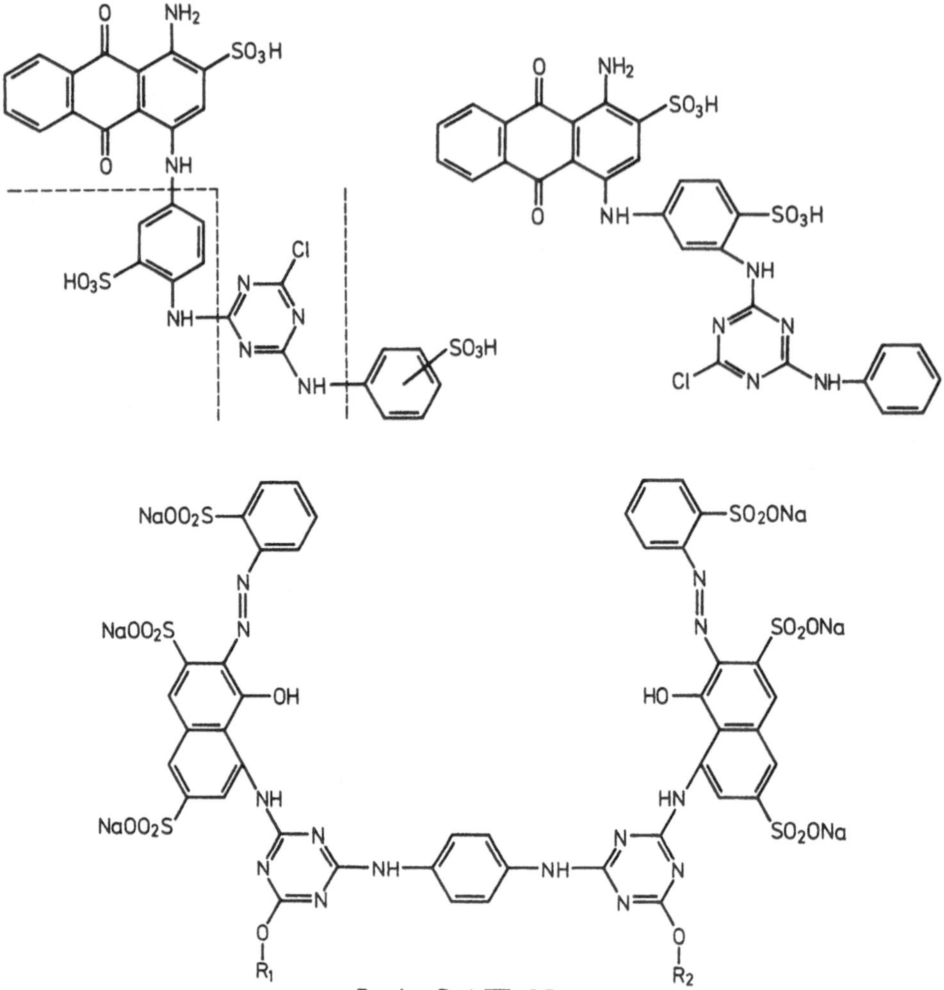

Procion Red III-3 B

Fig. 6. Formulae of Cibacron Blue and Procion Red dyes [300] Cibacron Blue 3 G-A (left); Cibacron Brilliant Blue BRP (right); Procion Red III-3 B (below)

specific for the dinucleotide fold than Cibacron Blue and it therefore can be expected to be a better affinity sorbent in some cases.

A lot of other dyes are used for protein labeling in order to detect them in histochemistry [304-306], after electrophoretic and chromatographic separation, as well as in protein determination (7.6).

Specific labeling of proteins can be achieved by the use of reagents which are "site directed" [307]. Alternatively, the presence of a particularly reactive group in the protein allows this group to be selectively modified (labeled) by stopping the reaction at an appropriate time. The reactivity of group-specific chromophoric reagents has been described by Birket et al. [308]. They paid attention to the kinetics of two chromophoric reagents for protein modification: 2,4,6-trinitrobenzene sulphonic acid (TNBS) which reacts mainly with NH_2 groups [309] and 7-chloro-4-

nitrobenzo-2-oxa-1,3-diazole chloride (NBD) which reacts specifically with SH-groups [310] under certain conditions. The degree of NH_2-groups labeling in proteins may be evaluated spectrophotometrically [309].

Selective photosensitized oxidation of some amino acids present in proteins can be also utilized for specific labeling with a chosen sensitizer (methylene blue and hematoporphyrin for methionine, crystal violet or cresol red for cysteine, proflavine for tryptophan, etc.). This topic is described by Scoffone et al. [311].

6.2 Labeling with Fluorophores

All proteins are "naturaly labeled" in some way or they may be labeled artificially with different types of fluorescent labels.

Natural fluorescent labeling of proteins is derived from their primary structure, i.e. mainly from the type, number and occurrence of amino acids having fluorescent properties. For native (intrinsic) fluorescence of proteins tryptophan and tyrosine are specially responsible, although some other amino acids (phenylalanine, histidine, arginine) are fluorescent, too. The fluorescence contribution of these "other" amino acids is, however, extremely low [312,313]. The fluorescence of tyrosine is normally greatly overshadoved by that of tryptophan. This is due to the quenching of phenol fluorescence by interaction with uncharged carboxyl groups [314,315] and peptide bonds [315,316]. Contrary to tyrosine, tryptophan fluorescence in proteins has been found to be higher, lower or about the same as that of the free amino acid [317]. This is a very brief description about the origin of native fluorescence in simple proteins.

The final fluorescence of conjugated proteins can be influenced by the non-protein component either by affecting the fluorescence of the aromatic amino acids bound in the peptide chain or by its own fluorescence. Among the most important nonprotein components of conjugated proteins, which significantly contribute their own fluorescence are mainly pyridine and flavine coenzymes (NAD, NADP, FAM, FAD), and pyridoxal phosphate with its derivatives. (Natural fluorescence of cofactors is, however, considered as a source of extrinsic fluorescence [313].) Even nonfluorescent protein components may influence the native fluorescence of protein by increasing or decreasing the energy transfer between the residues. Besides the above-mentioned cofactors, there are other protein components, mainly different pigments (e.g. chlorophyll), which are fluorescent. The fluorescent properties of the above compounds together with some other fluorogenic substances occurring in organisms are the reason for so-called "autofluoroscence" of a number of tissues and cell organelles. For instance in the case of lysosomes, flavine nucleotides seem to play the most important role [318,319].

In addition to native fluorescence, proteins may be labeled with three different types of fluorescent labels. The most important examples are given in Table 8 and the formulae of some of them are shown below (Fig. 7).

The first group comprises "fluorescent probes". The compounds listed in this group attach to protein (or other target substance) without forming covalent bonds. They are applied mainly for studying protein structure.

The second group consist of the "real fluorescent labels" (fluorescent dyes) which are characterized by forming covalent bonds with the labeled protein. The third and

Table 8. Fluorescent labels

Fluorescent probes	Fluorescent markers (labels)	Fluorogenic markers (labels)
characterized by not forming covalent bonds with the target substance	characterized by forming covalent bonds with the target substance	not fluorescent, but forming fluorescent products by reacting with target substance
2-Anilinonaphthalene-6-sulfonic acid 1-Anilinonaphthalene-8-sulfonic acid Auramine 0 4,4'-Bisdimethylamino-benzhydrol 3,3-Dihexyloxacarbocyanine iodide 3,3-Dioctadecyloxacarbo-cyanine p-toluene sulfonate Ethidium bromide 8-Hydroxy-1,3,6-pyrene-trisulfonic acid 6-(2-Imidazolin-2-yl)-2-[4-(2-imidazolin-2-yl)-phenyl]indole (DIPI) Merocyanine 7-(p-Methoxybenzylamino)-4-nitrobenzo[α]-21,3-oxa-diazole 2-n-Octadecylamino-naphthalene-6-sulfonic acid 1-Pyrenebutyric acid 2-Toluidinonaphthalene-6-sulfonic acid 1-Toluidinonaphthalene-8-sulfonic acid	Acetamido-4'-isocyanostilbene- 2,2'-disulfonic acid 9-Acridine isothiocyanate N-(p-Benzoxazolylphenyl)- maleinimide Bisbenzidine (H 33258) 4-Bromoethyl-7-methoxycoumarin 4-Chloro-7-nitrobenzo[α]-2, 1,3-oxadiazole Dansylchloride 4',6-Diamidino-2-phenylindole (DAPI) 5-[(4,6-Dicholorotriazin-2-yl)amino]- fluorescein 4-Dimethylamino-4'-isothiocyano- stilbene 5-Dimethylamino-1-naphthalene sulfonylaziridin (Dansylazaridin) 5-Dimethylamino-1-naphthylamine- sulfonyl hydrazide (Dansylhydrazide) 4-Dimethylamino-1-naphthyliso- cyanate Fluram 4-Hydroxybenzoic acid hydrazide 5-(N'-Iodoacetylethylenediamiwo)- naphthalene-1-sulfonic acid N-(Iodoacetylaminoethyl)-5- naphthylamine-1-sulfonic acid N-(Iodoacetylaminoethyl)-8- naphthylamine-1-sulfonic acid 7-(p-Methoxybenzylamino-4-nitro- benz-2-oxy-1,3-diazole N-3-Pyrene maleimide Quinacrine mustard Rhodamine B isothiocyanate	2-Aminobiphenyl 2'-Azido-2'-deoxyuridine N-(1-Anilinonaphthyl-4)- maleimide N-(p-2-Benzimidazolyl- phenyl)-maleimide N-(7-Dimethylamino-4- methyl-3-coumarinyl)- maleimide N-(3-Fluoranthyl)- maleimide o-Phthaladehyde

last group includes those compounds which are not fluorescent, but become fluorescent after the reaction with the labeled protein.

The compounds of the second group will be of primary concern in the following lines. The most important advantage of the application of these fluorescent dyes (fluorophores) in comparison with the ordinary ones (chromophores) is a considerable increase in the sensitivity of protein detection. The introduction of fluorescent dyes in order to increase the sensitivity of the dye techniques is ascribed to Creech and Jones [320] who treated proteins with benzantryl isocyanate to form the corresponding carbamido conjugates. Approximately at the same time, Coons et al. [321] started to apply the fluorescence antibody technique. Since that time this technique is generally applied everywhere when proteins need to be localized and identified within cells and tissues.

Nowadays fluorescent labeling is widely applied in studying protein structure and composition, protein interaction with other proteins or small molecules, protein tracing (immunochemistry, histochemistry, etc.), separation techniques and in the determination of protein concentration. The fluorescent protein tracing has been reviewed by a number of authors, e.g. [322].

The general procedure for fluorescent protein labeling, and, the requirements demanded of the labeled protein are similar to those when other labels are used. The fluorescent dye is chemically combined with protein under conditions leading to a high yield of fluorescent protein derivative which is subsequently purified from unreacted dye. It is expected that the labeled protein will have approximately the same or at least very similar properties to the unlabeled one.

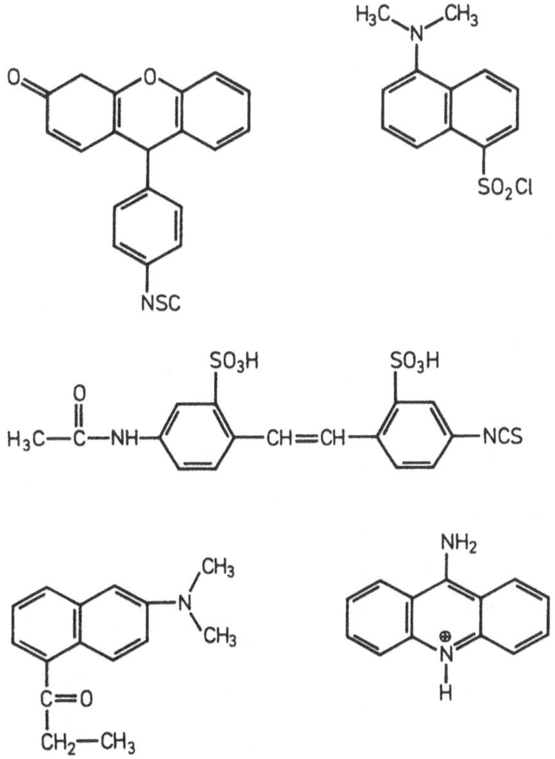

a

Fig. 7.a–c. Formulae of fluorescent labels
a) Fluorescein isothiocyanate (FITC top left)
Dansylchloride (1-dimethylamino-1-naphalene-5-sulfonylchloride, top right)
4-Acetamido-4′-isocyanostilbene-2,2′-disulfonic acid (SITS, middle)
2-dimethylamino-6-propionyl-naphthalene (bottom left)
Neuhoff et al.: Hoppe-Seyler's Z. Physiol. Chem. 360, 1657 (1979)
9-Aminoacridine (9-AA, bottom right)

The most frequently used fluorogenic substance for protein labeling is fluorescein-isothiocyanate, most likely owing to its availability and easy way of labeling. Detailed procedure for protein labeling (and especially antibody labeling) are described elsewhere [312, 323, 324].

Other isothiocyanate derivatives are also used for protein labeling, e.g. isothiocyanate of tetramethylrhodamine (TMRITC). Different fluorescent dyes can be used in double tracing methods. For instance, Wild [325] has described the use of FITC and TMRITC or Lissaminerhodamine RB 200 for double tracing of proteins. FITC conjugates give an apple green fluorescence and TMRITC and Lissaminerhodamine conjugates give an orange-red fluorescence. Where both green and red fluorescing conjugates occur together an intermediate yellowish fluorescence results.

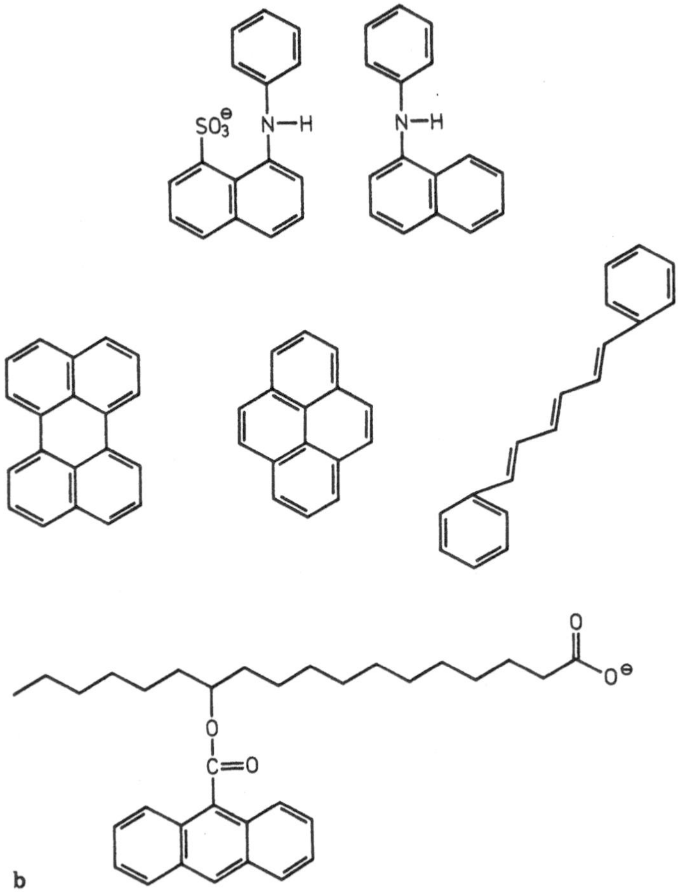

$SO_3^{\ominus}$ N—H N—H

O

$O^{\ominus}$

O

C=O

b

b) 1-Anilinonaphthalene-8-sulfonic acid (ANS, top left)
N-phenyl-1-naphylamine (NPN, top right)
Perylene (centre left)
Pyrene (centre)
Trans-diphenylhexatrien (DPH, centre right)
12-(9-antroyl)-stearic acid (AS, bottom)

New fluorogenic substance simplify the method of labeling. Very fast labeling of proteins (mainly antigens) can be achieved using fluorescamin (4-phenylspiro) furan 2 (3H),1-phtalan (3,3-dione) known commercially as Fluram. This substance condenses quantitavely with primary amino groups yielding a product with high fluorescence. The unreacted reagent is hydrolysed to a product with non-fluorescence and therefore the separation after labeling can be omitted. Labeling with Fluram is undoubtly advantageneous compared with fluorescein or even ^{125}I [326].

Fluorescamin has found application in diverse fields (protein detection and determination [327] etc.) in the same way as similary like Hoechst Dye No. 2495 (see formula below) [328, 329]. Recently a new fluorescent label called formycin appeared [330]. It is an analog of adenosine which differs structurally by the reversal of the carbon atom at position 8 and the nitrogen at position 9 of the purine ring. Only 2 picomoles of this compound can be fluorometrically detected. Such high sensitivity give these compounds potential for much wider applications in the future.

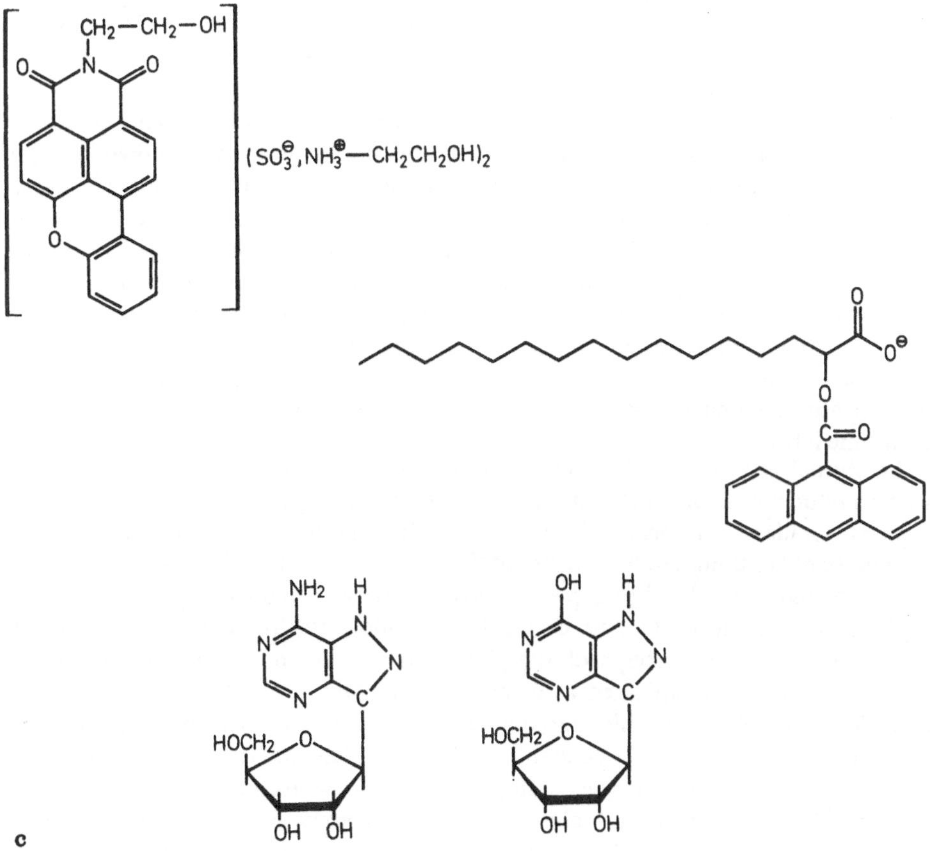

CH2—CH2—OH

$(SO_3^{\ominus}, NH_3^{\oplus}—CH_2CH_2OH)_2$

O

C=O

NH2 H

N

HOCH2 O

OH OH

OH H

N

HOCH2 O

OH OH

c

c) Hoechst dye 2495-40/70 (top left, according to Neuhoff et al.: Hoppe-Seyler's. Z. Physiol. Chem. 360, 1657 (1979))
2-(9-antraoyl)-palmitic acid (top right)
Formycin A (bottom left)
Formycin B (bottom right)

The use of rare-earth elements as fluorescent labels, e.g. europium seems to be very promising. Their application in immunochemical techniques permits assays of high sensitivity which are not troubled by non-specific short-lived fluorescence since this can be eliminated by time-resolution fluorometers [331].

Increasing attention is paid to chemiluminiscent and bioluminiscent labels because they offer much higher sensitivity. These advantages have recently been exploited in immunochemical techniques. In this respect, mainly luminol, isoluminol and their derivatives have been applied [332, 333]. Owing to the availability of suitable instruments for luminiscence measurements (a lot of analytical applications such as enzyme assays etc.) much wider application of luminiscent labels can be expected in the near future.

6.3 Labeling with Enzymes

The first scientists who started to label proteins with enzymes were histochemists. They labeled different antibodies and antigens for easier detection by light and electron microscopy of certain proteins in histochemical preparations [334, 335]. Further application of proteins labeled with enzymes is closely associated with the development of enzyme immunoassay (EIA) [336–338].

The conjugation of enzymes to proteins (usually antigens and antibodies) is performed in many different ways, in principle the same as are known for the immobilization of enzymes or preparation of immobilized affinity ligands [5, 339]. For the conjugation of enzyme labels to different proteins all reactive groups of the proteins (amino, sulphydryl, phenolic, imidazol, hydroxyl and carboxyl groups), as well as those of nonprotein components within the conjugated proteins, can be used. In common practice, however, only a limited number of binding reactions is applied. Free amino groups (of lysyl residues, or of amino acids in terminal positions) and carbohydrate components of glycoproteins (e.g. peroxidase, glucose-oxidase) are included in the majority of methods used for protein coupling with an enzyme label.

The mutual attachment of protein and enzyme label can be also achieved using different bifunctional or polyfunctional reagents; of these glutaraldehyde has become the most popular. Avrameas [340] introduced it for the purpose of EIA. To avoid a high degree of crosslinking, two-step modification is prefered [341, 342]. Glutaraldehyde is firstly bound to the label (e.g. peroxidase), its excess is removed by gel chromatography and then the protein to be labeled is added. Its free amino groups react with the free aldehyde groups of the glutaraldehyde bound to the enzyme label. This reaction is often explained as a formation of Shiff bases between amino and aldehyde groups. However, this coupling reaction cannot be convincingly explained in this simple way. The remarkable stability of glutaraldehyde cross-linked proteins is contary to the well-known reversibility of Schiff base formation. Richards and Knowles [343] have shown that glutaraldehyde is largely polymeric and contains α,β-unsaturated aldehydes which most likely participate in the coupling reaction. The detailed mechanism of the crosslinking reaction is still under discussion. For everyday daily practice, it is important to remember that the yield of the coupling reaction (usually about 30%) is affected very much by the quality of glutaraldehyde (batch, age, method of storing, etc.).

Two-step conjugation procedures are also preferred when other bifunctional reagents are applied. Kato et al. [342] applied the same method for coupling β,D-galactosidase to merkaptosuccinylated insulin by means of N,N-o-phenylendimaleimide. Other bifunctional reagents (bis-diazobenzidine, bis-maleinimide,N,N-1,3-phenylene, 3,6-bis(mercurimethyl dioxen, bisoxirane, etc.) may be used for the same purpose.

Apart from the above-mentioned bifunctional reagents, different polyfunctional (heterofunctional) reagents have been applied in coupling reactions in recent years (Table 9). The coupling reaction is divided into several steps. First, the heterofunctional reagent (i.e. N-succinimidyl-3-/2-pyridyldithiopropionate, SPDP) is bound both to the enzyme label and protein to be labeled in two separate reaction mixtures. Secondly, both preparations are purified by means of gel chromatography. Thirdly, disulphide groups of SPDP bound to one of coupled proteins is reduced to the sulphydryl group. Fourthly, the reduced derivative is separated by means of gel chromatography on Sephadex G-25. Finally, both modified proteins are mixed. The mutual interaction of sulphydryl and disulfide bonds of mixed modified proteins leads to a conjugate formation (enzyme labeled protein). The yields of this coupling reaction are very high, almost 100%. The course of all reaction steps can be controled photometrically on the basis of released N-hydroxysuccimide (at 280 nm) and pyridin-2-thion (at 343 nm) [344].

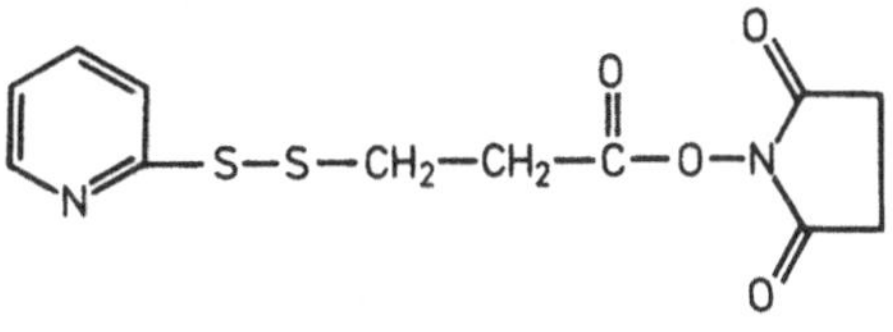

Fig. 8. N-Succinimidyl 3-(2-pyridyldithiol)propionate (SPDP). According to Carlsson, J., Drevin, H., Axen, R.: Biochem. J. 173, 723 (1978)

Because a lot of enzymes are glycoproteins, the carbohydrate moiety may be used for the coupling reaction after preliminary periodate oxidation. Periodate specifically splits —C—C— bonds bearing free vicinal hydroxyl groups; two aldehyde groups are formed. Aldehyde groups of modified glycoprotein enzyme react with free amino groups of the protein. The Schiff base formed is stabilized by sodium borohydride. In the original method of Nakane [345] the free amino groups of the glycoprotein are blocked with dinitrofluorobenzene before periodate oxidation. Other methods for labeling antibodies can be found in literature [346,347].

Apart from the general application of enzyme reactions which can be monitored on the basis of coloured reaction products, other methods of enzyme activity assay have been employed in EIA. For instance, in the case of urease potentiometric detection using an ammonia gas-sensing electrode [348] is possible and lysozyme activity can be assayed potentiometrically [349], by means of a TMPA$^+$ (trimethylphenylammonium ion) ion-sensitive electrode [356], or turbidimetrically [351].

Human choriogonadotropine (HCG) labeled with catalase has been used by Aizawa et al. [352] for the determination of HCG. Their method employs a sensor consisting of a Clark oxygen electrode covered with a cellulose membrane which contains a covalently bound antibody (against HCG).

Table 9. Bifunctional and heterobifunctional reagents

Reagent	Abbreviation	Cleavable (C) Non-cleavable (NC)
Homobifunctional:		
glutaraldehyde		
p-phenylenediisothiocyanate	DITC	NC
4,4′-difluoro-3,3′-dinitrodiphenylsulfone		NC
1,5-difluoro-2,4-dinitrobenzene		NC
4,4′-diisothiocyano-2,2′-disulfonic acid stilbene	DIDS	NC
erythritol-bis-carbonate		C
carbonyl-bis(L-methionine p-nitrophenyl ester)		C
4,4-dithio-bis-phenylazide		C
disuccinimidyl suberate	DSS	NC
bis-[2-(succinimidooxycarbonyloxy)ethyl]sulfone	BSOCOES	C
dithio-bis-(sussinimidyl propionate)	DSP	C
ethylene glycol-bis-(succinimidyl succinate)	EGS	C
disuccinimidyl tartarate	DST	C
disuccinimidyl(N,N′-diacetyl homocystine)	DSAH	C
dimethyl pimelimidate · 2 HCl	DMP	NC
dimethyl adipimidate · 2 HCl	DMA	NC
dimethyl suberimidate	DMS	NC
2-iminothiolane · HCl	(Traut's reagent)	C
dimethyl-3,3′-dithio-bis-propionimidate · 2 HCl	DTBP	C
Heterobifunctional:		
N-succinimidyl-3-(2-pyridyldithio)propionate	SPDP	C
N-(4-azidophenylthio)phtalimide	APTP	C
N-succinimidyl(4-azidophenyldithio)propionate	SADP	C
N-succinimidyl-6(4′-azido-2′-nitrophenylamino)hexanoate	(Lomant's reagent II)	NC
succinimidyl 4-(p-maleimidophenyl)butyrate	SMPB	NC
succinimidyl 4-(N-maleimidomethyl)cyclohexane-carboxylate	SMCC	NC
N-hydroxysuccinimidyl-4-azido-benzoate	HSAB	NC
p-nitrophenyl 2-diazo-3,3,3-trifluoropropionate		NC
methyl-4-azidobenzoimidate	MABI	NC
m-maleimidobenzoyl N-hydroxysuccinimide ester	MBS	NC
4-fluoro-3-nitrophenyl azide	FNPA	NC
p-azidophenylglyoxal		NC
N-5-azido-2-nitrobenzoyloxysuccinimide		NC
p-azidophenacyl bromide		NC

For details see:
Das M, Fox, C. F.: Ann. Rev. Biophys. Bioeng. 8, 165 (1979)
Peters, K., Richards, F. M.: Ann. Rev. Biochem. 46, 523 (1977)
Freedman, R. B.: TIBS-September, 193 (1979)
Double-Agents technical bulletin of Pierce Eurochemie, Holland

Apart from the covalent bonding of an enzyme label, its attachment on the principle of immunochemical interaction has been utilized in immunocytochemistry by Sternberger [353] (see also 7.5).

6.4 Biospecific Labeling

Biospecific labeling, which takes place in vivo, can be also utilized in experiments in vitro. In the case of natural biospecific labeling a reversible complex is always formed as a result of the mutual interaction between a given protein and a counterpart compound (sometimes also protein). The interaction between an antigen and corresponding antibody can serve as the most common example. Such interactions have important consequences in vivo. In vitro, they are used in separation techniques (affinity chromatography), in detection, localisation, identification and determination of certain proteins (immunochemical and histochemical methods). From this point of view, the main groups of substances having ability to form complexes with definite proteins, and which are at present used in practice, will be briefly characterized.

In some cases one of these two counterparts is additionally labeled with a radionuclide, chromophore, fluorophore, enzyme, etc. to facilitate the detection and to make the method more sensitive. Leucoagglutinin labeled with tritium [354] and protein A labeled with FITC can serve as examples.

Protein A is a specific protein isolated from the cell wall of *Staphylococcus aureus* whose characteristic property is the ability to react and to form precipitates with a variety of IgG molecules from several species. This interaction is reminiscent of the formation of antigen-antibody complexes, and has been used to study different aspects of immune response as well as cell surface structure and function [355-357]. For easier detection, Protein A is covalently coupled to fluorescein isothiocyanate (FITC). The commercial preparations of FITC-Protein A contain an avarage of 6 FITC substituent groups per molecule of protein. Such a degree of labeling does not affect the biological properties of the native protein.

Protein A binds to the F_c region of IgG and does not interfere with the antigen binding capacity of IgG, thus FITC-Protein A can form tertiary complexes with antigen and antibody (IgG) [358]. FITC-labeled protein A has been applied in different immunochemical studies [359,360] and separations of cells bearing IgG or other surface receptors. In general, it is suitable for investigating cell surface antigens and immunochemical reactions in vitro.

Another group of substances which offer themselves as labels for a lot of metabolic, morphological and physiological investigations, and also for separation purposes, are lectins. Lectins is a term generally used for carbohydrate-binding proteins which have been isolated from a variety of plants (i.e. beans, nuts, grains) and invertebrates (i.e. *Helix pomatia*). They may be distinguished not only by their origin and structure, but mainly by the differences in their affinities for different carbohydrate moieties. In the whole range of separation techniques they are applied preferably in an immobilized form for isolating a variety of glycoproteins and cells. For instance, snail (Helix pomatia) lectin bound to an unsoluble carrier (e.g. Sepharose 6MB Pharmacia) may be applied for the purification of human, bovine and mouse peripheral lymphocytes [361,362], rat and mouse spleen cells [361]. The desorption of cells under mild conditions enables them to keep their viability and functions. The carbohydrate specificity of the other known lectins and examples of their applications are given in the catalogues and literature of many commercial producers (e.g. Serva, Pharmacia, P.L. Biochemicals, Inc.).

6.5 Affinity Labeling

As we have already shown in the introduction, biochemical processes, taking place in living objects, are catalyzed and regulated by the specific interactions between corresponding molecular species. This phenomenon of mutual affinity of specific compounds has been manipulated in order to determine structural and chemical features; to study active sites both on enzymes and antibodies; to localize proteins of different functions on membranes, cell and organelle surfaces; to study intracellular transport and other events occurring in living organisms. For these purpose the methods of affinity labeling have been developed [363–366].

They differ from the biospecific labeling described in the preceeding paragraph by introducing a chemically reactive group (e.g. bromacetyl, aminoacyl) which enables a firm covalent bond (irreversible complex) to be formed. The appropriate substrates, inhibitors or haptens become fixed labels.

Affinity labels (RL) consist of a biologically active moiety (R), capable of forming a reversible complex with a given protein (P) and a properly positioned, chemically reactive leaving group (L) [367]. The affinity label interacts with the protein counterpart resulting in an irreversible protein-ligand complex (see Fig. 9)

$$P + R - L \underset{k_2}{\overset{k_1}{\rightleftharpoons}} P \ldots R - L \xrightarrow{k_3} P - R + L^-$$

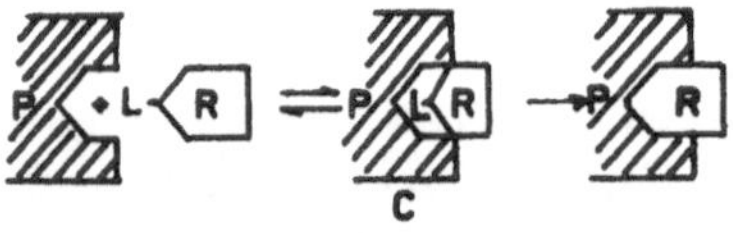

Fig. 9. Affinity labeling
P ... protein, R—L ... affinity label (R ... biologically active moiety, L ... chemically reactive leaving group) (Reproduced by kind permission of Academic Press [367])

The inhibitory capacity of an affinity label is related to its ability to form the reversible complex and to the reactivity of the leaving group. The reversible protein ligand complex dissociates to an active protein (e.g. enzyme), while the affinity labeled protein is either partly or totally inactivated. The degree of inactivation is dependent on the importance of the blocked functional group to the biological (e.g. enzyme) activity of the affinity labeled protein.

The experimental criteria for affinity labeling, as well as the general rules for selection of an appropriate affinity labeling reagent, can be found in a paper by Wilchek and Bayer [367].

Since 1961, when the affinity labeling was introduced into research practice, a lot of different affinity labels have been applied [368]. The phenomenon of mutual affinity of certain proteins to the other definite compounds has been successfully applied in affinity cytochemistry [369], affinity chromatography [6, 370, 371] and affinity therapy [372].

There is no doubt that affinity labeling, as well as the associated approaches mentioned above, will have increasing importance in studying the structure, function and mechanism of action of proteins which play significant roles in vital functions of organisms and in many practical aspects of life (medical diagnosis, therapy, etc.).

6.6 Miscellaneous Labeling

Apart from the various labeling methods described above some other labeling techniques are applied in certain cases. Labeling with enzymes can be replaced by labeling with certain other proteins having properties suitable for this purpose. This requirement is fulfilled mainly by various chromoproteins and antibodies. Their practical applications lie in histochemistry [304–306] and immunochemistry [373,374]. For the detection of antibodies in electron microscopy, ferritin was introduced as a label by Singer in 1959 [375]. The coupling can be performed with xylylene diisocyanate or toluene diisocyanate [376]. This procedure was later further modified [377] and some methods were even published as patents [378].

The advantage of iron labeling for electron microscopy has inspired the use of some other metals and iron in the form of other compounds, for instance ferrocene, e.g. bis-pentadienyl iron [379]. Mercury [380], uranium [381] and osmium [382] are also widely used in electron microscopy for protein labeling. Because they are easy to detect and because they attach themselves readily to antibodies and antigens, metals are very promising labels in immunochemical methods. Voller [383] forecasts the development of a new generation of simple metaloimmunoassays.

Metals are used not only as labels in histochemistry and immunochemistry, but also for studying protein structure and properties. Some metal ions can serve as valuable probes by replacing the original ions in different metaloenzymes or other metaloproteins. For instance, cobalt can replace zinc on the active side of carboxypeptidase, aldolase, carbonic anhydrase, phosphatase or yeast alcohol dehydrogenase.

Cobalt can act as an electron spin resonance, absorption spectral and nuclear resonance probe. It can also be used as a proton probe because its paramagnetism perturbs the proton resonances of neighbouring organic groups. The probe properties of cobalt are described by Foster et al. [384].

Spin label tempone (2,2,6,6-tetramethylpiperidone-N-oxyl) has been used by Wagner et al. [385] to label metal ion chelation and complex formation with cellular components such enzymes, ribosomes and membranes. The assay is directly applicable only for paramagnetic ions such as nickel, iron, cobalt and copper. It is based on the broadening of nitroxide spin-label absorption lines due to electron spin exchange between the free radical and the paramagnetic ions. The nitroxide spectral broadening is diminished when a paramagnetic ion forms coordinate bonds with another species. The observed competition between the paramagnetic ion and other nonparamagnetic ions for the sites of complex formation results in the paramagnetic ion being released into solution thus restoring its spin-broadening capacity. The formation of a coordination bond between the metal ion and the protein can be either essential for its activity or it can modify or inactivate the protein.

A very interesting method of labeling is the subject of an U.S. patent [386]. Protein to be labeled is coupled to methacrylate microspheres containing entrapped particles of a ferromagnetic metal (iron). This labeling technique is suitable for membrane mapping.

For certain purposes two different labels can be applied. A suitable example is the surface labeling of erythrocytes with ^{125}I and dijodfluorescein isothiocyanate which makes it possible to keep hemoglobin inside the cells almost unlabeled [387].

A special case of labeling, is indirect labeling with ^{125}I-labeled Triton X-100 [388]. Covalent coupling of soluble polymers (in fact also a kind of labeling) to pharmacologically active substances (hormones, vitamins, drugs) enables their activity in vivo to be prolonged. Such prolongation of the pharmacological activity has been achieved with insulin [389,390], thyroxin [390], dimethofrine [391] and cobalamins [392] bound to dextran. Similar treatment decreases the loss of trypsin activity due to autolytic digestion [393].

In a broad sense, pulse radiolysis can be also regarded as a labeling technique. It causes, by its oxidative or reductive attack, certain changes (i.e. OH-adducts of aromatic and heterocyclic rings or radical intermediates of cystine) which label definite places in protein molecule. This technique is useful for studying the role of functional groups of enzymes [394].

Some other labels, for instance bacteriophage [395], can be found in the literature, but they are of no great importance at present. The range of labels which can be used is so broad and their applications are so diverse that it is very difficult to review this topic fairly from all angles.

7 Application of Labeled Proteins

7.1 Enzymology

Labeled proteins are used in enzymology especially as substrates for determining the activity of proteolytic enzymes, for studying structural properties of enzyme molecules (7.2), and for some analytical purposes. Here, attention is paid to the determination of proteolytic activities when proteins labeled with radionuclides, chromophores and fluorophores are used as substrates.

Radiochemical methods for assaying enzyme activities are nowadays very widely used owing to their high sensitivity and accuracy. A few reviews concerning this topic can be recomended [396–398]. Among the enzymes whose activities may be determined using radioactively labeled substrates proteases (i.e. hydrolases acting on peptide bonds) should not be left out. Many different proteins, labeled in an suitable way, may serve as substrates for their activity assay. The assessment of proteolytic activity includes in general the application of a gently denatured labeled protein in a medium containing a buffer of suitable pH (according to the type of protease assayed), and activators (if necessary). The reaction is started by the addition of an enzyme solution, carried out at standardized conditions and interrupted, after certain reaction time, with a denaturing reagent (e.g. trichloroacetic acid). The precipitated protein is removed (often by centrifugation) and the radioactivity of the protein fragments in the supernatant liquid is determined by an appropriate method (e.g. Fig. 10).

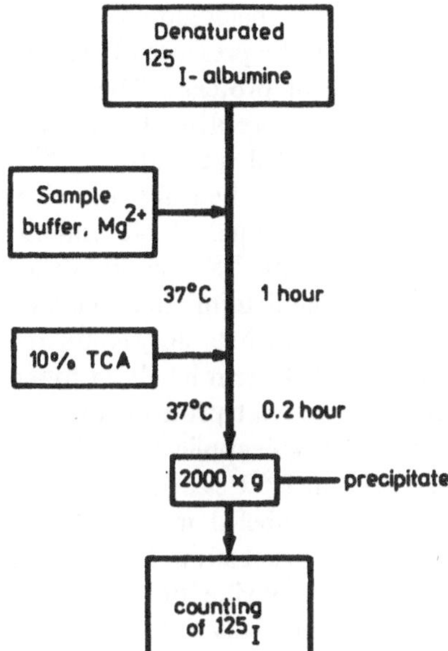

Fig. 10. Schematic diagram of a radioisotopic method for assaying proteolytic activity

Serum albumin labeled with an iodine radionuclide was firstly used as a substrate for determining protease activity by Absolon [399]. This method was later on modified several times and applied for assaying various proteolytic activities in different materials. Mego et al. [400] injected denaturated ^{125}I-human serum albumin into the tail vein of rats and measured the rate of intralysosomal proteolysis on isolated lysosomes containing endocytosed substrate. This method was also used for the determining the intralysosomal pH on the basis of differences found in the rate of ^{125}I-albumin breakdown in intact and lysed lysosomes [401]. ^{14}C-bovine serum albumin, ^{125}I-casein or ^{131}I-albumin have been alternatively used as substrate for measuring the activity of trypsin, chymotrypsin and papain [402–404].

The authors of this review have used ^{125}I-human serum albumin for determining proteolytic activity in various types of foodstuffs, enzymatic preparations (determination of the contaminating proteolytic activity), pharmaceuticals and other materials [405–409]. Proteolytic activity assays using radioactively labeled substrates are very sensitive, about hundred times more than the classical colorimetric ones (e.g. Anson's method [410]). The sensitivity, however, varies in relation to the protease assayed, the substrate used (origin and quality) and other circumstances. In general, among the other advantages of the assays using radioactively labeled substrates, we should remembered that they can be used with cloudy and coloured samples and that there are no problems with the interfering background of peptides and amino acids, usually present in the investigated natural samples (extracts of animal and plant materials, body fluids, etc.). There are, of course, also some disadvantages of these methods such as the limited period of usage of such substrates (in respect of radionuclide label used), the relatively expensive instrumentation required for the radioactivity measurement, and the need for safety precautions. In some rare cases, unfavourable environmental effects can also appear, similar to that described when Orcein or Congo Red dyed proteins are used as protease substrates in the presence of serum protein [295]. Roth and Losty [411] advise that ^{14}C-hemoglobin (labeled with $K^{14}CNO$) should not be used as a protease substrate in the presence of cysteine because the radioactivity is released even in the absence of protease. The use of proteins labeled in some other way (e.g. with ^{125}I) solves this problem. The activity of collagenase has been often studied using radioactively labeled substrates [412–416]. Collagen for proteolytic activity assay can be labeled biosynthetically (in vivo) either with ^{14}C-proline [413,414] or with glycine [415]. The technique of biosynthetic labeling has also been used by Čikalo [417] to label protein with ^{35}S. Casein labeled with ^{51}Cr has been employed by Ritz and Rubin [418] for the determining α-chymotrypsin activity. The main advantage of this substrate is its higher stability compared with those using other radioactive labels. ^{14}C—N,N-dimethylcasein has been found to be a suitable substrate for assaying the activity of bacterial protease thermolysine [419], and ^{131}I-casein for assaying plasmin activity [420]. The application of solid phase substrates is also advantageous. Such a preparation (^{125}I-casein-Sepharose) has been used by Sevier [403]. Tritiated bovine fibrinogen labeled in the fibrinopeptide region has opened up the possibility of applying this substrate for selective investigations on the enzymatic phase of clotting [421]. ^{3}H-bovine fibrinogen for assaying trombin and plasmin activity is available commercially (New England Nuclear), as are a lot of other substrates (^{3}H-collagen and methyl ^{13}C-collagen for assaying the activity of mammalian and bacterial collagenases, ^{125}I and methyl ^{14}C, albumin bovine and human, methyl ^{14}C-casein, ^{125}I-gelatine for various proteolytic activity assays, etc.).

The application of substrates labeled with various chromophores has become more and more popular in recent years. It has been already shown that proteolytic activity assays using dyed proteins have a long tradition (see 6.1). Their enforcement has not, however, been simple. There have been objections (often correct) against a discutable correlation between the colour intensity and the number of split peptide bonds. This disadvantage can be overcome by specific binding of a certain dye to the amino acid involved in the proteolytic breakdown of the protease being assayed.

On the other hand, such a relationship is not always required and very high sensitivity is preferred. The assay sensitivity is affected, in the same way as with other labels, by the degree of labeling. Higher labeling usually causes a higher "amplifying effect". However, it must be taken in account that the degree of labeling also affects the accessibility of the substrate and in this way the rate of proteolysis. For instance, strong labeling reduces the accessibility and also the specificity of elastin substrate to elastase [422]. On the other hand, Rinderknecht et al. [295] found that a tenfold increase in label content (from 1.33 to 14 molecules of dye to 1000 amino acid residues) caused a twentyfold increase in the accessibility of Remazolbrilliant Blue labeled hide powder to trypsin. The increased accessibility of the substrate cannot however, be ascribed only to the higher number of label molecules bound to the protein, but also to the degree of protein denaturation, sterical changes in the protein molecules, charge changes and some other phenomena accompanying the labeling procedure. Thus the only conclusion that can be drawn is that the degree of labeling is very important both for the sensitivity and real applicability of dyed protein for the proteolytic activity assay, but that the most suitable degree of labeling cannot be predicted in advance.

Great competitors of dyed proteins in proteolytic activity assays are various low-molecular weight chromogenic substrates, including in particular p-nitroanilides and β-naphthylamides of amino acids and peptides which, release by the action of proteases coloured p-nitroaniline or β-naphthylamine. A very high number of derivatives is commercially available and widely used for assaying many proteolytic enzymes. These derivatives imitate the natural protein substrate only to a slight extent and thus the rate of their splitting is different from that of natural protein substrates. More complicated synthetic chromogenic substrates, as for example S-2160 (Bz—Phe—Val—Arg—NH —⟨◯⟩— NO_2) of the Swedish firm AB Bofors, Nobel Division, are very expensive and therefore not suitable for routine analysis.

The dyed proteins, though also not very cheap, are much more suitable from an economical point of view. The use of insoluble dyed proteins (e.g. hide powder azure) is very simple and handy. After the chosen reaction time the insoluble protein is removed by filtration or centrifugation and the absorbance of the filtrate is immediately measured. For instance, Rinderknecht [295)] proved that it was possible to determine trypsin activity with HPA (hide powder azure) in ng quantities per ml and proteolytic activities in biological materials, tissue extracts, serum, urine, faeces, etc. Other authors applied this substrate for assaying proteolytic activity in beer stabilized with chillproofing preparations containing proteases (mainly papain) [423, 424, 404)].

The application of these substrates for proteolytic activity determination is suitable when very low activities are expected. The main risk in routine application of dyed proteins in proteolytic activity assay is the possibility of transmigration of dye from substrate to other proteins occurring in the reaction mixture. This environmental effect on the dyed protein can be demonstrated using Orcein or Congo Red as labels for elastine substrates. Albumin and, to a lesser extent, other plasma proteins, because of their affinity for acid dyes, release Orcein or Congo Red from the insoluble elastin-dye conjugate and in this way simulate enzymatic breakdown of both dyed substrates [425)]. Such substrates are thus unsuitable for proteolytic activity assay in the presence of albumin or other serum proteins. To avoid erroneous results the conditions of dye-labeled proteins must be discreetly chosen. Salts can also, under certain circumstances, affect the binding capacity of the protein and contribute to the colour intensity of the blank. These obstacles can be overcome by changing the reaction conditions and thoroughly checking the blanks in all experiments.

Fluorogenic substrates could be applied in similar cases as the chromogenic or radioactive ones. The fluorogenic substances released by proteolysis are, in general, detectable in lower concentrations than the coloured substances and, therefore, higher sensitivity can be achieved in this way. Nevertheless, this advantage cannot be always utilized owing to the presence of different substances in natural materials which affect the fluorescence, mainly by quenching. This is the main reason why fluorogenic substrates are used in assaying proteolytic activity less often than would be expected. Fluorescein-labeled fibrin has been applied by Strassle [426)]. Lűscher and Kăser-Glasszmann [427)] have developed a method in which the fluorescent dye Lissamine Rhodamine B 200 is covalently bound to fibrinogen by treating the protein with the sulfonyl chloride of the dye. To measure the fibrinolytic activity of a tissue, the dye-labeled protein is incubated with it and other components required for clot formation. The fluorescence liberated into the fluid on clot lysis is proportional to fibrinolytic activity[427,428]. Rinderknecht[422,429] utilized fluorogenic elastin for assaying elastase activity. A simple assay for proteases based on the fluorescent labeling of insoluble proteins (e.g. fibrin) or of soluble casein by 2-methoxy-2,4-diphenyl-3-(2H)furanone (MDPF) has been developed recently by Wiesner and Troll[430]. Fluorescence of the peptide-fluorophores liberated as a result of enzymatic hydrolysis is easily measured in the supernatant after unreacted fluorescent fibrin has been separated by centrifugation or unreacted casein-fluorophor has been separated by acid precipitation. Nanogram quantities of trypsin, chymotrypsin and elastase can be assayed by this method.

The properties of protein fragments can be also used for the fluorometric assay of proteolytic activity. Curoff[431] proposed that the increase in trichloroacetic acid-soluble tyrosine (both free and bound in small peptides) can be fluorometrically assayed after the condensation with nitrosonaphthol.

7.2 Protein Structure and Mechanism of Enzyme Action

Various techniques of protein labeling help to solve a wide variety of problems which have to be considered when studying protein structure and mechanism of enzyme action. In spite of the complexity of this problem, only a few general approaches can be given. In this paragraph we shall pay attention mainly to the application of fluorescent labels and affinity labeling. Some other approaches, for example exchange of metal ions, pulse radiolysis (see 6.6), labeling by metabolic conversion, etc. are also useful.

The rapid development of fluorescence spectroscopy in the last twenty years has accumulated knowledge concerning fluorescence, polarization, energy transfer, quenching mechanisms, and spectral shifts. This increased understanding of fluorescent phenomena — together with technical inovations such as lasers, computers and new electronic equipment — has brought fluorescent techniques to the point where one can start using them in biochemical research.

Circular polarization of luminiscence, stopped-flow fluorescence, fluorescence-monitored chemical relaxation, the evaluation of relative orientation by polarized excitation energy transfer, time-resolved fluorescent polarization ("nanosecond polarization"), and other new techniques have become valuable means for studying protein structures, their interactions and structural changes in relation to various treatments (e.g. denaturation). New fluorescent probes and quenchers have enabled the research field to expand from isolated proteins to more complicated systems such as membranes, muscle and nerve components and other subcellular structures (see also 7.3).

In studying protein structure, two main approaches which utilize fluorescent labels are most frequently realized.

Firstly, the evaluation of changes in the fluorescent properties of the fluorescent dye absorbed or covalently bound to protein.

Secondly, the technique of fluorescent depolarization which was introduced into protein chemistry by Weber in 1952[432].

When the changes in fluorescent properties of a given dye are followed, after either its absorption or covalent attachment, the fluorescent labels are called "fluorescent probes" [433] or "reporter groups" [434]. Such a probe (label) undergoes changes in absorption characteristics and these are a function of enviromental perturbations. Among the most important changes in fluorescence properties, which have been already studied, the following can be mentioned:

1. Spectral shift of the fluorescent band [435].
2. Enhancement or quenching of fluorescence in relation to:
   (a) protein nativity (or degree of denaturation) [436,437];
   (b) probe binding, e.g. certain dyes (ANS, TNS — see 6.2) are nearly non-fluorescent in aqueous solutions but highly fluorescent when absorbed onto certain proteins [437,439–441];
   (c) solvent properties (polarity, etc.) [438,439].
3. Both spectral shift and changes in fluorescence intensity, when antibodies against the probe (e.g. TNS, DNS, fluorescein) are prepared and combined with the probe [442,443].
4. Changes in life-time and fluorescence intensity of the attached probe [444].

The technique of fluorescence depolarization uses in principle fluorescent dyes covalently bound to proteins. The label is excited by polarized light; the fluorescence is also polarized to a degree that is inversely related to the amount of Brownian motion occurring during the interval between absorption and emission of the light. The information about the amount of motion helps to form an idea on the size and shape of the protein being investigated.

The most frequently used probes for studying of protein structure seem to be ANS, TNS, NPN, 9-aminoacridine, umbelliferone (7-hydroxycoumarin), auramine 0 and dansylchloride and in membranology they are pyren, perylen, AS, AP, ANS and NPN. Among the main criteria for selecting an appropriate probe are the character of the interaction, its polarity and hydrophobicity (Table 10). For details see for example [445–451].

From many studies on this subject only a few examples will be given. The interaction of fluorescent probe ANS with proteins to which they are bonded has been studied by Streyer [438]. The mutual interaction of the same probe with bovine serum albumin has been investigated in detail by Daniel and Weber [452]. ANS has also been used for evaluating the thermal denaturation of human serum albumin [453] and for studying its interaction with a model lipid bilayer [454]. Ziegler and Rogus [455] investigated the effect of temperature on the fluorescence of an ANS-probe in a suspension of biomembranes of sarcolema.

Affinity labeling (6.5) is a very valuable technique for obtaining information about the active sites of enzymes and certain other biopolymers. For instance, small phosphate esters containing chemically reactive substituents have been used as

Table 10. Examples of polar and nonpolar fluorescent probes

Polar probes	Nonpolar probes
12-(9-anthroyl)-stearic acid	perylene
2-(9-anthroyl)-palmitic acid	pyrene
1-anilino-8-naphthalene sulfonate	trans-diphenylhexatriene

active-site-specific probes for certain enzymes of carbohydrate metabolism which have an affinity for phosphate esters. Hartman et al. [456] have described such labels while investigating properties of active sites in triosephosphate isomerase (E.C. 5.3.1.1), phosphoglycerate mutase (E.C. 2.7.5.3) and ribulose-bis-phosphate carboxylase (E.C. 4.1.1.39).

Wilchek and Bayer [367] have described in detail the application of bromacetylated substrates, inhibitors and haptens. These compounds are advantageous owing to their easy preparation, stability under various conditions and common applicability. The haloacetyl functional groups react with all the nucleophile side chains present in the protein molecule and thus have the potentional to modify many of the amino acids in the neighborhood of the active site. Homologous series of such labels make it possible to achieve various spacings of the reactive group from the binding site and thus enable the stereochemical arrangement of amino acid chains in the active site to be elucidated.

Bromacetyl derivatives have been applied in affinity labeling of staphylococcal nuclease [367], for studying the binding of thyroxin to human serum prealbumin [457], and for investigating the antibody structure-function relationship (using homologous series of bromacetyl derivatives of 2,4-dinitrophenyl haptens) [458]. Knorre and Lavrik [459] have described the reactive derivatives which have been used for affinity-labeling of aminoacyl-t-RNA synthetase (E.C. 6.1.1) for studying the localisation and interaction of the active site of the enzyme. They have also stated the requirements for the analogs that are used for affinity labeling of an enzyme:

a) the analog has to be a competitive inhibitor or substrate;
b) covalent binding of the analog has to inactivate the enzyme;
c) covalent attachment must be prevented in the presence of a sufficient excess of substrate.

The catalytic aspect of enzymes has prompted the development of new types of enzyme inactivators, which have been called "suicide inhibitors". These inhibitors are substrates which are modified by masking the reactive functions. If they are unmasked by the catalytic action of the target enzyme, reaction of the reactive group with an active site or cofactor is made possible and this leads to the inactivation of the enzyme. Since the chemically reactive group is only liberated at the active site of the target enzyme, reactions with foreign molecules cannot occur. A number of examples of suicide inhibitors (or K_{cat} inhibitors) can be found in a paper by Rando [460].

Information which can be obtained from many types of spectrometric analysis of compounds of hematological interest can be enhanced when measurements are made on series of isotopically labelled compounds [231]. The labeling of the protein of infected cells which have been grown in the presence ^{14}C-acetate offers the possibility of detecting particular regions in some proteins (e.g. adenovirus hexon protein) [226]. The glycosylation of the viral glycoproteins has been studied by means of radioactively labeled sugars [155].

The examples given above demonstrate a variety of possibilities for labeling protein when studying their structure and function.

7.3 Studies on Cell Morphology and Organisation (Cells, Membranes, Cell Receptors, etc.)

Radioactive labeling of proteins is very useful for studying the interactions and binding of these molecules to cells, membranes, as well as to molecular structures. For studying such interactions, radioactive ricin can serve as an example [461].

Labeled proteins make it possible to study the surface topography of organelles (e.g. ribosomes) and the molecular organisation of mammalian and bacterial cell membranes, e.g. [462–465]. The techniques of labeling and solubilisation of cell-surface proteins have been studied in relation to these problems. An evaluation of the techniques used for labeling the surface proteins of cultured mammalian cells is the subject of a paper by Juliano [463]. A number of studies deal with the problem of specific and differential labeling of membrane components.

For the selective labeling of the acetylcholine receptor Saitoh et al. [466] used the covalently bound non-competitive blocker 5-azido-^{3}H-trimethioquin. Asymmetric labeling of the proteins in the kidney microvillar membrane by lactoperoxidase-catalysed radioiodination and by photolysis of 3,5-di-^{125}I-4-azidobenzenesulphonate is a new approach which may be applied to the topological investigation of complex membranes [76].

Combined labeling of B lymphocytes with ^{125}I-anti-Ig-antibodies and E-rosetting permits simultaneous detection of B- and T-lymphocytes [72]. Different types of "immunobead reagents" (an analogy of "Enzymobeads") makes it possible to label various cells by quite a simple technique, which opens up wide possibilities for studying cell functions. [28]

An abundance of techniques is available for membrane labeling, e.g. biosynthetic incorporation of labeled amino acids [176]; absorption of labeled glycoproteins [137]; specific and nonspecific modification of membrane proteins using ^{125}I-iodosulphanilic acid [75], tritium labeled N-hydroxysuccinimide esters [134], ethyl-1-^{14}C-acetimidate [255], etc. Ethyl-1-^{14}C acetimidate hydrochloride labels protein under physiological conditions, penetrates cells without impairing membrane function. The methods described by Packer enable to label inner membrane surfaces to be labeled with diazobenzene sulphonate or p-mercuriphenylsulphonate [467].

The problem of the interaction of different biologically active compounds which bind to specific proteins for different reasons (transport, storage, etc.) is also a very large one.

The properties of membranes can also be used for analytical purposes. A radioreceptor assay for somatomedin A has been developed using human placental membrane. This assay is more sensitive and simpler than the chick bioassay and requires only 40 μl of serum. It is not species-specific [468].

Besides the common fluorescent probes used for studying protein structure (ANS, TNS, etc.), other labels suitable especially for the modification of cell surfaces and the membranes of intracellular structures have been introduced in research. ANDS, for instance, is such a label (3-azido[1,7]napthalene disulphonate) [469]. It is used for nonspecific covalent modification of hydrophilic cell surfaces. Another is FDNB (1-fluoro-2,4-dinitrobenzen) [470] which is used for identifying membrane proteins associated with glucose transport in human erythrocytes. More detailed information can be found elsewhere [445,446,471].

Affinity labeling was originally developed for studying isolated enzymes, but it has been applied to other structures occurring in living cells (see also 6.5). Affinity labeling agents can be also used for labeling receptors. Effector analogs containing chemically reactive functional groups are usually used as labels. Upon binding to the receptor, the reaction between the affinity labeling reagent and an active-site amino acid causes irreversible inactivation of the binding site. The selectivity of the label is based on how tightly it binds the target receptor. The lower the $K_{I(diss)}$ for the binding of the label is, the greater is the selectivity [438].

Affinity labels, in this case antibodies against the individual ribosomal proteins, have been successfully applied for recognizing antigenic sites of these proteins on the surface of the ribosomes [472,473].

Other affinity labels (derivatives of aminoacyl-t-RNA, antibiotics, messenger RNA or of GTP) were also used in studying specific binding sites of ribosomes (see review papers [474,475]).

The application of the affinity labeling technique for labeling intact cells in order to study various transport processes is more complicated owing to the loss of activity after such treatment. Nevertheless, this approach has been used too, for instance, for studying the biotin transport complex in yeast [476].

Besides many other labels, ribosomes bind specifically various antibiotics which interfere with some of their functions. The studies with chloramphenicol and bromamphenicol brought a lot of interesting findings [477,478].

This is only a brief look at the exciting prospects offered by labeling techniques in membranology and cell research. Labeling serves here not only for identifying proteins but mainly for elucidating their role in membrane processes and also for separation purposes.

7.4 Metabolic Studies

Labeled proteins are very important tools for studying various metabolic processes which are to be evaluated from different viewpoints. This section therefore presents only a few examples of such studies. Its aim is to show the various possibilities in this field.

One of the most thoroughly studied problems is the turnover of proteins in the bodies of animals, plants and man, in cells and subcellular structures, in the blood stream, the digestive tract and so on. Different approaches have been chosen for such studies and among them the use of labeled proteins or their labeling in vivo plays an important role (e.g. [44,160,211,479]).

Double-labeled proteins from rat liver cytosol (^{14}C in long-lived, ^{3}H in short-lived proteins after in vivo labeling) have been used as substrates for proteinases in vitro. The differences in the degradation rates of short-lived and long-lived proteins in vitro by different proteinases in the presence or absence of different effectors enabled conclusions to be drawn concerning their role in in vivo turnover. The main activity of lysosomal proteinases at pH values of 6.1 and 6.9 was found to be caused by thiol proteinases which decompose short-lived cytosol proteins preferentially. Autolysis of double-labeled cell fractions showed a remarkably faster breakdown of short-lived substrate proteins only in the soluble part of lysosomes [480].

The method of double labeling has been applied in other studies of protein turnover. Jung[481] has used the radionuclides ^{3}H and ^{14}C. A similar approach was chosen by Konno et al.[482] to investigate the relative turnover rate of soluble proteins in rat liver.

The utilization of proteins from the fodder of growing poultry and laying hens has been investigated with ^{15}N-wheat (proteins)[483] and ^{15}N-labeled casein[484].

Similar problems are with other domestic animals, mainly ruminants, currently being tackled. In relation to this problem, the hydrolysis of selected plant protein has been investigated by non-growing and growing cells of Bacterium *Bacteroides ruminicola*[485] and the digestibility of microbial protein in buffalo calves[486].
neuronal structures has been studied by means of an in vivo labeling with radio-
and distribution in rat brain[490]. Questions concerning the axonal transport in

The methods for measuring the total body protein turnover in man with ^{15}N labeled compounds are reviewed in a paper by Garlick and Waterlow[487] and James et al.[488].

Protein labeling in vivo has been used for the measurement of protein turnover[489]
active amino acids[491]. Other neurochemical problems have been studied in similar way[492].

A review of methods and metabolic studies associated with the labeling of lipoproteins can be found in a paper by Fidge[493]. The factors affecting the secretion of insulin and glucagon in lean and genetically obese mice have been studied by Beloff-Chain et al.[494] with ^{125}I-labeled compounds. ^{3}H-prolin has been used to prove that collagen could have a stabilizing effect on the differentiation of the ameloblasts[495].

7.5 Immunochemistry

7.5.1 Immunochemical Analysis (IA)

One of the most important contributions of immunochemistry to daily practice is the development of immunoanalysis. The interest in immunochemical methods increased when sensitive labels for detecting the complex antigen-antibody were introduced. The labeling of antibodies or antigens with radionuclides increased the sensitivity of immunochemical methods so remarkably that the determination of specific compounds at ng levels is quite common and under certain circumstances even much lower concentrations may be detected.

The first technique to use such labeling was radioimmunoassay which is based on the observations of Berson and Yalow[496–498] who showed that low concentrations of antibodies to the antigenic hormone, insulin, could be detected by their ability to bind radiolabeled ^{131}I-insulin. Since that time (about 1956–1959) other radionuclides have been applied as labels and a high number of different assays of hormones, drugs, pharmaceuticals, pollutants and other substances are used daily in many laboratories, mainly in clinical ones.

Detailed information on radioimmunoassay can be found in an outstanding review containing 591 references by Skelley et al.[499] and in many monographs and handbooks, e.g.[500,501].

Besides radionuclides (detected on the basis of beta and gamma radiation), labels detected by the measurement of magnetic resonance have been proposed (spin labels → spin immunoanalysis, abr. SIA). Other labels are fluorescent dyes already mentioned (fluoroimmunoanalysis, abr. FIA), chemiluminiscent compounds, and bacteriophage, which is, however, a rather exotic label [397].

Labeling with enzymes has become very popular in last ten years. About 50 commercial enzyme-immunoassay kits are now available. The first papers concerning enzyme immunoanalysis appeared in 1971 and 1972 (Engval and Perlmann [336], Avrameas and Guilbert [502], Van Weemen and Schuurs [337], Miedema et al. [503] and Rubenstein et al. [504].

Since that time a several types of enzyme immunoanalysis have been developed. They can be classified in following way (the usual abbreviations for these methods are given in brackets):

Enzyme immunoanalysis

1. heterogenic. Enzyme-linked immunosorbent assay (ELISA) or solid phase enzyme immunoassay (SOPHEIA). It can be competitive or uncompetititive.
2. homogenic. Enzyme multiplied immunoassay technique (EMIT)

ELISA and SOPHEIA are two names for the same technique. The principles are described elsewhere [338, 346, 505–507]. Both types of ELISA (or SOPHEIA), i.e. competitive and uncompetitive, are called heterogenic because the labeled protein bound into the complex of the antigen-antibody must be separated from unbound protein. Both these methods are in principle similar to radioimmunoassay (RIA). They differ only in the label applied. In the competitive ELISA the antigen labeled with enzyme competes with the unlabeled one, present in the sample, by binding to a limited number of binding sites of the antibodies. A larger amount of antigen in the sample causes a smaller amount of bound labeled antigen and vice versa. The method is mainly applied for determining hormones and pharmaceuticals.

Uncompetitive (sandwich) ELISA is divided into two steps. First, antigen from the sample under examination is bound to the immobilized antibody. Secondly, after preliminary washing, antibodies labeled with enzyme are bound on to fixed antigen. This method is applied only for polyvalent antigens. Of course, the reverse procedure can be applied for assaying antibodies, too. These methods are used nowadays for the detection of infectious hepatitis and have found a number of applications in microbiology. EMIT is mainly used for the estimations of low molecular weight substances (haptens).

All these methods use as the most important "reagent" proteins (antigens or antibodies) labeled with different enzymes. According to Engval [336] the enzyme used in an immunoassay should have certain properties:

a) Enzyme determination should be performed with a simple spectrophotometric technique.
b) The enzyme should be detectable in small amounts. This means that the enzyme should have a high activity and the product of the enzyme reaction should have a high molar extinction coefficient.
c) The enzyme should be stable.
d) The enzyme should be commercially available in a purified form.
e) It may also be an advantage if the enzyme is not abundant in the biological materials being used as samples in the assay. Neither should the test material contain a potent inhibitor of the enzyme being used.

The enzymes most frequently used for protein labeling are: horse radish peroxidase, alkaline phosphatase, glucose oxidase, glucoamylase, β-D-galactosidase, lysosym, malate dehydrogenase, glucose-6-phosphate dehydrogenase, pyruvat kinase and urease.

EIA is gaining ground and competes with its older sister RIA. Commercially available EIA kits for the determination of hormones, pharmaceuticals, toxins, bacterial and virus antigens, specific antibodies against pathogenic microorganisms and protozoa, cancer markers, etc. offer wide application of this sensitive technique. With this technique it is possible to detect proteins or haptens even in pico or femtomols. Apart from other factors, the sensitivity of assay depends on the activity of the enzyme label (turnover number) and the sensitivity of the enzymatic assay. The high activity of the enzyme label is in relation to the properties of the chosen enzyme, its purity and stability in the conjugate form. The sensitivity of the enzymatic assay depends on the substrate used and the method applied for detecting the reaction product. In general, the fluorogenic substrates allow the sensitivity of enzymatic assay to be increased about hundred times compared to the photometric ones. However, the limited application of these substrates, owing to widespread occurrence of quenching agents must be taken into account.

Those who try to make publicity for EIA usually make a strict comparison with RIA (which is not necessary) and give the following advantages of EIA: it is less expensive than RIA, has longer expiration time (about one year compared with a few weeks, for RIA) and it can be performed in the majority laboratories equipped with common instrumentation. People dealing with RIA have objections to such an evaluation. The price is always of very relative consideration and it can be changed drastically in a few years. The expiration time is a question of the stability of the labeled preparation and the chosen label. Progress should be made on both sides (radionuclide and enzyme labeled compounds). Of more serious concern seems to be the objection to the official regulations imposed upon staff and laboratories using isotopes and this limits the wider use of RIA in microbiological laboratories [383]; the same may be expected of food processing laboratories, too.

A serious disadvantage of enzyme labeling is the introduction of a further macromolecule into the system; it is a time, milieu and temperature dependent extra step [508]. It is, of course, not necessary to decide beforehand which method of labeling is more suitable for IA. Both RIA and EIA proved to be very valuable analytical tools and will surely compete in more convenient analytical approaches in the particular cases in future. Besides RIA and EIA, immunological techniques utilizing other labels are also steadily progressing. The sensitivity of fluorescent immunoassays (FIA), direct and indirect, and chemiluminiscent immunoassay (LIA) is increasing owing to the application of new types of labels and the introduction of new instrumentation. LIA becomes comparable with RIA in terms of sensitivity, precision and specificity [509, 510]. For instance, a detection limit for testosterone of 0.5 fmol has been attained using a Berthold Biolumat LB 9500 as a luminometer and ABEI (6-(N-)4-aminobutyl(-N-ethylamino-2,3-dihydrophtalazine-1,4-dione)aminobutyl-ethyl-isoluminol) as a chemiluminiscent label [511].

Other promising labels for immunochemical methods seem to be metals [376].

7.5.2 Immunocytochemistry

Immunocytochemistry employs the specificity of antibodies for the detection of cell components bearing antigenic determinants. A lot of methods have been developed to label antibodies in order to visualize the antigen-antibody reaction in situ. At first, only covalently bound labels were used [512] (since 1961). Later on, labels attached to the antibody via immunological bonds only were introduced by Sternberg in 1969 [514]. The labeling technique has to be adapted to the special requirements of electron microscopy, which differ from those of light microscopy. Fluorescein [513], ferritin [375,377], mercury [380], uranium [381], osmium [382] and ferrocene [379] (bis-pentadienyl iron) are most frequently used firmly bound labels which regularly react with antibodies in a stoichiometric relationship. Enzymes can be bound to antibodies either by covalent linkages [334,335] or simply on an immunological basis [514]. Independently on the way in which the enzyme is attached to the antibody, the enzyme reaction, in the presence of substrate, yields a high number of product molecules per one molecule of antibody. This "amplifying effect" of the enzyme reaction is the reason for the high sensitivity of the methods using an enzyme label (EIA etc.). High sensitivity is advantageous only in those cases where the specificity of the reaction is not decreased by high background, nonspecific binding, quality of the antiserum and other factors. A detailed description of these methodological problems is out of the scope of this review, but it can be found elsewhere, e.g. [515].

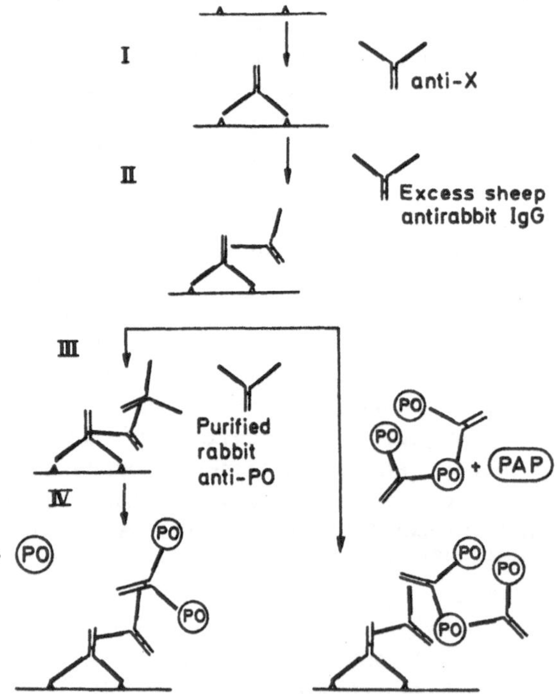

Fig. 11. Detection of an antigen-antibody complex using a enzyme label (peroxidase) attached on an immunological basis (according to Sternberger [515])
PO ... peroxidase, PAP ... antiperoxidase complex (Reproduced by kind permission of Van Nostrand Reinhold Company)

The method of using an enzyme as a covalently bound label to an antibody appears at the first sight to be simpler, but the conjugate formation is usually accompanied by some drop in specific reactivity and yields a diminished effect compared to the original label. Attention has been therefore paid to keeping the enzyme activity of the label unchanged, or at least less affected. This is the final goal in two versions of Sternberg's approach. Both modifications start with the same reaction between an antigen (X = cell component to be detected) with a rabbit unlabeled antibody (anti-X). In the following step an antibody (prepared by immunization of another species, e.g. sheep) (anti-IgG) against the "first antibody" reacts with the complex X-anti-X (see Fig. 11).

The third and fourth steps differ according to the modification applied. In the first version, purified rabbit antiserum against peroxidase (anti-PO) reacts with above complex as an antigen with antigenic determinants of anti-rabbit-IgG. Then peroxidase (PO) reacts as antigen with bound anti-PO. In the last step bound peroxidase reacts with diaminobenzidine (DAB) and hydrogen peroxide. The excess of hydrogen-peroxide is detected with the osmium tetroxide.

In the second version, the separate application of anti-PO and PO was combined into a single step. For this purpose soluble peroxidase-antiperoxidase PO-(rabbit)-anti-PO has been prepared. Details can be found elsewhere [515]. This method is one example of labeling in which covalent binding is not necessary.

7.6 Analytical Applications

Apart from assays of proteolytic activity and immunochemical methods (7.1 and 7.5), protein labeling is exploited in other areas of analysis. Noteworthy among them are protein determination in complex mixtures or even directly in cells, protein separation (affinity chromatography and electrophoresis) and protein detection following various separation techniques. For this purpose different labels are applied.

The labeling of proteins with dyes, radionuclides, etc. is often used for the determination of proteins. A lot of methods were worked out for assaying proteins in samples of various origin. The general interest in these methods is increasing owing to their simplicity, speed and possibility of automation. The choice of a suitable dye makes it possible to regulate the sensitivity of the method and, to some extent, even the type of protein to be assayed (e.g. if the dye is combined with amino groups of lysyl residues then proteins with high lysine content can be preferentially determined). Basically two methods can be applied: measurement of dye losses in the solution after the absorption of dye on the protein (e.g. Orange G) or, more frequently, direct measurement of dye absorbed on the protein.

A lot of applications are at present used in food analysis, as well as in clinical biochemistry. Examples of the dyes used are given in Table 11.

For instance, Brilliant-Orange is bound to basic amino acids (histidine, arginine and lysine), and it is therefore used for assaying proteins in corn. Amido Black is prefered for determining milk proteins. From the available commercial kits "Pierce's microprotein rapid stat kit" and "Bio-Rad Protein Assay Kit" can be shown as examples. Examples of protein analysers based on this principle are also given in Table 11.

Table 11. Dye binding methods used for assaying proteins

Dye	Assaying proteins in	Commercial device or kit
Acid Orange 12	ice creams	—
Orange G	milk	—
Amido Black	milk	Pro-Milk MK II[a] Pro-Milk Automatic[a]
Briliant Orange	milk, cereals	Udytec[a] Pro-Metr Mk III[a]
Commasie Blue	beer, wort	
Ruthenium Red	plant materials	
Commasie Briliant Blue G 250	body fluids	Pierce Microprotein Rapid Stat Kit
?	serum fluids and various	Bio-Rad Protein Assay Kit

a A/S N. Foss Electric, Denmark

The direct labeling of proteins in microbial cells (e.g. yeast cells) makes it possible to determine the protein content by the method of flow cytometry [516]. For this purpose, the fluorescent label sulforhodamine 101 was used. Very sensitive determination of proteins (10–50 μg) can be performed using another fluorescent label, fluorescamin [517].

The determination of the quantity of protein bound to the insoluble carrier sometimes causes difficulties. The methods usually applied are laborious or somewhat inaccurate. Labeling of assayed protein, for instance with ^{14}C-acetanhydride, makes it possible to carry out a very fast and exact determination of immobilized protein [518]. The determination of bound enzyme ^{14}C-labeled aldolase after its immobilization on polyacrylamide can serve as an example [519]. The concentration measurements of certain proteins are based on their ability to bind certain ligands. Radiolabels such as ^{14}C or ^{3}H-biotin have been used for the determination of avidin by direct binding [520] or for biotin assay by isotopic dilution [521]. Cofactor and fluorescent labeled ligands [522] have been also used for the monitoring of specific protein binding reactions.

Gebauer and Rechnitz [523] applied the enzyme labeling technique, tillnow used in immunoassay, for assaying specific binding proteins and their ligands. In this case, hydrolytic enzyme lysosyme was used as a label. Instead of a common coloured reaction, the activity of enzyme label was followed potentiometrically by observing the release of trimethylphenylammonium ion or turbidimetrically using Micrococcus lysodeikticus cells as a substrate.

Various types of protein interactions (mainly biospecific and those which resemble them in some way due to similar spatial conformations) have been utilized in very effective separation techniques which will be here only very briefly characterized.

Affinity chromatography [4] has been defined as a separation technique for biologically active substances, generally called ligands or affinity ligands (Lowe and Dean) or simply affinants (Reiner and Walch). If the method is understood in this way it would be better to call it bioaffinity of biospecific affinity chromatography. However, a lot of mutualy interacting pairs of substances cannot be considered as biospecific because they do not occur in nature. There are, for instance, different dyes which interact with proteins and can thus be used for their separation on the basis of affinity chromatography. They usually have affinity for a group of substances and this phenomenon enables one to use them to a variety of separation problems. In this respect, they are advantageous to the biospecific sorbents. The most successfully applied dye, as a specific group adsorbent, is Cibacron Blue, which is commercially available bound to solid supports (e.g. Blue Sepharose CL-6B from Pharmacia, which has been applied for the separation of a number of enzymes and selective removal of albumin from other plasma proteins [524]). This very promising technique is steadily progressing. Other dyes (Procion Red, Congo Red, etc.) and certain substances (lectins, microbial proteins, etc.) which reversible complexes with some proteins may be used for separation purposes. The key problem regarding the successful application of dye-ligand chromatography is to choose the most appropriate triazine dye. This strategy is described in a paper by Hey and Dean [300]. They routinely screen as many as 65 immobilized dyes (a dye screening kit is available commercially, e.g. from Amicon Corporation). This approach makes it possible to distinguish: first, dyes which bind protein to be separated; secondly dyes which do not bind it; and thirdly, dyes with intermediate behaviour (the majority). The first group of dyes is used for optimizing the separation conditions with regard to the most important parameters affecting the yields of the purified protein (for details see [300]). The dyes which do not bind the purified protein are also useful for the application of the so-called "tandem system" (7.8). The dyes showing intermediate behaviour are, of course, excluded.

Why is such attention paid to dyes as affinity ligands? It is necessary to remember the wide range of possible applications. The conformation of Cibacron Blue mimics the space conformation of NAD, NADP or ATP [525, 526]. Two of the sulphate groups of the dye are similar in orientation to the pyrophosphate group of NAD. The binding to the dinucleotide fold region probably involves both electrostatic forces and hydrophobic interaction [527, 528]. The ratio in which they share in binding the individual proteins can vary greatly. For instance, the binding of interferon to Cibacron Blue seems to be largely hydrophobic because for eluting the protein ethyleneglycol is required [529].

The specific interaction of Cibacron Blue and its derivatives to dinucleotides, mainly to NAD, NADP and ATP offers the possibility of purifing all enzymes which are dependent on these coenzymes. According to Mosbach [530, 531] there are 163 enzymes requiring NAD, 141 enzymes requiring NADP, about 40 enzymes requiring NADP or NAD and 225 enzymes dependent on ATP. Besides these specific interactions non-specific interactions of Cibacron Blue and its derivatives with proteins can also be applied for separation purposes. The non-specific interaction of Cibacron Blue with human serum albumin, for instance, enables albumin to be removed from transferrin, ceruloplasmin or other plasma proteins [532, 533], in order to purify human and mouse interferons [534]. Also some proteolytic enzymes [531] and blood coagulation factors [535, 536] interact with Cibacron Blue.

A number of other group-specific affinity ligands have been employed in the affinity separation of various proteins. The question of which affinity ligand to choose for separating proteins of different structure and function, as well as cells, viruses and a number of non-protein compounds, has been reviewed elsewhere, e.g. [4].

The affinity principle is also utilized in the novel technique of affinity electrophoresis [537].

Protein labeling is quite a common procedure following various separation techniques. The methods of labeling are adapted to the conditions of the appropriate separation technique and are described in the corresponding literature, e.g. following electrophoresis [538].

Apart from the wide range of staining techniques, fluorescent and radioactive labels are often used. FITC is used for protein detection following SDS-electrophoresis, e.g. [539]. Rat hemoglobin fractions have been assayed by labeling with ^{131}I [540]. Double labeling with ^{14}C and ^{3}H makes it possible to detect proteins from two different cultured cell lines, after electrophoretic separation, by means of a single two-dimensional gel [541].

Protein molecular weight markers in the form of ^{14}C-methylated proteins are very useful for gel electrophoresis. They are available commercially either individually or as mixtures with a wide range of molecular weights (e.g. from the Radiochemical Centre, Amersham or New England Nuclear).

7.7 Medical Applications

Labeled proteins are widely used in medicine and related disciplines. Proteins labeled with various radionuclides are more and more applied in scintigraphic methods, which are rapidly developing (Tab. 12). The research is concentrated on searching for suitable proteins or some other compounds which are specifically concentrated in the tissue being examined. Suitable labeling (preferably radiolabeling) makes it possible to visualize the indicator compound which then serves as an imaging agent for diagnostic purposes.

The problems in question are mainly: specificity in localisation; distinguishing between healthy and injured tissue (i.e. normal and pathological); none or, at least, low toxicity; the convenient form of administration; the short time of distribution in the body; the quality and reliability of imaging at low levels of labeling, the most suitable forms of labeling, especially with regard to the health risk; and a lot of technical problems.

Scintigraphic methods help physicians not only to distinguish the morphological changes, tumor formation and different kinds of injuries, but also to follow and to evaluate the function of various organs. They have become very important tool of modern diagnostic methods. Table 12 shows examples of scintigraphic methods both in applications and research.

Several radioactively labeled proteins are also used for the measurement of blood volume. Yang et al. [542] compared human serum albumins labeled with ^{131}I and ^{99m}Tc. They obtained a correlation coefficient of 0.987; nevertheless ^{99m}Tc-HSA was found to be a superior agent for repeated plasma volume determinations because of its shorter half-life and the reduced radiation doses to the subject being examined. Alternatively ^{68}Ga-labeled transferrin can be used for measuring of blood volume, too [543]. Frolov et al. [544] measured orthostatic changes in the blood volumes of the lower extremities in practically healthy humans by means of erythrocytes labeled

Table 12. Radiolabeled proteins in scintigraphic methods

Examination of	Indicator	Ref.
trombi	^{99m}Tc-labeled Trasylol	548)
	^{99m}Tc-labeled plasmin	115)
	^{99m}Tc-labeled fibrinogen	112,550)
	^{131}I-soluble canine fibrin	82)
	^{125}I and ^{131}I-fibrinogen	42,47,549)
	^{123}I-fibrinogen, ^{123}I-soluble fibrin	14,15,50)
	^{131}I-anti-human fibrin antibody	551)
	^{123}I and ^{99m}Tc-plasminogen	552)
tumors	^{99m}Tc-labeled Trasylol	548)
	^{99m}Tc-labeled plasmin	115)
	^{111}In-human serum albumin and bovine fibrinogen	48,187)
	^{99m}Tc-anti-human chorionic gonadotropin	553)
	^{99m}Tc-labeled liposomes	554)
	^{67}Ga-^{125}I-transferin (changes in ^{67}Ga/^{68}Ga ratio)	555)
	^{99m}Tc-concavaline A	556)
	^{125}I-protein A	557)
lung	^{68}Ga-albumin microspheres	173)
	^{99m}Tc-albumin microspheres	90,95,108,558,561,562,563)
	^{111}In and ^{113}In-albumin microspheres	95,188,191,194,
	^{99m}Tc-HSA, HSA macroaggregates, phytate etc.	118,559,560)
	^{131}I-serum albumin, MAA	79,563)
liver	^{99m}Tc-HSA, HSA macroaggregates (MAA), microagregates (MiAA)	83,99,559,565)
	^{113m}In-HSA	83,565)
	^{131}I-HSA	78,
	^{131}I-asialo-acid glycoprotein	566)
	^{99m}Tc-fibrinogen	98)
bone	^{99m}Tc-HSA and HEDSPA/1-hydroxyethylidene-1,1-disodiumphosphate	95,103,567)
bladder	^{99m}Tc-microspheres	570)
kidney	^{131}I-autologous fibrinogen	81)
	^{99m}Tc, ^{60}Co, ^{125}I, ^{131}I-bovine orgotein	563,564)
myocardium	^{99m}Tc and ^{113m}In-albumine microspheres and HSA	83,88,568,569,571)
	^{111}In-antibody to canine cardiac myosin	190)

with ^{51}Cr and albumin labeled with ^{99m}Tc. ^{99m}Tc-phytate colloids formed in vitro and in vivo were examined as radioindicators for estimating the volume of third-space fluid in an ovarian ascites model using C3HeB/FeJ mice. ^{99m}Tc-phytate colloids have clinical use for the simultaneous imaging of radiotracer migration in order to assess potential occlusion of diaphragmatic lymphatics by neoplastic cells, and for dilution analysis to estimate the volume of ascitic fluid [545)].

Red blood cells labeled with ^{99m}Tc were used for the measurements of the red cell mass in newborn infants [546)]. In the medical applications of labeled proteins different metabolic studies must be included: mainly the measurements of the half-life of industrially produced plasma components (e.g. HSA); analytical applications used in clinical laboratories (mainly RIA and ELISA); and studies on the blood coagulation effects (e.g. with ^{131}I or ^{125}I-fibrinogen [547)] or ^{3}H-fibrinogen [132)]).

Radiolabels are also important tools in pharmacokinetic studies for detecting various immune modalities [165], and also for different kinds of external injuries. Toth [572] studied the distribution of ^{125}I-labeled fibrinogen in rats during endotoxin shock.

It can be concluded that the application of labeled proteins is at present the standard method used in medical practice and research.

7.8 Possible Industrial Applications

An increasing interest in biotechnology can be followed all over the world. The lower requirements on energy sources, the diminished extent of enviromental pollution, the completly new attempts to solve various problems are some of the main factors which favour biotechnology and forecast it to be a widely used technology in the future. In "biotechnology news", frequently discussed in journals and books or at different symposia, congresses and meetings, proteins, especially those having biological activity (enzymes, hormones, etc.) play a significant role. Nevertheless, it is no secret, that the serious retardation of such projects lies very often in the uneconomical purification of these proteins. These obstacles, however, can be overcome with new, economically acceptable, separation techniques which will make repeated purification possible in a one-step procedure. Theoretically affinity chromatography offers such advantages. Unfortunately the first attempts to apply affinity chromatography on an industrial scale were not too successful owing to the high price of affinity ligands, the necessary prepurification procedures, the unfavourable absorption effects which usually limited the number of repeated separations and so on.

The introduction of dyes in affinity chromatography seems to remove a lot of these problems. The advantages of dye-ligand chromatography summarized by Hey and Dean [300], are given in the Table 13. One of the most important merits of this technique is undoubtly the relatively low cost of the dyes (they were introduced by ICI in 1956 for colouring in textile industry), the stability and re-usability of the columns and the possibility of applying a "tandem operation". This technique [300]

Table 13. Advantages of dye-ligand chromatography (according to Hey and Dean [300])

Ease of preparation
Economy
Lack of complex organic ligand synthesis
Avoids CNBr chemistry
Stable both enzymatically and chemically
Dry storage of gels possible
Readily available
Group-specific nature of dyes
Large-scale potencial
Reusability
High capacity of dye matrices
High flow rates
Chromatography unaffected by detergents

enables (i) the balast of proteins on the column to be removed with immobilized dye, which does not adsorb isolated protein and (ii) this partly purified protein to be then further purified and concentrated on the immobilized dye which binds it, as far as possible, specifically. It is preferable that the elution of the bound protein is performed unspecifically (i.e. by ionic strength alternations, buffer pH or solvent changes), which is the cheapest way. In some cases, however, the biospecific elution should also be suited, as far as the degree of purification is concerned, to the cost of the biospecific agent applied, e.g. the 6590-fold purification of orotidylate decarboxylase in one step on a Cibacron Blue 3G-A column using UMP for elution can serve as an example [574].

Given below are examples of preparative and large-scale applications of dye-ligand chromatography which indicate that this type of affinity chromatography may be anticipated as an industrial separation technique in near the future.

Immobilied Cibacron Blue 3G-A has been found very efficient for the isolation of human serum albumin [575], interferons produced in tissue cultures [576], phosphoglycerate kinase from Saccharomyces cerevisiae [577], and alcohol dehydrogenase from equine liver. Immobilized Procion Red HE-3B and Procion Blue MX-4GD have been successfully applied for the isolation of 3-hydroxybutyrate and malate dehydrogenases [578].

8 Concluding Remarks

The aim of this study was to show wide spectrum of possible labels for proteins and the techniques for incorporating them into protein molecules. The choice of label used and the method of labeling is closely related to the purpose for which the labeled protein is to be employed. Although there are many different approaches, they can often fulfil the same final goal (e.g. various protein labels in immunoanalysis).

The required properties of labeled proteins are sometimes contradictory. In the majority of applications involving labeled proteins, preservation of the original properties of the protein before labeling is the longing wish. Labeling should be gentle and, if possible "inconspicuous". A labeled protein is expected to behave in the same manner as an unlabeled one. On the other hand, under certain circumstances, only the specific changes in properties of the proteins (e.g. loss of enzymatic activity) which result from the labeling (e.g. with certain affinity labels) enables one to study the arrangement of the binding and activity sites of biologically active proteins, or to modify the course of metabolic processes in order to study the course of various cellular events. Sometimes even denatured labeled proteins are used for assaying proteolytic activity or in scintigraphy (protein macroaggregates).

Concerning label attachment, all types of binding (covalent, adsorption, biospecific interactions, etc.) are included in the labeling techniques used.

Taking into account all these facts, the following definition of a protein label is proposed:

A label can be any natural part of a protein molecule or (more frequently) an artificially introduced component which can be easily detected by a suitable analytical technique or which causes certain typical change in some of protein properties.

The present trends in protein labeling may be briefly summarized as follows:

a) The search for the most suitable labels for a given application.
b) The search for the simplest method of labeling (mainly eliminating the separation steps) which will produce the properties required of the labeled protein.
c) To combine the lowest degree of labeling with the highest efficiency by sensitive analytical techniques combined with steadily improving instrumentation.
d) To diminish all unfavourable side effects, including the health risk, mainly in medical applications and routine analytical procedures.
e) The search for new types of specific labels (e.g. affinity ones) which will help us to obtain completely new information on the structure, function and role of proteins in different cellular processes. This is the most difficult, but also most promising field of labeled protein application.
f) Wide utilization of labeled proteins in analysis and medical diagnosis.
g) "Temporary labeling", based on the different specific protein interactions, offers effective separation procedures, which could be applied in the large-scale purification of biologically active proteins.

9 References

1. Stenesh, J.: Dictionary of Biochemistry, John Wiley and Sons, New York 1975, p. 88
2. Hughes, R. C., Butters, T. D.: Trends in Biochemical Sciences (TIBS) *6*, 228 (1981)
3. Montreuil, J.: Adv. Carb. Chem. *37*, 157 (1980)
4. Schlesinger, P. H. et al.: Biochem. J. *176*, 103 (1978)
5. Turková, J.: Affinity chromatography, Elsevier, Amsterdam 1978 (extended edition is in preparation)
6. Chapman, A., Kornfeld, R.: J. Biol. Chem. *254*, 816 (1979)
7. Sandmeier, E., Christen, P.: ibid. *255*, 10284 (1980)
8. Sonderegger, P. et al.: Biochem. Biophys. Res. Commun. *94*, 1256 (1980)
9. Aten, A. H. W., Hevesy, G.: Nature *142*, 111 (1938)
10. Banks, T. E. et al.: Biochem. J. *40*, 734 (1946)
11. Loiseleur, J.: Compt. Rend. *201*, 1511 (1935)
12. Bolton, A. E.: Rev. Radiochem. Cent. *18*, 77 (1977)
13. Report IAEA-SM—210/308, Vienna, Austria
14. Harwig, J. F. et al.: Int. J. Appl. Radiat. Isot. *28*, 157 (1977)
15. Harwig, J. F. et al.: J. Nucl. Med. *17*, 397 (1976)
16. Rijk, P. P. van.: Int. Nuclear Information System (INIS)-mf—3500, 25 Jan. 1977
17. Caro, R. A. et al.: Int. J. Appl. Radiat. Isot. *26*, 527 (1975)
18. Doran, D. M., Spar, I. L.: J. Immunol. Methods *39*, 155 (1980)
19. Stevens, A. M. et al.: Int. J. Nucl. Med. Biol. *5*, 83 (1978)
20. Stahr, G., Stansbury, M., Sammon, P. J.: Int. J. Appl. Radiat. Isot. *30*, 359 (1979)
21. Koch, N., Haustein, D.: J. Immunol. Methods *41*, 163 (1981)
22. Tatum, J. L., Briner, W. H., Goodrich, J. K.: Invest. Radiol. *14*, 185 (1979)
23. Krueger, D., Maelicke, A.: Labelled, Immunoassay Horm. Drugs, 81, Proc. Int. Symp. (1978)
24. Sammon, P. J., Stansbury, M., Stahr, G.: Int. J. Appl. Radiat. Isot. *30*, 359 (1979)
25. Johannsen, R., Syhre, R.: Zentralinstitut für Kernforschung, ZfK—280, Dresden, GDR (1974)
26. Valeyre, J., Deltour, G., Delisle, M. J.: 18. French language symp. on nuclear medicine, Reims, 9–12 June 1977
27. Catalogue fy Pierce Eurochemie B.V., Febr. 1982
28. Catalogue fy Bio-Rad Laboratories, Technical Bulletin 1071G, May 1981
29. Bolton, A. E. et al.: Clin. Chem. *25*, 1826 (1979)

30. Knight, L. C., Budzynski, A. Z., Olexa, S. A.: J. Labelled Compd. Radiopharm., 3. Int. symp. on radiopharmaceutical chemistry, St. Louis, USA, 16–20 June 1980
31. Karonen, S. L. et al.: Anal. Biochem. *67*, 1 (1975)
32. Tixhon, C.: Rev. IRE *3*, 13 (1978)
33. Wood, W. G., Wachter, C., Scriba, P. C.: Fresenius' Z. Anal. Chem. *301*, 119 (1980)
34. Jilg, W., Hannig, K., Zeiller, K.: Hoppe-Seyler's Z. Physiol. Chem. *361*, 389 (1980)
35. Linhart, J., Beneš, J.: Radioisotopy *14*, 871 (1973)
36. Tuszynski, G. P., Knight, L. C., Piperno, J. R.: J. Labelled Compd. Radiopharm. *18*, 206 (1981)
37. Berkes, I., Jovanovič, V., Bzenic, J.: Isotopenpraxis *11*, 18 (1975)
38. Bonetto,O.et al.: Comision Nacional de Energia Atomica, CNEA—428, Buenos Aires, Argentina 1976
39. Caro, R. A. et al.: Comision Nacional de Energia Atomica, CNEA—403, Buenos Aires, Argentina 1975
40. Welch, M. J., Krohn, K. A.: Int. symp. on radiopharmaceuticals, Atlanta, Georgia, USA, 12 Febr. 1974
41. Almeida, M. R. H., Abreu Maffei, F. H.: Cienc. Cult. (Sao Paulo) Supl. *32*, 873 (1980)
42. Hawker, R. J., Hawker, L. M.: J. Clin. Pathol. *29*, 495 (1976)
43. Rockmann, T. M. B. et al.: Cienc. Cult. (Sao Paulo) *27*, 559 (1975)
44. Moza, A. K., Kumar, M., Sapru, R. P.: J. Lab. Clin. Med. *87*, 169 (1976)
45. Loberg, M., Miller, I., Cooper, M.: Int. symp. on radiopharmaceuticals, Atlanta, Georgia, USA, 12 Febr. 1974
46. Moza, A. K. et al.: Bull. Postgrad. Inst. Med. Educ. Res. *8*, 121 (1974)
47. Harwig, J. F. et al.: J. Nucl. Med. *16*, 756 (1975)
48. Simon, J. et al.: Nouv. Presse Med. *4*, 3063 (1975)
49. George, E. A. et al.: J. Nucl. Med. *17*, 175 (1976)
50. Charlton, J. R., Gravett, D. L.: US Pat. 3,933,996 (1976)
51. Mozhajskij, A. M. et al.: Isotopenpraxis *11*, 305 (1975)
52. Berghem, L. et al.: Scand. J. Clin. Lab. Invest. *36*, 849 (1976)
53. Schaffalitzky de Muckadell, O. B., Fahrenkrug, J.: ibid. *36*, 661 (1976)
54. Hemmaplardh, D., Morgan, E. H.: Int. J. Appl. Radiat. Isot. *27*, 89 (1976)
55. Baur, S., Bacon, V. C., Rosenquist, G. L.: Anal. Biochem. *87*, 71 (1978)
56. Delpech, M., Pasquier, C.: Centre de Recherches du Service de Sante des Armees, CRSSA-RA—1978, p. 291, 92-Clamart (France)
57. Kim, Y. S., Park, K. B.: J. Korean. Nucl. Soc. *10*, 1 (1978)
58. Taylor, D. M.: Int. J. Appl. Radiat. Isot. *31*, 192 (1980)
59. Cavalieri, R. R., McMahon, F. A., Castle, J. N.: J. Clin. Invest. *56*, 79 (1975)
60. Cloix, J. F. et al.: Pathol.-Biol. *23*, 827 (1974)
61. Caro, R. A. et al.: Comision Nacional de Energia Atomica, CNEA—389, Buenos Aires, Argentina 1975
62. Bjorklid, E., Gierchsky, K. E., Prydz, H.: Br. J. Haematol. *39*, 445 (1978)
63. Kaplan, S. L.: Symp. on nuclear medicine in pediatrics, San Francisco, California, USA, 16 Febr. 1973
64. Schaik, F. W., Lequin, F. M., Thyssen, J. H. H.: 18th Dutch Federative Meet., abstracts symposia and communications, Leiden, The Netherlands, 13–15 April 1977
65. Gens, J.: Zentralinstitut für Kernforschung, ZfK—294, p. 82, Rossendorf bei Dresden, GDR (1975)
66. Hauger-Klevene, J. H., Quihillalt, E. L., Mitta, A. E. A.: CNEA—386, Buenos Aires, Argentina 1975
67. Hutchinson, H. D., Ziegler, D. W.: Appl. Microbiol. *28*, 935 (1974)
68. Schmidt, H. E. et al.: Isotopenpraxis *10*, 401 (1974)
69. Dambuyant, C., Sizaret, Ph.: J. Immunol. Methods *8*, 289 (1975)
70. Roll, F. J., Madri, J. A., Furthmayr, H.: Anal. Biochem. *96*, 489 (1979)
71. Fainaru, M., Havel, R. J., Felker, T. E.: Biochim. Biophys. Acta *446*, 56 (1976)
72. Jarošková, L., Kovářů, F.: J. Immunol. Methods *22*, 253 (1978)
73. Ramlau, J., Bjerrum, O. J.: Scand. J. Immunol. *6*, 867 (1977)
74. Wood, F. T., Wu, M. M., Gerhart, J. C.: Anal. Biochem. *69*, 339 (1975)
75. Yoshioka, T. et al.: Proc. Jpn. Acad. *54-B*, 310 (1978)

76. Booth, A. G., Kenny, A. J.: Biochem. J. *187*, 31 (1980)
77. Chowdhary, S. Y. et al.: Isotopenpraxis *11*, 281 (1975)
78. Kutas, V., Kocsar, L., Holland, J.: Int. J. Appl. Radiat. Isot. *26*, 31 (1975)
79. Chavez, L. C., Sison, B.: Philipp. J. Cardiol. *2*, 35 (1974)
80. Frisbie, J. H. et al.: J. Nucl. Med. *16*, 393 (1975)
81. Hagan, P. et al.: J. Nucl. Med. *15*, 974 (1974)
82. Coleman, R. E. et al.: Circ. Res. *37*, 35 (1975)
83. Rhodes, F. A., Bolles, T. P.: Int. symp. on radiopharmaceuticals, Atlanta, Georgia, USA, 12 Febr. 1974
84. Colombetti, L. G., Moerlien, S., Pirsky, S.: Int. J. Nucl. Med. Biol. *2*, 180 (1975)
85. Lin, M. S.: Int. Symp. on Radiopharmaceuticals, Atlanta, Gerogia, USA, 12 Febr. 1974
86. Kaepfer, I., Nissler, K., Schneider, G.: Radiobiol. Radiother. *19*, 728 (1978)
87. Stanko, V. I. et al.: Isotopenpraxis *14*, 6 (1978)
88. Senda, K., Imaeda, T.: Rinsho Hoshasen *21*, 481 (1976)
89. Bruno, G. A., Haney, T. A., Rajamani, P. N.: US pat. 3,862,299 (1975)
90. Villa, M. et al.: Int. J. Nucl. Med. Biol. *5*, 51 (1978)
91. Richards, P., Steigman, J.: Int. symp. on radiopharmaceuticals, Atlanta, Georgia, USA, 12 Febr. 1974
92. Yokoyama, A. et al.: Int. J. Appl. Radiat. Isot. *26*, 291 (1975)
93. Billinghurst, M. W., Rempel, S., Westendorf, B. A.: J. Labelled Compd. Radiopharm. *18*, 651 (1981)
94. Pettit, W. A., DeLand, F. H., Bennett, S. J.: Int. J. Appl. Radiat. Isot. *29*, 344 (1978)
95. Alvarez Cervera, J.: Report IAEA-R—1048-F, Vienna, Austria 1975
96. Rehani, M. M., Sharma, S. K.: J. Nucl. Med. *21*, 676 (1980)
97. Dewanjee, M. K., Brueggemann, P., Wahner, H. W.: Radiopharm 2, Proc. Int. Symp. 2nd, p. 435 (1979)
98. Wong, D. W., Mishkin, F. S.: J. Nucl. Med. *16*, 343 (1975)
99. Bunko, H., Tatsuno, I., Kato, S.: Radioisotopes *23*, 706 (1974)
100. Narasimhan, D. V. S., Mani, R. S.: Radiochem. Radioanal. Lett. *20*, 307 (1975)
101. Ligny, C. L., Dekker, B. G.: ibid. *34*, 7 (1978)
102. Pettit, W. A. et al.: J. Nucl. Med. *21*, 62 (1980)
103. Union Carbide Corp., New York: Netherlands patent document 7613339(A), (1976)
104. Wong, D. W., Mishkin, F., Lee, T.: Int. J. Appl. Radiat. Isot. *29*, 251 (1978)
105. Darte, L., Persson, B. R. R.: Lund Univ. Radiofysiska Inst. LURI—1975-07, Sweden
106. Cífka, J.: Report IAEA-R—1021-F, Vienna, Austria 1975
107. Darte, L., Persson, B. P., Scederbom, L.: Nucl.-Med. *15*, 80 (1976)
108. Chervu, L. R., Vallabhajosula, S. R., Blaufox, M. D.: Int. J. Nucl. Med. Biol. *4*, 201 (1977)
109. Wong, D. W. et al.: J. Nucl. Med. *20*, 967 (1979)
110. Paik, C. H. et al.: Proc. of Korea symp. on science and technology, p. 332, Seoul, Korea 1980
111. Wong, D. W., Huang, J. T.: Int. J. Appl. Radiat. Isot. *28*, 719 (1977)
112. Harwig, J. F. et al.: ibid. *27*, 5 (1976)
113. De Ligny, C. L., Gelsema, W. J., Beunis, M. H.: ibid. *27*, 351 (1976)
114. Janoki, Gy., Kocsar, L., Spett, B.: Izotoptechnika *19*, 90 (1976)
115. Persson, B. R. R., Darte, L.: Int. J. Appl. Radiat. Isot. *28*, 97 (1977)
116. Hebestreit, H. P., Pryss, C.: Nucl.-Med. *14*, 172 (1975)
117. Marqversen, J.: Scand. J. Clin. Lab. Invest. *35*, 50 (1975)
118. Gil, M. C., Palma, T., Radicella, R.: Int. J. Appl. Radiat. Isot. *27*, 69 (1976)
119. Winchell, H. S. et al.: French patent document 2187307(A), 1973
120. Soini, E.: Science Tools *25*, 38 (1978)
121. Vobecký, M.: Tables of isotopes, in Nuclear chemistry tables (ed. Jakeš, D. et al.) p. 14, Praha, SNTL 1964
122. Ehrenkaufer, RL. E., Wolf, A. P., Hembree, W. C.: U.S. 4,162,142, 24 July 1979
123. Tschesche, H., Behr, W., Wick, R.: Insulin: Chem. Struct., Funct. Insulin Relat. Horm., Proc. Int. Insulin Symp., 2nd 1979, (ed. Brandenburg, D., Wollmer, A.) de Gruyter, Berlin, 1980
124. Wessels, B. W. et al.: Radiat. Res. *74*, 35 (1978)

125. Report COO—3105-6, Columbia Univ., Coll. of Physicians and Surgeons, New York, USA 1977
126. Fromagect, P., Hung, L. T., Morgat, J. L.: U.S. pat. 3,828,102, 6 Aug. 1974
127. Moran, T. F., Powers, J. C., Lively, M. O.: U.S. pat. document 2044772(A), 22 Oct. 1980
128. Hukkanen, V., Frey, H., Salmi, A.: Brain Res. *205*, 207 (1981)
129. Ascoli, M., Puett, D.: Biochim. Biophys. Acta *371*, 203 (1974)
130. Lefevere, M. F., Slegers, G. A., Claeys, A. E.: Clin. Chim. Acta *92*, 167 (1979)
131. Marsh, J. B.: J. Lipid Res. *19*, 107 (1978)
132. Wegrzynowicz, Z., Kloczewiak, M., Kopec, M.: J. Lab. Clin. Med. *86*, 380 (1975)
133. Mueller, G. H.: J. Cell Sci. *43*, 319 (1980)
134. Fink, D. J., Gainer, H.: Brain Res. *177*, 208 (1979)
135. Toribara, T. Y.: Int. J. Appl. Radiat. Isot. *29*, 597 (1978)
136. Kummer, U. et al.: J. Immunol. Methods *42*, 367 (1981)
137. Ginsel, L. A., Onderwater, J. J. M., Daems, W. T.: Virchows Arch., Abt. B. *30*, 245 (1979)
138. Seproedi, J., Teplan, I., Medzihradszky, K.: Int. symp. on radioactive isotope in clinacal medicine and research, Bad Gastein, Austria, 12 Jan. 1976
139. La Llosa, P. de, La Llosa-Hermier, M. P. de.: FEBS Lett. *45*, 162 (1974)
140. Marche, P.: FRNC-TH—589, Paris-7, Univ., 75 (France), 1975
141. Kolodziejczyk, A., Arendt, A.: Rocz. Che. *51*, 659 (1977)
142. Montelaro, R. C., Rueckert, R. R.: J. Biol. Chem. *250*, 1413 (1975)
143. Menzel, J.: J. Immunol. Methods *18*, 257 (1977)
144. Straatmann, M. G., Welch, M. J.: J. Nucl. Med. *16*, 425 (1975)
145. Marche, P. et al.: Radiochem. Radioanal. Lett. *21*, 53 (1975)
146. MacKeen, L. A., Diperi, C., Schwartz, I.: FEBS Lett. *101*, 387 (1979)
147. Kraal, B., Hartley, B. S.: J. Mol. Biol. *124*, 551 (1978)
148. Ottesen, M., Svensson, B.: Compt. Rend. Trav. Lab. Carlsberg *38*, 445 (1971)
149. Jentoff, N., Deaborn, D. G.: Anal. Biochem. *106*, 186 (1980)
150. Koch, G. K., Heertje, I., Van Stijn, F.: Radiochim. Acta *24*, 215 (1977)
151. Lier, J. E. van, Sanche, L.: J. Radioanal. Chem. *53*, 273 (1979)
152. Lier, J. E. van, Sanche, L.: 11. Int. symp. on chemistry of natural products, v. 3, Golden Sands, Bulgaria, 17–23 Sept. 1978
153. Sanche, L., Lier, J. E. van: Nature *263*, 79 (1976)
154. Laengstroem, B., Sjoeberg, S., Ragnarsson, U.: J. Labeled Compd. Radiopharm. *18*, 479 (1981)
155. Schwarz, R. T. et al.: J. Virol. *23*, 217 (1977)
156. Jakubick, V., Diehl, J. F.: Z. Ernaehrungswiss. *19*, 33 (1980)
157. Heise, K. H., Bubner, M.: Isotopenpraxis *12*, 125 (1976)
158. Holt, L. A., Milligan, B., Rivett, D. E.: 5. Int. symp. on wool textile research, Aachen, FRG, 2 Sept. 1975
159. Menzel, J.: J. Immunol. Methods *15*, 77 (1977)
160. Mitrius, J. C., Morgan, D. G., Routtenberg, A.: Brain Res. *212*, 67 (1981)
161. Schmidt, G., Billwitz, H., Lindigkeit, R.: Acta Biol. Med. Ger. *39*, 21 (1980)
162. Amir-Zaltsman, Y. et al.: FEBS Lett. *122*, 166 (1980)
163. Litske Petersen, J. G.: Carlsberg Res. Commun. *44*, 395 (1980)
164. Uemura, I., Murakami, H.: Radioisotopes (Tokyo) *24*, 318 (1975)
165. Nanba, H.: Naika Hokan *26*, 137 (1979)
166. Bjoernson, J.: Scand. J. Haematol. *13*, 252 (1974)
167. Malcolme-Lawes, D. J.: Int. J. Appl. Radiat. Isot. *31*, 190 (1980)
168. Dauchin, A.: Biochimie *55*, 17 (1973)
169. Whitaker, T. R. et al.: J. Lab. Clin. Med. *84*, 879 (1974)
170. Decsy, M. I., Sunderman, F. W. Jr.: Bioinorg. Chem. *3*, 95 (1974)
171. Wagner, S. J., Welch, M. J.: J. Nucl. Med. *20*, 428 (1979)
172. Hnatowich, D. J., Schlegel, P.: ibid. *22*, 623 (1981)
173. Hnatowich, D. J.: ibid. *17*, 57 (1976)
174. Maziere, B. et al.: 20. French language symp. on nuclear medicine, Bordeaux, France, 20–22 Sept. 1979
175. Karpatkin, S.: Br. J. Haematol. *39*, 459 (1978)
176. Dosseto, M. et al.: J. Immunol. Methods *41*, 145 (1981)
177. Knight, L. et al.: Int. symp. on radiopharmaceuticals, Atlanta, Georgia, USA, 12 Febr. 1974

178. McElvary, K. D., Knight, L. C., Welch, M. J.: Mar. Algae Pharm. Sci. 429 (1979) after C.A. *92*, 71924g (1980)
179. Pettit, W. A., Deland, F. H., Blanton, L.: Radiopharm., Proc. Int. Symp., 2nd, (ed. Sorenson, J. A.), Soc. Nucl. Med. Inc. New York 1979, p. 33
180. McElvary, K. D., Welch, M. J.: J. Nucl. Med. *21*, 953 (1980)
181. McElvary, K. D. et al.: Univ. St. Louis, Mo, USA, report 1980
182. McElvary, K. D., Welch, M. J.: Int. J. Appl. Radiat. Isot. *31*, 679 (1980)
183. Kselikova, M. et al.: Radiobiol. Radiother. *20*, 701 (1979)
184. Kselikova, M. et al.: Industrial and Environmental Xenobiotics, Proc. of an Int. Conf. held in Prague, Czechoslovakia, 27–30 May 1980, (ed. Gut, I., Cikrt, M., Plaa, G. L.), Springer-Verlag 1981, p. 93
185. Kolchakov, K., Vnekov, L.: Compt. Rend. Acad. Bulgare Sci. *18*, 219 (1965)
186. Goodwin, D. A. et al.: Int. symp. on medical radionuclide imaging Los Angeles, Calif., USA, 25–29 Oct. 1976
187. Goodwin, D. A. et al.: Nucl.-Med. *14*, 365 (1975)
188. Hagan, P. L. et al.: J. Nucl. Med. *19*, 1055 (1978)
189. Scheffel, U., Tsan, M. F., McIntyre, P. A.: ibid. *20*, 524 (1979)
190. Khaw, B. A. et al.: Science *209*, 295 (1980)
191. Csetenyi, J. et al.: Radiobiol. Radiother, *15*, 55 (1974)
192. Choudhary, S. Y., Mari, T. S.: Proc. of the chemistry symp., Aligarh, India, 21–23 Dec. 1972
193. Caro, R. A. et al.: Int. J. Appl. Radiat. Isot. *25*, 501 (1974)
194. Caro, R. A. et al.: Comision Nacional de Energia Atomica, Buenos Aires, Argentina, report n. 385 (1975)
195. Allen, D. R. et al.: J. Nucl. Med. *15*, 821 (1974)
196. Petriyev, V. K., Stepchenkov, V. I.: Isotopenpraxis *16*, 202 (1980)
197. Saha, G. B., Girschek, P. K., Farrer, P. A.: Int. J. Nucl. Med. Biol. *4*, 92 (1977)
198. Hofmann, H. D., Kadenbach, B.: Eur. J. Biochem. *102*, 605 (1979)
199. Friedman, A. M., Zalutsky, M. R., Appelman, E. H.: Int. J. Nucl. Med. Biol. *4*, 219 (1977)
200. Vaughan, A. T. M.: Int. J. Appl. Radiat. Isot. *30*, 576 (1979)
201. Aaij, G. et al.: ibid. *26*, 25 (1975)
202. Fremlin, J. H., Vaughan, A. T. M.: Int. J. Nucl. Med. Biol. *5*, 229 (1978)
203. Vaughan, A. T. M., Bateman, W., Cowan, J.: J. Radioanal. Chem. *64*, 33 (1981)
204. Anon., Argonne National Lab., Illinois, USA, report n. 8096, April 1974
205. Woodworth, R. C., Dobson, C. M., Brown-Mason, A.: J. Biol. Chem. *256*, 1506 (1981)
206. Kranc, M. C.: Memphis State Univ., Tenn., USA, report, 1976
207. Giorgi, P. P., Dubois, H.: Biochem. J. *196*, 537 (1981)
208. Wickramasinghe, S. N., Hughes, M.: Br. J. Haematol. *38*, 179 (1978)
209. Innocenti, A. M., Floris, C.: Biochem. Physiol. Pflanz. *174*, 404 (1979)
210. Forman, D. S., Berenberg, R. A.: Brain Res. *156*, 213 (1978)
211. Hakarrainen, J.: Acta Vet. Scand., Suppl. *59*, 1 (1975)
212. Vivekanandan, R., Singh, L. N.: Symp. on use of radiations and radioisotopes in studies of animal production, Izatnagar, India, 16–18 Dec. 1975
213. Paik, W. K. et al.: Anal. Biochem. *90*, 262 (1978)
214. Vinogradova, R. P., Litvinenko, A. R.: Radiobiologiya *17*, 567 (1977)
215. Petersen, J. G. L., Pinon, R.: Carlsberg Res. Commun. *44*, 395 (1979)
216. Sinichkin, A. et al.: J. Neurochem. *29*, 425 (1977)
217. Zaidenberg, M. A., Korotkina, R. N., Konikov, A. S.: Int. J. Appl. Radiat. Isot. *25*, 333 (1974)
218. Sovová, V., Černá, H., Bosch, V.: Folia Biol. (Prague) *24*, 199 (1978)
219. Karjalainen, S., Soederling, E.: Arch. Oral Biol. *24*, 909 (1979)
220. Hatakeyama, T., Inaba, S., Sugimura, K.: Tokyo Nogyo Daigaku Nogaku Shuho *23*, 67 (1978)
221. Koide, N.: Kobe J. Med. Sci. *24*, 17 (1978)
222. Anttonen, O. et al.: Biochem. J. *185*, 189 (1980)
223. Maccioni, A. H. R. et al.: Brain Res. *187*, 247 (1980)
224. Fallon, R. D., Pfaender, F. K.: Chesapeake Sci. *17*, 292 (1976)
225. Freyssinet, G.: Biol. Cell. *30*, 17 (1977)
226. Joernvall, H., Bahr-Lindstroem, H. von, Philipson, L.: FEBS Lett. *88*, 237 (1978)
227. Triphaus, G. P., Schmidt, A., Buddecke, E.: Hoppe-Seylers Z. Physiol. Chem. *361*, 1773 (1980)
228. Bisseling, T.: Report INIS-mf—5993, Austria 1980

229. Stuik, E.: Report INIS-mf—5902, Vienna, Austria 1979
230. Rocijen, N. van: J. Immunol. Methods *15*, 267 (1977)
231. Lapidot, A., Irving, C. S.: 2. Int. Conf. on Stable Isotopes, Oak Brook, Illinois, USA, 20 Oct. 1975
232. Fourie, P. J.: School on chemistry in nuclear technology, Johannesburg, South Africa, 24–26 Oct. 1977
233. Gavaret, J. M. et al.: Horm. Metab. Res. *7*, 166 (1975)
234. Uehleke, H., Poplawski-Tabarelli, S.: Arch. Toxikol. *37*, 289 (1977)
235. Kasche, W.: Radiochem. Radioanal. Lett. *3*, 51 (1970)
236. Paus, E.: Int. Atomic Energy Agency-SM-259/18, 35 (1982)
237. Fracker, P. J., Speck, J. C.: Biochem. Biophys. Res. Commun. *80*, 849 (1978)
238. Markwell, M. A., Fox, C. F.: Biochemistry *17*, 4807 (1978)
239. Salacinski, P. R. P. et al.: Anal. Biochem. *117*, 136 (1981)
240. O'Leary, M. H., Westheimer, F. H.: Biochemistry *7*, 913 (1968)
241. Ostrowski, K. et al.: J. of Histochem. and Cytochem. *18*, 490 (1970)
242. Heinegard, D. K., Hascall, V. C.: J. Biol. Chem. *254*, 921 (1979)
243. Brems, D. N., Rilling, H. C.: Biochemistry *18*, 860 (1979)
244. Gersten, D. M., Goldstein, L. S.: Int. J. Appl. Radiat. Isot. *30*, 469 (1979)
245. Reinwald, E., Risse, H. J., Saelker, R.: Hoppe-Seyler's Z. Physiol. Chem. *359*, 939 (1978)
246. Rudinger, J., Ruegg, U.: Biochem. J. *133*, 538 (1973)
247. Bolton, A. E., Hunter, W. M.: ibid. *133*, 529 (1973)
248. Gonyea, L. M.: Clin. Chem. *23*, 234 (1977)
249. Fang, V. S., Cho, H. W., Meltzer, H. Y.: Biochem. Biophys. Res. Commun. *65*, 413 (1975)
250. Kasamatsu, H., Wu. M.: Proc. Natl. Acad. Sci. USA *73*, 1945 (1976)
251. Morrod, P., Clark, D. G.: FEBS Lett. *59*, 316 (1975)
252. Rosano, T. G., Kenny, M. A.: Clin. Chem. *23*, 69 (1977)
253. Airhart, J. et al.: Anal. Biochem. *96*, 45 (1979)
254. Davies, G. E., Stark, G. R.: Proc. Nat. Acad. Sci. USA *66*, 651 (1970)
255. Whiteley, N. M., Berg, H. C.: J. Mol. Biol. *87*, 451 (1974)
256. New Product News 1, March 1982 (catalogue NEN — New England Nuclear)
257. Gerber, G. B., Remy-Defraigne, J.: Anal. Biochem. *11*, 386 (1965)
258. Tolleshaug, H., et al.: Biochim. Biophys. Acta *585*, 71 (1979)
259. Bregman, M. D., Cheng, S., Levy, D.: ibid. *539*, 489 (1978)
260. Catalogue Amersham Radiochemical Centre, England 1982
261. Butler, P. J. G. et al.: Biochem. J. *112*, 679 (1969)
262. Wood, F. T., Wu, M. M., Gerhart, J. C.: Anal. Biochem. *69*, 339 (1975)
263. Callewaert, G. L., Vernon, C. A.: Biochem. J. *107*, 728 (1968)
264. Kumarasamy, R., Symons, R. H.: Anal. Biochem. *95*, 359 (1979)
265. Keul, V. et al.: J. Biol. Chem. *254*, 843 (1979)
266. Frist, R. H. et al.: Virology *26*, 559 (1965)
267. Fink, D. J., Gainer, H.: Science *208*, 303 (1980)
268. Galardy, R. E. et al.: J. Biol. Chem. *249*, 3510 (1974)
269. Schwartz, M. A., Hynes, R. O.: ibid., in press (1982)
270. Fanger, M. W., Hettinger, T. P., Harbury, H. A.: Biochemistry *6*, 713 (1967)
271. Gerwin, B. I.: J. Biol. Chem. *242*, 451 (1967)
272. Velick, S. F.: ibid. *203*, 563 (1953)
273. Waterman, M. R.: Biochim. Biophys. Acta *371*, 159 (1974)
274. Fischer, E. P., Thomson, K. S.: J. Biol. Chem. *254*, 50 (1979)
275. Hendee, W. R.: Radioactive Isotopes in Biological Research, J. Wiley and Sons, London 1973, p. 83
276. Jirousek, L.: On the stability of Radioiodine Labeled Compounds, p. 5, New England Nuclear brochure, April 1979
277. Opresko, L. et al.: Proc. Natl. Acad. Sci. USA *77*, 1556 (1980)
278. Linde, S. et al.: Anal. Biochem. *107*, 165 (1980)
279. Seidah, N. G. et al.: ibid. *109*, 185 (1980)
280. Jurčovičová, J., Klimeš, I.: Radiochem. Radioanal. Lett. *34*, 187 (1978)
281. Anbar, M., Neta, P.: Int. J. Appl. Radiat. Isot. *18*, 493 (1967)
282. Ramsey, N. W. et al.: Br. J. Radiol. *53*, 357 (1980)

283. Kutas, V. et al.: Izotoptechnika *17*, 406 (1974)
284. Grützner, P.: Arch. Physiol. *8*, 452 (1874)
285. Scharvin, V.: Z. Angew. Chem. *26*, 254 (1913)
286. Rakuzin, M. A.: Bull. Inst. Sci. St. Petersbourg *1*, 171 (1920)
287. Gilbert, G. A.: Proc. Roy. Soc. A *183*, 167 (1944)
288. Nelson, W. L., Ciacio, E. I., Hess, G. P.: Anal. Biochem. *2*, 39 (1961)
289. Shackell, L. F.: J. Biol. Chem. *56*, 987 (1923)
290. Sabin, F. R.: J. Exptl. Med. *70*, 67 (1939)
291. Pechman, E. V.: Biochem. Z. *321*, 248 (1950)
292. Nelson, W. L. et al.: Anal. Biochem. *2*, 39 (1961)
293. Sachar, L. A. et al.: Proc. Soc. Exptl. Biol. Med. *90*, 323 (1955)
294. Hall, D. A.: Biochem. J. *101*, 29 (1966)
295. Saletan, L. T. et al.: Wallerstein Lab. Commun. *26*, 147 (1963)
296. Rinderknecht, H. et al.: Clin. Chem. Acta *21*, 197 (1968)
297. Stam, O. A.: Helv. Chim. Acta *46*, 3008 (1963)
298. Kun, L. J., Yoa, Ch. J.: Anal. Chem. *47*, 1634 (1975)
299. Hess, R., Pearse, A. G. E.: Nature *183*, 260 (1959)
300. Hey, Y., Dean, P. D. G.: Chemistry and Industry 726 (1981)
301. Stam, O. A. et al.: Helv. Chim. Acta *44*, 1123 (1961)
302. Bohme, H. J. et al.: J. Chromatogr. *69*, 209 (1972)
303. Edwards, R. A., Woody, R. W.: Biophys. Res. Commun. *79*, 470 (1977)
304. Gabe, M.: Histological Techniques, Springer-Verlag, Heidelberg 1976
305. Lilie, R. D., Fullmer, H. M.: Histopathologic Technic and Practical Histochemistry, McGraw-Hill Book Company, New York 1970
306. Culling, C. F. A.: Handbook of Histopathological and Histochemical Techniques, Butterworths, London 1974
307. Singer, S. J.: Adv. Prot. Chem. *22*, 1 (1967)
308. Birkett, D. J. et al.: The Reactivity of Group-Specific Chromophoric Reagents, in: Chemical Reactivity and Biological Role of Functional Groups in Enzymes (ed. Smellie, R. M. S.), Academic Press, London and New York 1970, p. 147
309. Freedman, R. B., Rada, G. K.: Biochem. J. *108*, 383 (1968)
310. Birkett, D. J. et al.: FEBS Lett. *6*, 346 (1970)
311. Scoffone, E., Jori, G., Galiazzo, G.: Selective photo-oxidation of amino acids in protein, in: Chemical Reactivity and Biological Role of Functional Groups in Enzymes (ed. Smellie, R. M. S.) Academic Press, London and New York 1970
312. Udenfriend, S.: Fluorescence Assay in Biology and Medicine, Academic Press, New York 1964[3], p. 191
313. Chen, R. F.: Extrinsic and Intrinsic Fluorescence of Protein, 467, in: Practical Fluorescence, Theory, Methods and Techniques (ed. Guibault, G. G.) Marcel Dekker, New York 1973
314. Feitelson, J.: J. Phys. Chem. *68*, 391 (1964)
315. Feitelson, J.: Photochem. Photobiol. *9*, 401 (1969)
316. Edelhoch, H. et al.: Biochemistry *7*, 3893 (1968)
317. Teale, F. W. J.: Biochem. J. *76*, 381 (1960)
318. Koening, J.: Histochem. Cytochem. *11*, 556 (1963)
319. Tapel, A. L.: Lysosomal enzymes and other components, in: Lysosomes in Biology and Pathology (ed. Dingle, J. T., Fell, H. B.) North-Holland Publ. Co., Amsterdam 1975, p. 207
320. Creech, H. J., Jones, R. N.: J. Am. Chem. Soc. *62*, 1970 (1940) and *63*, 1660 (1941)
321. Coons, A. H. et al.: Proc. Soc. Exptl. Biol. *47*, 200 (1941)
322. Nairn, R. C.: Fluorescent Protein Tracing, Livingstone, E. S., London 1969[3]
323. Riggs, J. L. et al.: Am. J. Pathol. *34*, 1081 (1958)
324. Reisher, J. I., Orr, H. C.: Anal. Biochem. *26*, 178 (1968)
325. Wild, A. E.: Fluorescent protein tracing in the study of endocytosis, in: Lysosomes in Biology and Pathology, (ed. Dingle, J. T.) North-Holland, Publ. Co., Amsterdam 1973, p. 511
326. Udenfriend, S. et al.: Science *378*, 871 (1972)
327. Felix, A., Jimenez, M. H.: J. Chromatogr. *89*, 361 (1974)
328. Lalande, M. E. et al.: Proc. Natl. Acad. Sci. *78*, 363 (1981)
329. Neuhoff, V. et al.: Hoppe-Seyler's Z. Physiol. Chem. *360*, 1657 (1979)
330. Rossomando, E. et al.: Anal. Biochem. *116*, 80 (1981)

331. Voller, A.: IAEA-SM—259/109, Vienna, Austria 1982
332. Schroeder, H. R. et al.: Methods in Enzymol. *57*, 424 (1978)
333. Fricke, H. et al.: J. Clin. Chem. Clin. Biochem. *20*, 91 (1982)
334. Avrameas, S., Uriel, J.: Compt. Rend. *262*, 2543 (1966)
335. Nakane, P. K., Pierce, G. B.: J. Histochem. Cytochem. *14*, 929 (1966)
336. Engvall, E., Perlmann, P.: Immunochemistry *8*, 871 (1971)
337. Van Weemen, B. K., Schuurs, A. H. W.: FEBS Lett. *15*, 232 (1971)
338. Engvall, E.: Enzyme-linked Immunosorbent Assay, ELISA, in: Biomedical Applications of Immobilized Enzymes and Proteins (ed. Chang, T. M. S.) Plenum Press, New York and London 1977, p. 87
339. Goldstein, L., Manecke, G.: The Chemistry of Enzyme Immobilization, in: Immobilized Enzyme Principles, (ed. Wingard, L. B., Katchalski-Katzin, E., Goldstein, L.), Academic Press, New York 1976
340. Avrameas, S.: Immunochemistry *6*, 43 (1969)
341. Avrameas, S., Ternynck, T.: ibid. *8*, 1175 (1971)
342. Kato, K. et al.: J. Biochem. *78*, 235 (1975)
343. Richards, F. M., Knowles, J. R.: J. Mol. Biol. *37*, 231 (1968)
344. Carlsson, J. H. et al.: Biochem. J. *173*, 723 (1978)
345. Nakane, P. K., Kawaci, J.: J. Histochem. Cytochem. *22*, 1084 (1974)
346. Schuurs, A. H. W. B., Van Weemen, B. K.: Clin. Chim. Acta *81*, 1 (1977)
347. Nakane, P. K.: Preparation and Standardization of Enzyme-Labeled Conjugates, in: Immunoassay in the clinical laboratory, (ed. Nakamura, P. M., Dito, W. R., Tucker, E. S.) Alan R. Liss. Inc., New York 1979, p. 81
348. Meyerhoff, M. E., Rechnitz, G. A.: Anal. Biochem. *95*, 483 (1979)
349. D'Orazio, P. et al.: Anal. Chem. *50*, 1513 (1978)
350. Meyerhoff, M. E., Rechnitz, G. A.: Science *195*, 494 (1977)
351. Henry, R. J., Cannon, D. C., Winkleman, J. W.: Clinical Chemistry. Principles and Technics, Chap. 21, Harper and Row, Hagerstown. Md. 1974
352. Aizawa, M. et al.: Anal. Biochem. *94*, 22 (1979)
353. Sternberger, L. A.: Enzyme Immunocytochemistry, in: Electron Microscopy of Enzymes. Principles and Methods. (ed. Hayat, M. A.) Van Nostrand Reinhold Comp., New York 1973, p. 150
354. Weber, T. H.: J. Labeled Compounds *8*, 449 (1972)
355. Grov, A. et al.: Acta Pathol. Microbiol. Scand. *61*, 588 (1964)
356. Sjoquist, J. et al.: Eur. J. Biochem. *30*, 190 (1972)
357. Björk, I.: ibid. *29*, 579 (1972)
358. Forsgren, A., Sjoquist, J.: J. Immunol. *97*, 822 (1966)
359. Ghetie, V. et al.: Immunology *26*, 1081 (1974)
360. Tokuda, S.: Nature *261*, 433 (1977)
361. Hellström, U. et al.: Scand. J. Immunol. *5*, Suppl. 5, 45 (1976)
362. Brochure: Helix Pomatia Lectin, Pharmacia Fine Chemicals, Uppsala, Sweden
363. Jakoby, W. B., Wilchek, M.: Affinity Labeling, in: Methods in Enzymology, Vol. 46, Academic Press 1977
364. Wofsy, L. et al.: Biochemistry *1*, 1031 (1962)
365. Rudnick, G. et al.: J. Biol. Chem. *250*, 1371 (1975)
366. Sonnenberg, N. et al.: Proc. Natl. Acad. Sci. USA *70*, 1423 (1973)
367. Wilchek, M., Bayer, E. A.: Affinity labeling from the isolated protein to the cell, in: Theory and practice in affinity techniques, (ed. Sundram, P. V., Eckstein, F.) Academic Press, New York 1978, p. 201
368. Baker, B. R. et al.: J. Am. Chem. Soc. *83*, 3713 (1961)
369. Bayer, E. A. et al.: FEBS Lett. *68*, 240 (1976)
370. Turková, J.: Affinity Chromatography, in: O. Mikeš: Laboratory Handbook of Chromatografic and Allied Methods, Ellis Horwood, Chichester 1979, p. 385
371. Jakoby, W. B., Wilchek, M.: Affinity techniques, in: Methods in Enzymology, Vol. 34, Academic Press 1974
372. Wilchek, M.: Enzyme Engineering, Vol. 4, Plenum Press, New York 1978
373. Roitt, I. M.: Essential Immunology, Blackwell Scientific Publications, Oxford—London 1980[4]

374. Kabat, E. A.: Structural Concepts in Immunology and Immunochemistry, Holt, Rinehart and Winston Inc., New York 1976
375. Singer, S. J.: Nature *183*, 1523 (1959)
376. Singer, S. J., Schick, A. F.: J. Biophys. Biochem. Cytol. *9*, 519 (1961)
377. Pierce, Jr., G. B., Ram, J. S., Midgley, A. R.: Int. Rev. Exp. Path. *3*, 1 (1964)
378. Zielinski, E.: Ger. (East) *129*, 216 (1978)
379. Franz, H.: Histochemie *12*, 230 (1968)
380. Kendall, P. A.: Biochim. Biophys. Acta *97*, 179 (1965)
381. Sternberg, L. A. et al.: J. Histochem. Cytochem. *14*, 711 (1966)
382. Sternberg, L. A. et al.: Exp. Mol. Path. Suppl. *3*, 36 (1966)
383. Voller, A.: IAEA-SM—259/109, p. 193 Vienna, Austria 1982
384. Foster, M. A. et al.: Cobalt as a Functional Group, in: Chemical Reactivity and Biological Role of Functional Groups in Enzymes, (ed. Smellie, R. M. S.) Academic Press, New York 1970, p. 187
385. Wagner, S. J. et al.: Anal. Biochem. *99*, 175 (1979)
386. Yen, Shiao-Ping, S. et al.: U.S. pat. 4,157, 323 (1979)
387. Gabel, Ch. A., Shapiro, B. M.: Anal. Biochem. *86*, 396 (1978)
388. Siviglia, G. et al.: ibid. *73*, 65 (1976)
389. Suzuki, F. et al.: Endocrinology *90*, 1220 (1972)
390. Armstrong, K. J. et al.: Biochem. Biophys. Res. Commun. *47*, 354 (1972)
391. Zambeletti SPA 04.04.73.IT.022595 (31. 07. 74) C07d, cited after Pharmacia booklet BHP-activated Dextran 70 (1977)
392. Zambeletti SPA 04.04.73.IT.022596 (31. 07. 74) A61k C07d, cited after Pharmacia booklet BHP-activated Dextran 70 (1977)
393. Marshall, J. J., Rabinowitz, M. L.: Arch. Biochem. Biophys. *167*, 777 (1975)
394. Adams, G. E. et al.: Pulse Radiolysis: An Oxidative-Reductive Probe in General Biochemical and Enzyme Studies, in: Chemical Reactivity and Biological Role of Functional Groups in Enzyme, (ed. Smellie, R. M. S.) Academic Press, New York 1970, p. 171
395. Kimlová, I. et al.: Čs. farmacie *27*, 460 (1978)
396. McCaman, R. E.: Isotopes and Radiation Technology *3*, 328 (1966)
397. Reed, D. J.: Methodology of radiotracer enzyme assays, in: Advances in Tracer Methodology (ed. Rotchild, S.) Plenum Press, New York 1968, p. 145
398. Oldham, K. G.: Radiochemical method of enzyme assays, review 9, The Radiochemical Centre, Amersham 1968
399. Absolon, K. B.: Proc. of Forum Sessions 40th Clinical Congr. of the American College of Surgeons, p. 543, Atlantic City, New York 1964
400. Mego, J. L., Queen, J. D.: Biochim. Biophys. Acta *100*, 136 (1965)
401. Reijngoud, D. J. et al.: ibid. *448*, 290 (1976)
402. Padayatty, J. D. et al.: J. Chromatogr. *34*, 259 (1968)
403. Sevier, E. D.: Anal. Biochem. *74*, 592 (1976)
404. Sneider, M. et al.: Proc. of the Soc. for Experimental Biology and Medicine *120*, 130 (1965)
405. Káš, J., Rauch, P.: Z. Lebensm. Unters. Forsch. *174*, 290 (1982)
406. Káš, J. et al.: ibid. *174*, 271 (1982)
407. Rauch, P., Fukal, L., Káš, J.: Proc. of the First Eur. Conf. on Food Chemistry (Vienna 17–20 Febr. 1981), Verlag Chemie, Weinheim 1982, p. 240
408. Káš, J., Rauch, P., Fukal, L.: Biotechnol. Lett. *5*, 219 (1983)
409. Fukal, L., Rauch, P., Káš, J.: J. Inst. Brewing (accepted), 1983
410. Anson, M. L.: J. Gen. Physiol. *22*, 79 (1938)
411. Roth, J. S., Losty, T.: Anal. Biochem. *42*, 214 (1971)
412. Nagai, Y. et al.: Biochemistry *5*, 3123 (1966)
413. Wood, J. F., Nichols, G.: Nature *208*, 1325 (1965)
414. Wood, J. F., Nichols, G.: J. Cell Biol. *26*, 747 (1965)
415. Lazarus, G. S. et al.: Science *159*, 1483 (1968)
416. Sakamoto, S. et al.: Proc. Soc. Exptl. Biol. Med. *139*, 1057 (1972)
417. Čikalo, I. I.: Lab. delo *6*, 52 (1960)
418. Ritz, N. D., Rubin, H.: J. Lab. and Clin. Med. *63*, 344 (1964)
419. Drucker, H.: Anal. Biochem. *46*, 598 (1972)
420. Henson, I. C.: J. Lab. Clin. Med. *54*, 284 (1959)

421. Bhatnagar, R., Decker, K.: J. Biochem. Biophys. Methods *5*, 147 (1981)
422. Rinderknecht, M. C. et al.: Clin. Chim. Acta *19*, 327 (1968)
423. Scriban, R., Stiene, M.: J. Inst. of Brewing *76*, 243 (1970)
424. Savage, D. J., Thompson, C. C.: ibid. *76*, 495 (1970)
425. Geokas, M. C. et al.: Clin. Biochem. *1*, 251 (1968)
426. Strassle, R.: Thromb. Diath. Haemorrhag. *8*, 112 (1962)
427. Lüscher, E. F., Käser-Glasszman, R.: Vix Sang. *6*, 116 (1961)
428. Pappenhagen, A. R.: J. Lab. Clin. Med. *59*, 1039 (1962)
429. Rinderknecht, M. C. et al.: Clin. Chim. Acta *19*, 89 (1968)
430. Wiesner, R., Troll, W.: Anal. Biochem. *121*, 290 (1982)
431. Curoff, G.: J. Biol. Chem. *239*, 149 (1964)
432. Weber, G.: Biochem. J. *51*, 145 (1952)
433. Edelman, G. M., McClure, W. O.: Accounts Chem. Res. *1*, 65 (1968)
434. Burr, M., Koshland, D. E. Jr.: Proc. Natl. Acad. Sci. U.S. *52*, 1017 (1964)
435. Lippert, E.: Z. Elektrochem. *61*, 962 (1957)
436. Gally, J. A., Edelman, G. M.: Biochim. Biophys. Acta *94*, 175 (1965)
437. Callaghan, P., Martin, N. H.: Biochem. J. *83*, 144 (1962)
438. Stryer, L.: J. Mol. Biol. *13*, 482 (1965)
439. Brodie, B. B., Udenfriend, S., Baer, J. E.: J. Biol. Chem. *168*, 299 (1947)
440. Alexander, B., Edelman, G. M.: Federation Proc. *24*, 413 (1965)
441. McClure, W. O., Edelman, G. M.: Biochemistry *5*, 1908 (1965)
442. Parker, C. W. et al.: ibid. *6*, 3408 (1967)
443. Parker, C. W. et al.: ibid. *6*, 3417 (1967)
444. Steiner, R. F., Edelhoch, H.: J. Amer. Chem. Soc. *83*, 1435 (1961)
445. Raymond, F., Chen and Harold Edelhoch: Biochemical Fluorescence: Concepts, Vol. 1, Marcel Dekker, New York 1975
446. Guibault, G. G.: Practical Fluorescence, Theory, Methods and Techniques, Marcel Dekker, New York 1973
447. Scatchard, G.: Ann. N.Y. Acad. Sci. *51*, 660 (1949)
448. Klotz, I. M. et al.: J. Am. Chem. Soc. *68*, 1486 (1946)
449. Radda, G. K.: Current Topics in Bioenergetics (ed. Sanadi, R. K.), Vol. 4, Academic Press, New York 1971
450. Slavík, J.: Biol. listy *44*, 10 (1979)
451. Kovář, J.: Chemické listy *73*, 617 (1978)
452. Daniel, E., Weber, G.: Biochemistry *5*, 1893 (1966)
453. Terada, H. et al.: Biochim. Biophys. Acta *622*, 161 (1980)
454. Lesslauer, W. et al.: Proc. Natl. Acad. Sci. U.S. *69*, 1499 (1972)
455. Ziegler, K., Rogus, E. M.: Biochim. Biophys. Acta *551*, 389 (1979)
456. Hartman, F. C. et al.: Reactive phosphate esters as affinity labels for enzymes of carbohydrate metabolism, in: Theory and Practice in Affinity Techniques, (ed. Sundaram, P. V., Eckstein, F.) Academic Press, London 1978, p. 113
457. Cutrecasas, P. et al.: J. Biol. Chem. *244*, 4316 (1969)
458. Givol, D. et al.: Biochemistry *10*, 3461 (1977)
459. Knorre, D. G., Lavrik, O. I.: Affinity labeling of aminoacyl-t-RNA synthetases, in: Theory and Practice in Affinity Techniques, (ed. Sundaram, P. V., Eckstein, F.) Academic Press, Lonson 1978, p. 169
460. Rando, R. R.: The design of highly specific enzyme inactivators, in: Theory and Practice in Affinity Techniques, (ed. Sundaram, P. V., Eckstein, F.) Academic Press, London 1978, p. 135
461. Lugnier, A. A. J.: FEBS Lett. *67*, 80 (1976)
462. Miller, H. M., Friedman, S. M.: Mol. Gen. Genet. *144*, 273 (1976)
463. Juliano, R. L.: Biochim. Biophys. Acta-Biomembranes *375*, 249 (1975)
464. Larraga, V., Munoz, E.: Eur. J. Biochem. *54*, 207 (1975)
465. Crumpton, M. J. et al.: Aust. J. Exp. Biol. Med. Sci. *54*, 303 (1976)
466. Saitoh, T.: FEBS Lett. *116*, 30 (1980)
467. Packer, L.: Methods Enzymol. *56*, 613 (1979)
468. Takano, K.: ISBN 917 222 133 X
469. Dockter, M. E.: J. Biol. Chem. *254*, 2161 (1979)

470. Shanahan, M. F., Jacquez, J. A.: Membrane Biochem. *1*, 239 (1978)
471. Beddard, G. S., West, M. A.: Fluorescent Probes, Academic Press, New York 1981
472. Wabl, M. R.: J. Mol. Biol. *84*, 241 (1974)
473. Lake, J. A., Kahan, L.: ibid. *99*, 631 (1975)
474. Kurland, C. G.: Ann. Rev. Biochem. *46*, 173 (1977)
475. Zamir, A.: Methods in Enzymology *46*, 621 (1977)
476. Rogers, T. O., Lichtstein, H. C.: J. Bacteriol. *100*, 557 (1969)
477. Sonnenberg, N. et al.: Proc. Natl. Acad. Sci. U.S. *70*, 1423 (1973)
478. Kuechler, E.: Affinity labels for t-RNA and m-RNA binding sites in ribosomes, in: Theory and Practice in Affinity Techniques, (ed. Sundaram, P. V., Eckstein, F.) Academic Press, London 1978, p. 151
479. Dinsart, C. et al.: Horm. Metab. Res. *8*, 140 (1976)
480. Bohley, P. et al.: Acta Biol. Med. Ger. *35*, 301 (1976)
481. Jung, K.: Isotopenpraxis *14*, 304 (1978)
482. Konno, S., Kimura, S.: J. Nutr. Sci. Vitaminol. *23*, 439 (1977)
483. Henning, A. et al.: IAEA-SM—205/31, Vienna, Austria 1976
484. Richter, G.: Arch. Tierernähr. *27*, 711 (1977)
485. Hazlewood, G. P. et al.: J. of General Microbiology *123*, 223 (1981)
486. Verma, D. N., Singh, U. B.: J. Nucl. Agric. Biol. *9*, 16 (1980)
487. Garlick, P. J., Waterlow, J. O.: IAEA, ISEN 92-0-011070-0, Vienna, Austria 1977
488. James, W. P. T. et al.: IAEA-SM—185/67, Vienna, Austria 1975
489. Chee, P. Y., Dahl, J. Z.: J. Neurochem. *30*, 1485 (1978)
490. Pavlík, A., Jakoubek, B.: Brain Res. *154*, 95 (1978)
491. Lasek, R. J.: Trends Neurosci *3*, 87 (1980)
492. Cancalon, P.: Brain Res. *161*, 115 (1979)
493. Fidge, N.: Clin. Chim. Acta *52*, 5 (1974)
494. Beloff-Chain, A. et al.: Horm. Metab. Res. *9*, 33 (1977)
495. Woltiers, J. M. L.: Arch. Oral Biol. *23*, 51 (1978)
496. Berson, S. A., Yalow, R. S.: J. Clin. Invest. *36*, 873 (1957)
497. Berson, S. A., Yalow, R. S.: Advan. Biol. Med. Phys. *6*, 349 (1958)
498. Berson, S. A., Yalow, R. S.: Nature *184*, 1648 (1959)
499. Skelley, D. S. et al.: Clin. Chem. *19*, 146 (1973)
500. Abraham, G. E.: Handbook of Radioimmunoassay, Dekker, New York 1977
501. Radioimmunoassay and Related Procedures in Medicine, Proc. of IAEA Symp., Berlin 1977
502. Avrameas, S., Guibert, B.: Compt. Rend. *273*, 2705 (1971)
503. Miedema, K. B. et al.: Clin. Chim. Acta *40*, 187 (1972)
504. Rubenstein, K. E. et al.: Biochem. Biophys. Res. Commun. *47*, 846 (1972)
505. Oellerich, M.: J. Clin. Chem. Biochem. *18*, 197 (1980)
506. Scharpé, S. L. et al.: Clin. Chem. *22*, 73 (1976)
507. Voller, A. et al.: J. Clin. Pathol. *31*, 507 (1978)
508. Hunter, W. M.: IAEA-SM—250/01, Vienna, Austria 1982
509. Pazzagli, M. et al.: Clin. Chim. Acta *115*, 277 (1981)
510. Pazzagli, M. et al.: J. Steroid. Biochem. *14*, 1005 (1981)
511. Pazzagli, M. et al.: IAEA-SM—259/13, Vienna, Austria 1982
512. Conns, A. H.: J. Immunol. *87*, 499 (1961)
513. Von Mayerstach, H.: Acta Histochem. Suppl. *7*, 271 (1967)
514. Sternberg, L. A.: Mikroskopie *25*, 346 (1969)
515. Sternberg, L. A.: Enzyme Immunochemistry, in: Electron Microscopy of Enzymes. Principles and Methods, (ed. Hayat, M. A.) vol. 1, Van Nostrand Reinhold Comp., New York 1973, p. 150
516. Hutter, K. J., Stoehr, M.: Microbios. Lett. *10*, 121 (1979)
517. Castell, J. V. et al.: Anal. Biochem. *99*, 379 (1979)
518. Lenhoff, H. M. et al.: Sequential and Cyclical Actions of Immobilized Enzymes that Require Pyridine Nucleotides, in: Biomedical Applications of Immobilized Enzymes and Proteins, (ed. Chang, T. M. S.) vol. 2, Plenum Press, New York 1977, p. 277
519. Bernfeld, P., Bieber, R. E., MacDonnel, P. C.: Arch. Biochem. Biophys. *127*, 779 (1968)
520. Heikki, A. E., Tuohimaa, P. J.: Methods in Enzymology *62*, 290 (1979)
521. Hood, R. L.: ibid. *62*, 279 (1979)

522. Carrico, R. J. et al.: Anal. Biochem. *72*, 271 (1976)
523. Gebauer, C. R., Rechnitz, G. A.: ibid. *103*, 280 (1980)
524. Blue Sepharose CL-6B, firm literature of Pharmacia, Uppsala, Sweden
525. Haff, L. A., Easterday, R. L.: Cibacron Blue-Sepharose: A tool for general ligand affinity chromatography, in: Theory and Practice in Affinity Technoques, (ed. Sundaram, P. V., Eckstein, F.) Academic Press 1978
526. Thompson, S. T. et al.: Proc. Nat. Acad. Sci. USA *72*, 669 (1975)
527. Jencks, W. P.: Catalysis in Chemistry and Enzymology, McGraw Hill, New York 1969
528. Epstein, H. F.: J. Teor. Biol. *31*, 69 (1975)
529. Jankowski, W. J. et al.: Biochemistry *15*, 5182 (1976)
530. Mosbach, K.: cited after ref. 531
531. Easterday, R. L., Easterday, I. M.: in Immobilized Biochemicals and Affinity Chromatography, (ed. Dunlap, R. B.) Plenum Press, New York 1974, p. 123
532. Travis, J., Pannel, R.: Behring Inst. Mitt. *54*, 30 (1974)
533. Travis, J., Pannel, R.: Clin. Chim. Acta *49*, 49 (1973)
534. Slatery, E. et al.: J. Biol. Chem. *253*, 598 (1978)
535. Swart, A. C. W. et al.: Haemostasis *1*, 237 (1973)
536. Swart, A. C. W., Henker, H. C.: Biochim. Biophys. Acta *222*, 699 (1970)
537. Hořejší, V., Kocourek, J.: ibid. *336*, 338 (1974)
538. Gaál, O., Medgyesi, G. A., Vereczkery, L.: Electrophoresis in the Separation of Biological Macromolecules, Akademiai Kiado, Budapest 1980[1]
539. Muramoto, K., Meguro, H., Tuzimura, K.: Agric. Biol. Chem. *41*, 2059 (1977)
540. Szyszko, A.: Acta Pol. Pharm. *30*, 539 (1973)
541. Choo, K. H., Cotton, R. G. H., Danks, D. M.: Anal. Biochem. *103*, 33 (1980)
542. Yang, S. S. L.: J. Nucl. Med. *19*, 804 (1978)
543. Maziere, B.: J. Labelled Compd. Radiopharm. *18*, 165 (1981)
544. Frolov, U. K. et al.: Med. Radiol. *24*, 9 (1979)
545. Kaplan, W. D. et al.: J. Nucl. Med. *19*, 1138 (1978)
546. Linderkamp, O. et al.: ibid. *21*, 637 (1980)
547. Ikeno, L. C., Bowen, B. M., Der. M.: J. Radioanal. Chem. *65*, 179 (1981)
548. Itoh, K. et al.: Kaku Igaku *13*, 459 (1976)
549. Kaufman, H. H., et al.: Acta Neurochir. *52*, 185 (1980)
550. Hale, T. I., Jucker, A.: Br. J. Radiol. *51*, 139 (1978)
551. Bosnjakovic, V. et al.: Lancet *1*, 452 (1977)
552. Roux, F. et al.: 18. French language symp. on nuclear medicine, 9–12 June, Reims, France
553. Danish patent document 2073/79/A, 21 May 1979
554. Richardson, V. J. et al.: J. Nucl. Med. *19*, 1049 (1978)
555. Terner, U. K. et al.: 2. Int. Symp. on Radiopharmaceuticals, Seattle, WA, USA, 18–23 March 1979
556. Itoh, K. et al.: Radioisotopes (Tokyo) *26*, 163 (1977)
557. Zeltzer, P. M., Seeger, R. C.: J. Immunol. Methods *17*, 163 (1977)
558. Yeates, D. B., Warbick, A., Aspin, N.: Int. J. Appl. Radiat. Isot. *25*, 578 (1974)
559. Persano, S. C. M., Wagner, J., da Silva, C. P. G.: Publicacao IAE, Sao Paolo, Brazil, no. 517 (1978)
560. Wolfangel, R. G.: Australian patent document 73/53375/A, 16 March 1973
561. Robbins, R. J., Feller, P. A., Nishiyma, H.: Health Phys. *30*, 173 (1976)
562. Raju, A. et al.: Isotopenpraxis *14*, 57 (1978)
563. UK patent document 1561048/A, 13 Febr. 1980
564. Huber, W., Saifer, M. G., Williams, L. D.: GFR patent document 2638657/A, 17 March 1977
565. Knop, G.: ZfK—280, Rossendorf bei Dresden, DDR 1974
566. van Rijk, P. P., Graaf, C. N., van de Hamer, C. J. A.: IAEA-SM—185/70, Vienna, Austria, 1974
567. Pasquier, J. et al.: 17. French language symp. on nuclear medicine, 2 June 1975, Paris, France
568. Thrall, J. H. et al.: J. Nucl. Med. *19*, 796 (1978)
569. Dahlstroem, J. A. et al.: Nucl.-Med. *18*, 271 (1979)
570. Castronovo, F. P. Jr. et al.: Health Phys. *39*, 112 (1980)
571. Matsuo, M. et al.: Radioisotopes (Tokyo) *28*, 500 (1979)
572. Toth, J. et al.: Izotoptechnika *21*, 377 (1978)

573. Hale, T. I., Jucker, A.: Nucl. Compact. *11*, 278 (1980)
574. Reyes, P., Sandquist, R.: Anal. Biochem. *88*, 522 (1978)
575. Dean, P. D. G., Watson, D. H.: J. Chromatogr. *165*, 301 (1979)
576. Janowski, W. J. et al.: Biochemistry *15*, 5182 (1976)
577. Kulbe, K. D., Schuer, R.: Anal. Biochem. *93*, 46 (1979)
578. Lowe, C. R. et al.: Int. J. Biochem. *13*, 33 (1981)

Author Index Volumes 101–112

Contents of Vols. 50–100 see Vol. 100
Author and Subject Index Vols. 26–50 see Vol. 50

The volume numbers are printed in italics